ECOLOGICAL PROCESSES HANDBOOK

Applied Ecology
and Environmental Management

A SERIES

Series Editor

Sven E. Jørgensen

Copenhagen University, Denmark

Ecological Processes Handbook
Luca Palmeri, Alberto Barausse, and Sven E. Jørgensen

Handbook of Inland Aquatic Ecosystem Management
Sven E. Jørgensen, Jose Galizia Tundisi, and Takako Matsumura Tundisi

Eco-Cities: A Planning Guide, *Zhifeng Yang*

Sustainable Energy Landscapes: Designing, Planning, and Development,
Sven Stremke and Andy Van Den Dobbelsteen

Introduction to Systems Ecology, *Sven E. Jørgensen*

Handbook of Ecological Models Used in Ecosystem and Environmental Management, *Sven E. Jørgensen*

Surface Modeling: High Accuracy and High Speed Methods
Tian-Xiang Yue

Handbook of Ecological Indicators for Assessment of Ecosystem Health, Second Edition
Sven E. Jørgensen, Fu-Liu Xu, and Robert Costanza

ADDITIONAL VOLUMES IN PREPARATION

ECOLOGICAL PROCESSES HANDBOOK

Luca Palmeri
Alberto Barausse
Sven Erik Jørgensen

CRC Press
Taylor & Francis Group
Boca Raton London New York

CRC Press is an imprint of the
Taylor & Francis Group, an **informa** business

CRC Press
Taylor & Francis Group
6000 Broken Sound Parkway NW, Suite 300
Boca Raton, FL 33487-2742

First issued in paperback 2017

Version Date: 20130711

ISBN 13: 978-1-138-07394-4 (pbk)
ISBN 13: 978-1-4665-5847-2 (hbk)

Visit the Taylor & Francis Web site at
http://www.taylorandfrancis.com

and the CRC Press Web site at
http://www.crcpress.com

Contents

Section II Physical Processes

Section III Chemical Processes

Preface

May God us keep from single vision and Newton sleep.

William Blake, 1802

Ecological Processes Handbook presents an overview of a large number of processes that take place in ecosystems. The processes are in most cases quantified as mathematical equations that have found wide use in the application of ecology in environmental management. The past 50 years—since Rachel Carson's book *A Silent Spring*—ecology has found an increasing use in environmental management. A number of ecological subdisciplines have been developed to facilitate the application of ecological considerations to solve environmental problems: ecological modeling, ecological engineering, ecological economics, application of ecological indicators and ecological services, ecotoxicology, and ecological informatics. These ecological subdisciplines all demand a quantification of ecological processes. These subdisciplines form a bridge between ecology and environmental management.

Ecological models consist of five components: state variables to describe the state of the ecosystems; forcing functions to describe the impacts on the ecosystems; ecological process equations; coefficients in these equations, so-called ecological modeling parameters; and physical and chemical constants. Ecological process equations are, therefore, an indispensable part of ecological models. Ecological engineering covers ecological restoration and the use of natural or constructed ecosystems to solve environmental problems. All these applications of ecological engineering require quantification and therefore the use of mathematical equations for the description of the focal ecological processes. Applications of ecological indicators and the assessment of the ecological services offered by ecosystems imply most often a quantification, which often must be based on mathematical equations of ecologically significant processes. Ecotoxicology is about the harmful effects of toxic substances in ecosystems, which entails a quantification of the processes in which the toxic substances participate in the ecosystems, and it is indispensable to assess the distribution of toxic substances in the ecosystems. All uses of ecological informatics are rooted in quantifications including quantifications of ecological processes. All in all, like a real bridge requires the use of iron and concrete, the virtual bridge from ecology to environmental management requires mathematical equations of ecological processes. In addition, a quantitative understanding of ecological processes is crucial for the basic ecological science, particularly of course for systems ecology that has focus on how ecosystems are working as (very effective) systems. We have, therefore, found that there was an urgent need for a comprehensive quantitative overview of ecological processes and we do hope that this book is able to meet this need.

The core material for this text comes from lecture handouts circulated in the last decade during the courses of "Modelling and Control of Environmental Systems" and "Environmental Impact Assessment" for master students of the curricula in environmental engineering, as well as during several seminars on the broad topics of ecological processes and systems ecology held in various international academic contexts.

The book consists of five sections. The first section presents some basic knowledge about ecosystems physics and may be considered an introduction to systems ecology. In this section, after introducing the fundamental concepts of classical and far-from-equilibrium systems thermodynamics, systems ecology is discussed with the perspective of information and systems theory. Here, the indicators emergy, exergy, and ascendency, commonly used to derive goal functions in ecological modeling, are presented. By defining these indicators on the common ground of thermodynamics of far-from-equilibrium systems, it is shown how all the goal functions proposed ultimately rely on the very generalization of the second principle of thermodynamics as proposed by Ilya Prigogine.

Sections II through V present the ecological processes and their mathematical formulation, focusing, respectively, on physical, chemical, biological/ecological, and landscape processes. The appendix shows how it is possible to make fast quantifications of several chemical processes simultaneously by graphical methods. Use of models is of course also possible, but the graphical methods are in most cases faster, providing results with a sufficiently accurate standard deviation as required. Therefore, it was chosen to include these methods in the handbook. This book contains many illustrations providing exemplifying quantifications. The examples demonstrate how the indicators and ecological process equations presented could be applied directly in the above-mentioned ecological subdisciplines.

Luca Palmeri and Alberto Barausse
University of Padova

Sven Erik Jørgensen
Copenhagen University

Authors

Luca Palmeri is a physicist. Although he initially specialized in environmental thermodynamics, his career has led him to acquire expertise in ecological modeling and systems ecology. His research deals with various topics in environmental science and related information technology. In particular, he is an expert in environmental GIS-based models as well as models for water resources management, water quality assessment, and environmental impact assessment. The multidisciplinary nature of his research is reflected by his continuous scientific outputs in these and other topics in applied environmental science and theoretical ecology. He has been a lecturer for more than a decade within the programme for masters of science in environmental engineering at the University of Padova (Italy), where his courses include modelling and control of environmental systems, environmental impact assessment, environmental chemical engineering, and laboratory of environmental monitoring and processes in treatment wetlands.

Alberto Barausse is an environmental engineer currently employed as a postdoctoral research fellow at the University of Padova (Italy), where he is working as a modeler on the practical implementation of the "Ecosystem Approach in Europe's Seas." His current research focus is on the combined effects of multiple human pressures and natural forcing on the functioning of large-scale marine ecosystems. More in general, he is interested in both applied and theoretical ecology and environmental science, with a focus on the aquatic environment. His research interests span diverse scientific fields such as marine ecosystem modeling, trophic network analysis, development of indicators of ecosystem status, statistical analysis of long-term ecological time series, models of treatment wetlands functioning, atmospheric dispersion of pollutants, and environmental impact assessment.

Sven Erik Jørgensen is an ecologist and chemist. He currently works for the Royal Danish School of Pharmacy in the Department of Analytical and Pharmaceutical Chemistry, University of Copenhagen, where he is a professor emeritus. Together with Professor William J. Mitsch, he was awarded the Stockholm Water Prize in 2004. He has authored more than 300 articles, most of them in peer-reviewed journals, and has written or edited more than 50 books, monographs, and compilations of articles.

Section I

Introduction

1

The Ecosystem as an Object for Research

1.1 Introduction

The term *ecology*,* which is an extension of the term *economy*, was introduced by E. Haeckel in 1866 to signify the study of the mechanisms of the exchange of goods and services between living organisms. In economics, human society is the ambit of the study; whereas in ecology, the subjects of the study are biological systems that are viewed as societies of organisms. Biological systems present extremely ordered and complex structures; they are able to store and transfer information in the form of structures and functions that are acquired during evolution.

The maintenance of life requires a continuous metabolism, the synthesis of organic macromolecules, and the regulation of these processes. These phenomena occur due to an uneven distribution of matter in living cells; for example, it may be observed that the average molar weight of a protein that is approximately 10^5 g is much greater than that of water, which is 18 g. Through a delimitation of space,† living systems build a protected environment in which the conditions that are necessary to realize thousands of simultaneous biochemical reactions are maintained; that is, those which are irreversible and which possess complicated feedback mechanisms.‡ We can, therefore, say that the evolution of a living organism toward a steady-state equilibrium§ takes place in accordance with a certain number of constraints that are imposed by the surrounding environment (forcing functions) and basic physical laws, such as the conservation of mass/energy. The exact nature of these constraints can be specified only in relation to the particular living system studied. The external constraints are forces that act on the system and support the flow of matter and energy that are necessary for the performance of biochemical processes. In particular, in the case of living organisms, external constraints are such as to bring and maintain the system

* The term comes from two Greek words: *oikos*, which means *house*, and *logos*—that is, the *study*.

† For example, by assembling membrane proteins.

‡ Essentially, enzymatic reactions.

§ State in which the body is in homeostasis. Under normal conditions, homeostasis is necessary for the sustenance of life. In the following chapters, we will indicate this state simply as stationary.

in a state that is significantly far from equilibrium, in which a continuous flow of matter and energy is simultaneously the result and engine for the metabolic activity.

By analogy, a natural tool for ecological research, particularly in the study of energy flows and of the evolution (or dynamic) of the system, appears to be *Thermodynamics*. If the static energetic analysis in ecology has started at the beginning of the 20th century (Lotka, 1922) and now has well-defined protocols (Odum, 1983b), the situation is different for the study of the dynamics.

In classical mechanics, the dynamics does not distinguish between past and future (Prigogine, 1996). Conservative mechanical systems follow reversible trajectories, which consist of a defined set of points in phase space. For the dynamics, even when appropriately generalized as in the case of quantum theory or relativity, both forward motion and backward motion in time are equally possible. However, it seems unconceivable to describe systems that evolve in a nontrivial way without establishing a preferred direction in time.

Independently of dynamics, the idea of evolution was introduced in physics in the nineteenth century along with the second law of thermodynamics. This principle distinguishes between reversible and irreversible processes and introduces a new quantity, the *entropy** *S*.

The second principle says that an isolated system evolves in time toward that state of thermodynamic equilibrium which corresponds to a maximum of the entropy. The processes occurring in nature are those that lead to a global increase of *S*. There are many equivalent formulations of this principle. The formulation provided by Clausius in 1864 can be synthesized with the intriguing sentence: "the entropy of the universe can only increase." In other words, the universe evolves toward a state of "heat death" in which the temperature is both homogeneous and uniform in time and space.

1.2 Holism versus Reductionism

Reductionism has dominated science until a few years ago, and it has produced significant results. This approach consists of decomposing the system into elementary parts, studying the behavior of each component, and consequently deriving the overall behavior of the system. For example, the human body is composed of subsystems; each subsystem is composed of organs; each organ is made up of tissues, which are composed of cells, which are formed by molecules, which, in turn, consist of atoms; and atoms are made up of elementary particles. Reductionism may be synthesized in a sentence by saying that "the whole is equal to the sum of the parts." This approach

* The entropy is often referred to as "the arrow of time," to emphasize its evolutionary character, that is, the ability to distinguish between past and future (Prigogine, 1991).

has produced classical scientific models based on simplifications that are necessary in order to reduce the number of degrees of freedom or state variables and the computational demand. However, often, this makes the model unrealistic. In particular, this epistemology reveals severe limitations when dealing with living objects. The human body is just the machinery: The animal, the human being is more than the body. Indeed, it would be impossible to describe the behavior of a chasing lion by only looking at the animal unassembled into its constituting parts.

Most recently, complementary to reductionism, the holistic approach has been introduced. Holism interprets the behavior of a system as the result of the relationships between its parts. For example, the behavior of a flock of birds cannot be explained by the simple description of the flight (position and velocity) of the birds, but rather by the result of the interaction of individual elements. In one sentence "the whole is greater than the sum of the parts." If we accept that ecosystems are complex systems that may persist only by performing several simultaneous ecological processes, and that such processes imply inevitably the instance of relationships, it becomes natural to adopt a systemic perspective. This approach is our only chance of producing some limited description of a complex system without getting lost in the jungle of details.

Occasionally, complex systems may present chaotic behaviors; that is to say, they may perform unpredictable actions or follow erratic trajectories. Actually, this should not be very surprising when considering living things that possess, at least to a certain extent, a very peculiar emergent property: free will.

1.3 Interpreting Ecosystem Reality

When dealing with ecological processes in a holistic perspective, it becomes immediately evident that estimates in the long term are practically impossible. Of course, the other side of the coin is that we can never know the world with an arbitrary accuracy. Something similar happened in physics when, in the beginning of the twentieth century, Heisenberg's principle of uncertainty was established as one foundation block of quantum mechanics:

$$\Delta x \cdot \Delta p \geq h$$

where x and p are complementary quantities that describe an object (e.g., position and velocity), and h is an arbitrarily small constant (in quantum mechanics, Plank's constant). This uncertainty is related to our interpretation of reality. It comes from the atomic world and is dramatically amplified while climbing the hierarchy of nested systems. At the topmost levels, researchers say that complex systems (therefore ecosystems) are inherently irreducible. Said differently, it is not always possible to interpret observations of system operation through basic principles. Indeed, we have to accept this condition of the modern systems ecology.

The way out from this deadlock seems to be by making use of empirical models. By changing our perspective, a new scientific paradigm for the interpretation of reality through modeling can be proclaimed. It is sufficient to accept that experimental observations do not produce new laws, but a new model of a part of reality.

Thanks to the recent developments in systems ecology and to the fast growth of computing power these days, the practice of modeling of ecological processes is possible. Thus, the interpretation of some part of the ecosystem reality can be done through modeling. Let us say that a question is posed with regard to the reality of some ecosystem aspects: An answer is found by simulation by using a model. From the model answer, which is based on a simulation with restricted validity, it becomes possible to infer theorizations with general validity (Figure 1.1).

Our interpretation of ecosystem reality is grounded on the iterative or recursive use of analysis and synthesis. Given that it will never be possible to understand the enormous complexity and interconnectivity of ecological processes with arbitrary completeness, new analytical results are used for the formulation of a synthesis (model). The latter is necessary to investigate the emergent properties of systems, and it indicates the new analyses that are required (Figure 1.2).

In conclusion, "the world equation" does not exist. Nature is simply too complex; is nonlinear, chaotic, and catastrophic;* and presents emergent properties. In the laboratory, where several internal and external conditions of the system may be maintained constant, everything is much easier. However, as soon as interactions or relationships with the surrounding environment are allowed, the living object starts using its free will to perform its role as an unpredictable modeler of its own reality.

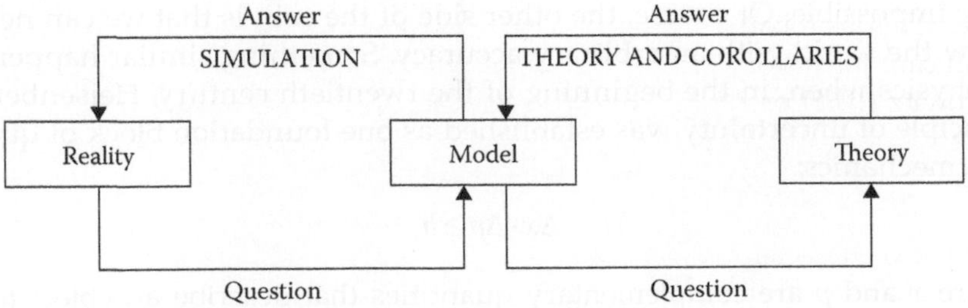

FIGURE 1.1
Interpretation of ecosystem reality throughout modeling.

* Refers to the sensitive dependence on initial conditions, in chaos theory the so-called "butterfly effect."

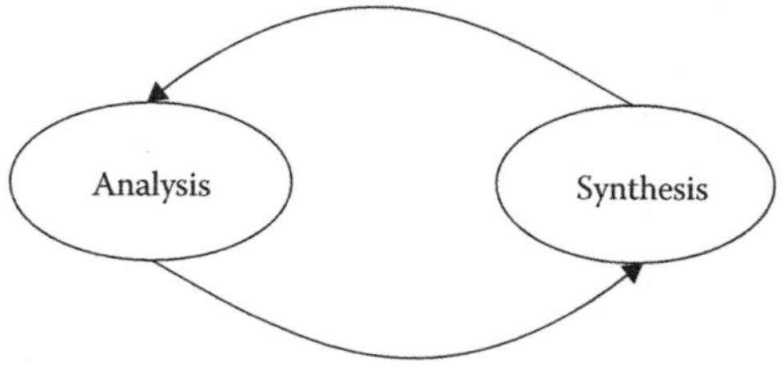

FIGURE 1.2
Analytical results are used for the formulation of a synthesis (model) that, in turn, indicates what new analyses are required.

1.4 Ecosystem Structure

A system can be generally identified by a set of items (elements) that are connected by relations. When dealing with ecosystems, it is usual to subdivide these items into two extensive basic classes: biotic and abiotic components. *Biotic components* refer to the populations of organisms; whereas *abiotic components* refer to state variables such as temperature, pressure volume, and a set of molecules in three phases: solid, liquid, and gaseous.

Jørgensen and Bendoricchio (2001) give the following definition: "An *ecosystem* is a biotic functional unit bounded in space and able to sustain the life of a community of organisms. It includes both biological and nonbiological variables" (Jørgensen and Bendoricchio, 2001) (see Figure 1.3). The elements of an ecosystem are, in turn, biotic functional subunits, as in a game of Chinese boxes. The spatiotemporal scales adopted in our description or model of the system depend on the objective of the model and determine the "dimensions of reference," that is, the boundaries of the system.

Therefore, meaningful ecosystem descriptions require the establishment of a specific level of organization, that is, the selection of appropriate spatiotemporal scales. This is probably the most general definition. Earlier definitions, which are significant for historical reasons, are those proposed by Odum and Morowitz. Morowitz (1968) proposed that "an ecosystem sustains life consistently to the actual conditions. It is a property of the entire system rather than of the individual species or populations that constitute it." Thus, well before the new epistemology was commonly accepted, the property of sustaining life had been recognized as a characteristic of the holistic system. On the other hand, E. P. Odum (1953) said that "an ecosystem is the set of physical–chemical–biological activities in a confined spatiotemporal unit." The fusion of both these definitions leads to the generalization proposed by Jørgensen and Bendoricchio (2001).

An ecosystem is a unit or functional biotic subsystem that sustains life and includes various biotic and abiotic components. Moreover, the scales are not defined a priori; instead, they depend on the objective of the study.

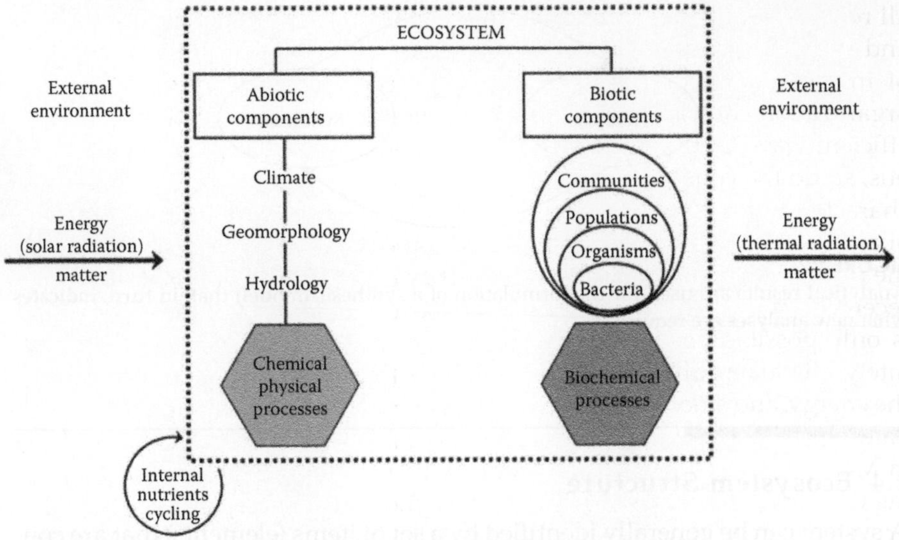

FIGURE 1.3
(See color insert.) Ecosystem structure, including biotic and abiotic components as well as the interactions (input and output) with the surrounding environment.

Obviously, the simplest description or model we can conceive will have to include the functional elements that are related to the interaction between the ecosystem and its surrounding environment (flows of matter and energy), and the functional elements of feedback (e.g., the internal cycling of nutrients). It is unconceivable to even partially describe an ecosystem while ignoring the context in which it is inserted and the internal recycling (Figure 1.3).

In practice, once the scope of the study has been identified, the researcher tries to assemble pieces of knowledge into an organic description through empirical studies. By means of comparative studies, descriptions of some functional and structural units of diverse ecosystems are included. In addition, experimental studies enable the manipulation of the entire system in an attempt to grasp relevant properties. Finally, a model or computer simulation proclaims the synthesis. All these steps are needed in order to obtain an adequate holistic description of the ecosystem structure.

1.5 Complexity and Self-Organization

Given these definitions, it is now useful to seek criteria that allow us to determine the manner and extent to which two ecosystems* are different, in short, classification criteria. Two fundamental ecosystem qualities that

* Or the same ecosystem at different stages of development.

all research in this field has confronted either earlier or later are complexity and self-organization. Both these concepts are intimately linked with that of information. More complex structures contain more information; self-organization contributes to the management of this information in a more efficient way. As the concept of information at first glance seems ambiguous, so do the concepts of complexity and self-organization. This equivocal character is due to the lack of a universal operational definition for these quantities. Actually, the informational content of a signal depends on both the reliability of the channel and subjective characteristics of the emitter and receiver* (Brillouin, 1962). Hence, a satisfactory definition of these quantities is only possible once the context has been established. Nevertheless, definitely, all biological processes seem to have the characteristic of "informing" the energy, and since these are processes of information exchange, along with Bateson (1972), we can call them *cognitive processes* ([Bateson, 1972], see also in Maturana and Varela [1980]). Indeed, Patten et al. (1997) give an interesting definition of life: Living objects make models of reality (i.e., act cognitive process), whereas nonliving ones do not (Patten et al., 1997).

The concept of order is closely linked to the concept of information (or its contrary disorder). More ordered systems contain more information than do disordered ones. This connection is made evident by the concept of entropy once this quantity has been appropriately defined (Boltzmann, 1896).

Incidentally, we may observe that highly ordered systems, for example, a solid crystal, as well as highly disordered ones, for example, a gas in equilibrium, are certainly noncomplex systems. In fact, for these systems, a limited set of numbers is enough for a satisfactory description. Truly complex systems, those that are capable of self-organizing similar to ecosystems, lie somewhere in between these two extremes (see Figure 1.4). While approaching the extremes, conservation of some potential and equilibrium are the distinctive traits; instead in the middle of this region (energy) dissipation and nonequilibrium states become predominant.

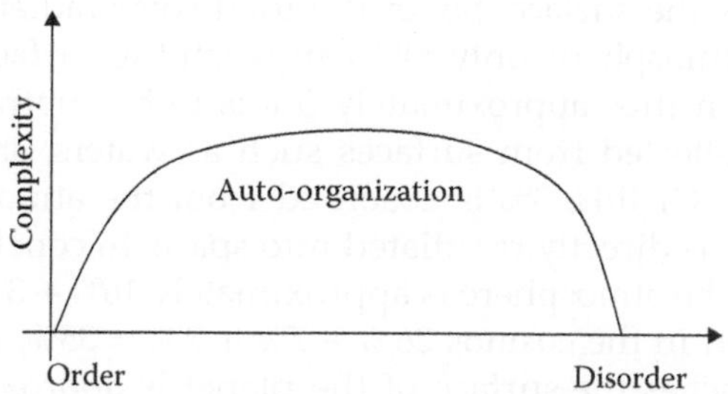

FIGURE 1.4
Complex auto-organizing systems lie somewhere in between a maximally ordered and maximally disordered system.

* In our terminology, depends on the scales and model adopted.

1.6 Energy Flows through the Biosphere

The biosphere may be considered a closed, nonisolated thermodynamic system that is traversed by a permanent 10^{22} kcal/year flow of solar energy. This flow of energy sustains what ecologists call the *thermal climatic machine*, as well as all biological processes. The earth's surface area is 510×10^6 km^2, of which 71% is covered by seas, and the remaining 29% is covered by emerged land. Of this, 16% is covered by sand and ice deserts; 2.7%, by continental water bodies; 38%, by forests (equatorial, tropical, temperate zone evergreen, and boreal), 34%, by savannas, scrubs, grasslands, tundras, and high mountain areas; and 9.3%, by croplands.

Satellite observations show that the areas of the planet that receive the majority of solar radiation (equatorial and tropical latitudes), which correspond to the continental areas covered by forests, are those in which the intensity of the emission in the infrared is greater.

The solar radiation intercepted by the earth above the atmosphere and at the average distance of the earth–sun (the astronomical unit) from a unit surface that is perpendicular to the rays per unit time is called the *solar constant*, C_s. The present value is $C_s = 1360$ W/m^2. This energy is not evenly distributed on the terrestrial semisphere, but depends on latitude, time, and day of the year. The quantity of this energy that will arrive on the surface of the planet depends on geographical (altitude, slope, etc.), atmospheric (transparency, humidity, etc.), and meteorological factors (cloud cover, turbulence, etc.).

The earth directly receives only 26% of the total radiation coming from the sun; of the remaining 74%, about 16% is absorbed by various atmospheric gases (vapor, CO_2, and O_3); whereas about 18% is reemitted by these same gases, dust, volcanic ashes, combustion products, and other substances; of this 18%, at least 11% is, in turn, diffused toward the surface. The remaining 40% is intercepted and reflected by clouds, but even of this portion, 14% indirectly reaches the surface. So, of the total solar radiation that reaches the limit of the atmosphere, only 51% can reach the surface either directly or indirectly. From this, approximately 5% is to be subtracted, because it is immediately reflected from surfaces such as waters, snow covers, deserts, and glaciers. Of this, 3% is absorbed from the atmosphere, whereas the remaining 2% is directly reradiated into space. In conclusion, the radiation absorbed by the atmosphere is approximately 16% + 3% = 19%, the one directly reradiated in the cosmos 26% + 7% + 2% = 35%, and the effective radiation that reaches the surface of the planet is approximately equal to 46% (Figure 1.5).

This radiation corresponds to the visible window (380–760 nm). Sunlight is composed by 9% of ultraviolet frequencies, 41% of visible light, and 50% of infrared frequencies. Infrared frequencies are almost completely

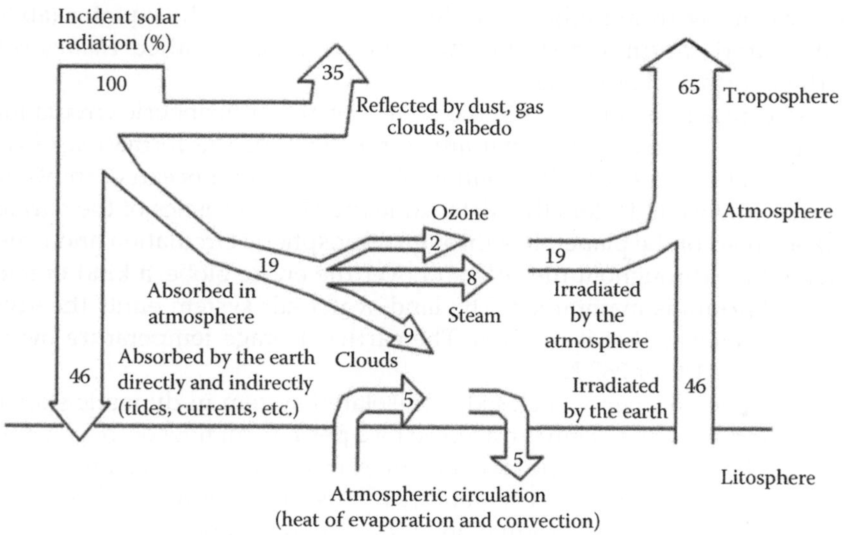

FIGURE 1.5
Solar radiation incident on the earth.

absorbed below a 10 km altitude by elements such as CO_2, N_2O, and they are then reradiated. Ultraviolet radiation is also almost totally absorbed in the high layers of the troposphere (20–100 km) by O_2, O_3, and N_2. Only a small part of this portion of the spectrum reaches the surface, that is, between 340 and 400 nm. The incident solar radiation is in the form of a short wavelength, whereas the earth reemits mostly infrared radiation at a low frequency. It is estimated that the atmosphere absorbs more than 90% of the radiation that is reflected from reflective surfaces when the sky is cloudy or foggy. Infrared radiation, which is not very penetrating, is then almost entirely absorbed by the atmospheric layers of the troposphere. It is as if the earth were surrounded by a partially thermo-insulating glass-like hood. The atmosphere has the same function as the glazing of a greenhouse: It is transparent to solar radiation at a high frequency, whereas it is opaque to the infrared one reflected by the earth. Despite the climatic changes of the last tens of thousands years, well documented at least in the northern hemisphere, the average temperature of the planet is roughly constant; polar regions lose more heat than they receive, whereas the opposite occurs for the equatorial ones. Along the equatorial belt, the soil warms, producing an intense up air blast that generates a band of high pressure and forcing air at high altitudes to flow to the surrounding lower pressure areas. Correspondingly, at the ground level, the air is attracted toward the equator to replace the ascending air. The latter, which is at some latitudinal distance from the equator, is replaced by the descending cold air coming from higher altitudes. On the sides of the equator (about 30° of latitude north and south), two convection cells called *Hadley cells* are, thus, formed (Figure 1.6).

These convection currents are the source of the atmospheric circulation, and they are the main ones that are responsible for the formation of different climatic zones. The distribution of continents and ocean currents are the other important factors that determine the characteristics of the various climatic zones of the planet. It is through atmospheric circulation and ocean currents that, throughout the year and over the entire globe, a kind of thermal equilibrium is maintained. The land–water–air system emits the same amount of energy that it receives. The earth's average temperature over a year is about 14°C or 287 K.

The biosphere is, hence, a closed, nonisolated system in dynamic stationary equilibrium. The system is traversed by a permanent flow of solar energy, with radiation entering in the form of a high-frequency infrared and exiting as a low-frequency infrared. The stability of this equilibrium is indeed maintained, thanks to this transformation of the flowing energy. For reasons that will be clarified in Chapters 2 through 4, we shall call this transformation *energy degradation* or *dissipation*, and say that dissipative processes maintain stability.

As proposed by Lovelock with "Gaia hypothesis" (Lovelock, 1979), the biosphere, as the container of all living systems, can be seen as an over-hierarchical living system in itself. Therefore, from a thermodynamic perspective, living systems are open systems that are far from maintaining thermodynamic equilibrium. By dissipating energy, they self-organize, accumulating local order (information) and transforming quantities of energy into energy of different qualities. Terrestrial vegetation and oceanic phytoplankton are the principal concentrators and transformers of the solar energy that passes through the biosphere. They are the so-called "autotrophic primary producers." Of the 1.2×10^{22} kcal/year arriving on the surface of our planet, these two elements use only 2.5×10^{20} kcal/year to carry out

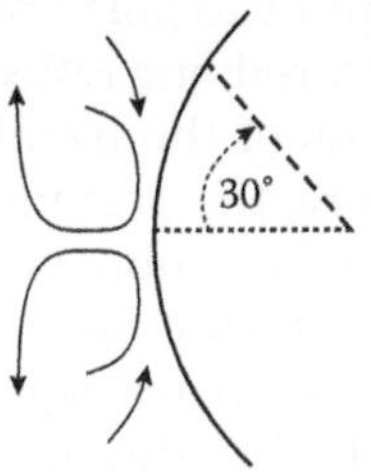

FIGURE 1.6
Hadley's convection cells are tropical atmospheric circulation cells that act as an engine for the entire earth's climatic machine.

various metabolic functions such as respiration, evapotranspiration, water and nutrients transport, and biomass growth. The last one consists of the production of 1.3×10^{18} kcal/year of vegetal organic matter, of which 60% is used for respiration (Svirezhev and Svirejeva-Hopkins, 1998). The net primary production of the entire biosphere is 5.2×10^{17} kcal/year. Thus, it is possible to estimate the production efficiency of the autotrophic component of the biosphere:

$$\eta = \frac{5.2 \times 10^{17}}{2.5 \times 10^{20}} \approx 0.2\%$$

Given that all organic matter has approximately the same composition, the energy content of the biomass can be calculated by the heat of combustion of dry material. The energy content of different classes of organisms and of different types of organic matter is as follows:

Plants	4.6 kcal/g
Algae	5.1 kcal/g
Invertebrates	5.5 kcal/g
Vertebrates	6.2 kcal/g
Carbohydrates	4.0 kcal/g
Proteins	5.0 kcal/g
Lipids	8.0 kcal/g

Primary production is realized by photosynthetic processes and is the autotrophs way to the synthesis of organic molecules, whereas production for heterotrophs is driven by consumption via fermentation and respiration processes. In all these production processes, energy is transported through the cells by molecules of adenosine triphosphate (ATP). All organisms from the simplest bacteria to humans use ATP as their primary energy currency.

Heterotrophic organisms consume only 3% of annual net primary production. On average, from one level to the next of the trophic chain, the efficiencies of energetic transfers are given by the following figures:

light → Primary production → Herbivores → Carnivores

0.2% 3% 1% ... 40%

At present, human societies consume about 6.9×10^{16} kcal/year (Svirezhev and Svirejeva-Hopkins, 1998). About 95% of this energy is extracted from fossil and nuclear fuels, and it corresponds to 10% of all net primary production. From the biological point of view, the annual energy needs per capita is about 10^6 kcal/year; therefore, since the current human population is approximately 7×10^9 individuals, for the entire human species, we have a biological energy requirement of 7×10^{15} kcal/year. Given that 3% of annual net primary production, 1.6×10^{16} kcal/year, is used by the entire animal kingdom, we find that human needs account for about 43% of the entire animal needs.

These considerations are indicative of the state of strong competition for resources between human activity and global biosphere. Pollution of the environment and reduction of biota diversity are the consequences of this competition.

1.7 Brief History of the Biosphere

Since it emerges from archeological investigations, the composition of the atmosphere, during the evolutionary history of the biosphere, has undergone drastic changes. There are strong indications that there were the first forms of life to change the conditions of the planet by making them more stable and favorable to the establishment of increasingly complex forms of life (Lovelock and Margulis, 1990). In this way, strong couplings were established, feedback mechanisms between the biota and the atmosphere, which allowed the maintenance of stable conditions in the biosphere. Research into the evolutionary history of the biosphere can, thus, help in understanding the importance of the feedback mechanisms that are required to maintain stability in biological systems.

The earth's history begins with the formation of a fireball of molten magma that appeared roughly 4.5 billion years ago, probably due to the condensation of gas and residues of the nebula of matter that originated when the sun was birth. The establishment of conditions that are favorable to life is due to the fortuitous combination of two factors: size and distance of the planet from the sun; these conditions allowed the formation of the atmosphere and oceans condensation. We can distinguish three main eras in the evolution of life on Earth, each of which extends for a period of 1–2 billion years.

The *prebiotic* era during which the conditions for the birth of life were established lasted from 4.5 to 3.5 billion years ago. From the very beginning, the earth began to cool. About 4 billion years ago, the condensation of atmospheric steam took place, thus forming early shallow oceans, where the first carbon-based compounds, catalytic rings, and membranes developed until the birth of the first living cell. The beginning of life can be traced back to about 3.5 billion years ago.

The second era extends for nearly two billion years, from 3.5 to 1.5 billion years ago, and is called the microcosm. During this period, bacteria and other microorganisms invented all the basic processes of life and established the complicated global feedback mechanisms for the self-regulation of the biosphere system. During the microcosm, bacteria have drastically changed the composition of the atmosphere, adjusting the levels of humidity, hydrogen, CO_2, oxygen (21%), and ozone (Lovelock and Margulis, 1990).

Finally, 1.2 billion years ago, the era of the macrocosm began, which saw the evolution of higher forms of life, and came up with the appearance of man, which is a very recent event in geological time.

In order to better understand the relative significance of evolutionary eras, it is instructive to linearly project the earth's history on a day (24 hours) time-scale (Table 1.1).

If it is assumed that the earth was born at midnight, we have to wait until noon to see the appearance of photosynthesis and eukaryotes. Actually, a whole day has to pass before mammals start wandering throughout the planet. Man is the very last creature to come on the stage some milliseconds before midnight of the next day (see Figure 1.7).

TABLE 1.1

Evolution in 1 Day Obtained by Linearly Projecting 4.6 Billion Years on a 24-Hour Period

10^6 years	h	min	s
4600 Birth of the earth	0	0	0
3500 Bacteria	5	44	21
2500 Water photosynthesis	10	57	23
2200 Eukaryotes (symbiosis)	12	31	18
2000 Aerobic respiration	13	33	55
570 Invertebrates	21	1	34
500 First vertebrates	21	23	29
395 First plants on land	21	56	21
385 Origin of reptiles, amphibians, fishes, and seeding plants	21	59	29
280 Insects	22	32	21
235 Superior reptiles	22	46	26
190 Conifers, mammals and birds, dinosaurs	23	0	31
135 Angiosperms, dinosaur extinction	23	17	44
38 Origin of modern mammals	23	48	6
7 Hominids	23	57	49
2.5 Glacial eras—Homo sapiens	23	59	13
0.01 History	23	59	60

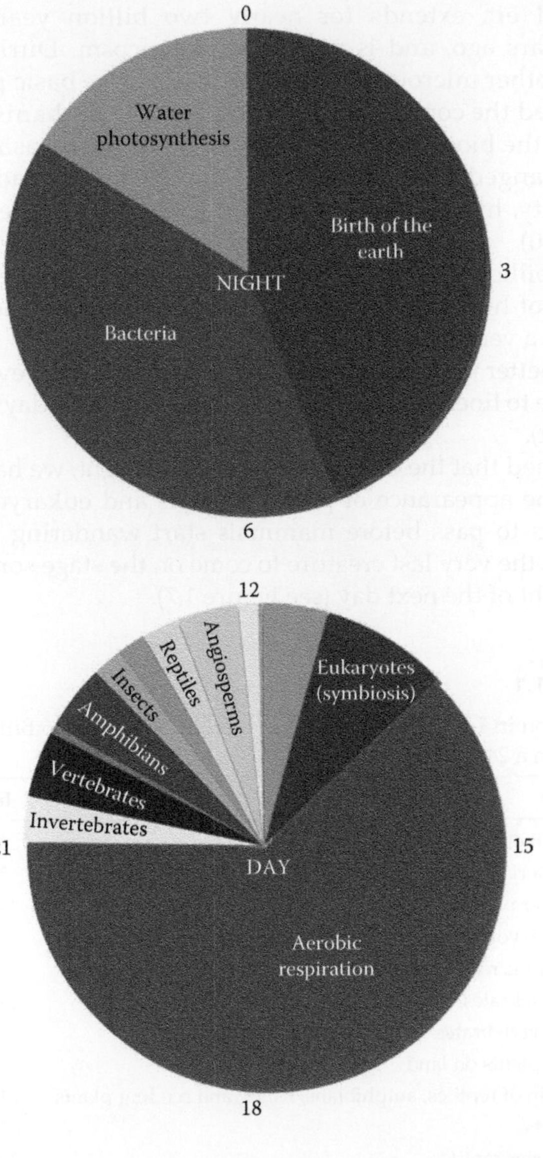

FIGURE 1.7
(See color insert.) Evolution on the watch. Astoundingly, a whole day has to pass before the appearance of mammals.

2

Conservation Principles

2.1 Introduction

Most significant laws in science are conservation laws; for example, we may recall the laws of conservation of mass, energy, momentum, angular momentum, and charge. In physics, a conservation law states that a particular measurable property of an isolated physical system does not change as the system evolves. One particularly important physical result with regard to conservation laws is that every conserved quantity corresponds to a particular symmetry: For example, the conservation of energy follows from the time invariance of physical systems.

In our holistic thermodynamic view, ecosystems are, by definition, far from equilibrium nonisolated systems. In recent years, Prigogine (1967) has generalized thermodynamics as referring to irreversible processes. The search for conserved quantities that are far from equilibrium systems is important, especially in ecology, because it may make explicit the effects of constraints, and reduce the degrees of freedom of the system description, thereby decreasing the computational demand.

Some conserved quantity is needed in order to use continuity equations in our descriptions of ecosystems. In this chapter, we will observe how the concept of conservation is seen in different perspectives: topological, physical, and ecological.

2.2 Space Conservation

There is certainly no standard concept known as "law of space conservation." In order to even consider such a law, one should specify the context, and what exactly is meant by that. Is conservation of space represented by a number or a mathematical structure? Trivially, we could delineate a law of conservation of Euclidean three-dimensional space: There is a three-dimensional space at a certain moment, and that space remains after some amount of time (i.e., it becomes neither a non-Euclidean space nor a two-dimensional one). In physics, this definition is not really useful. However, when looking at ecosystems, two forms of space conservation are immediately recognized: topological and ecological.

Topological space conservation is related to the ability of living objects to delimit an area by building membranes. For example, in the living cell, the delimitation of an area is associated with the persistence of boundaries, separating an internal secure and stable environment, in which the chemo-physical conditions are maintained stationary, from the external environment, which is stochastic and uncontrollable. Boundary conditions are required in order to maintain the homeostasis that is actually necessary for biochemical reactions in the cell to correctly proceed. So, in this context, the "law of space conservation" may be interpreted as the aptitude of living systems to persistently occupy a portion of space while maintaining stable conditions within it.

Ecological space conservation becomes apparent as soon as we begin looking at space as if it were a food-like resource. This is the case of fouling or biofouling when, for example, mussels compete for a place on the reef. In this latter case, once the total surface available for a given group of organisms has been defined, the carrying capacity of the ecosystem is defined by nutrient as well as space availability. The sum of available and occupied space is, thus, conserved.

2.3 Mass Conservation

In the late eighteenth century, Antoine Lavoisier discovered the general principle of mass conservation. This principle states that the mass of a closed system will remain constant over time. Given the mass/energy equivalence that we will discuss later on, this principle is theoretically equivalent to the conservation of energy.

The mass of a closed and isolated system (closed to matter and isolated to energy exchanges) cannot be changed as a result of the processes acting inside the system. Although matter may be rearranged in space and changed into different types of particles, this law implies that mass can be neither created nor destroyed. For any chemical process in a closed system, the mass of the reactants should equal the mass of the products.

This principle of mass conservation holds in ordinary interactions. This is true in classical chemistry or thermodynamics but does not hold when nuclear or relativistic processes are involved. If we want to look at the most general form, then it is the conservation of mass and energy that holds, as in many cases, mass and energy can be converted into one another.

Accepting the validity of this very general law, we may now write the continuity equation for an open nonisolated ecosystem. Let c be the concentration of a given element and V be the volume of the system (or portion of the system) considered; the continuity equation is

$$V\frac{dc}{dt} = \text{Input} - \text{Output} + \text{Formation} - \text{Transformation} \tag{2.1}$$

The first two terms in Equation 2.1 account for the interaction with the surrounding environment, whereas the last two are referred to as *net production:*

$$\text{Net production} = \text{Formation} - \text{Transformation}$$

The last term may also be expressed as

$$\text{net production} = \text{ingestion} - \text{respiration} - \text{excretion} - \text{unassimilated food}$$

Thus, the efficiency is

$$\text{Efficiency} = \frac{\text{Net production}}{\text{Ingestion}} \tag{2.2}$$

2.4 Energy Conservation: Thermodynamics' First Principle

The law of energy conservation, the classical conservation law that is soundly defined and has very general validity, is also known as *the first principle* in thermodynamics. This law states that the total amount of energy in an isolated system remains constant over time. For an isolated system, this means that energy can change its form and position within the system. For instance, chemical energy can become kinetic energy, but that energy can be neither created nor destroyed. In modern physics, mass/energy equivalence has been established by formally unexceptionable laws, whereas in ecological thermodynamics, this equality is accepted only once it is recognized that all organic matter has approximately the same chemical form (proteins, carbohydrates, lipids, etc.) and composition (e.g., fixed ratios of N, P, C, etc.—Redfield ratios [Redfield, 1958]).

In order to describe how solar energy is converted into organic matter and work, we shall be able to write a continuity equation by converting matter into energy and heat into work. This job is done by the first principle.

Let dq be the infinitesimal heat absorbed and dw be the infinitesimal amount of work done by a closed system during any thermodynamic process; then, the relation

$$dU = dq - dw \tag{2.3}$$

defines a thermodynamic potential, the internal energy U. Incidentally, one may note here that it is possible to define heat as the energy change in a transformation in which no mechanical work is done. From the fact that dU is an exact differential, it follows that the energy change for going from point a in the state space to point b

$$\Delta U_{a,b} = \int_a^b \mathrm{d}U \qquad (2.4)$$

does not depend on the trajectory followed by the system, in particular for closed trajectories, which is $a \equiv b$, we have $\Delta U_{a,b} = 0$. Note that dU is an exact differential, whereas dq and dw are not, if taken, separately. Let us consider a system described by n state variables; then, for a closed trajectory, we can write

$$\Delta U_{a,b}(x_1, x_2, \ldots, x_n) = 0 \qquad (2.5)$$

where x_1, x_2,..., x_n are state variables. Equation 2.5 identifies an $n - 1$ dimensional subspace, in which the trajectory of the system is constrained to lie. The existence of a conserved quantity U enables the diminishing of the degrees of freedom.

The experimental foundation of the first principle is the famous Joule (1850) experiment, which demonstrated the equivalence between heat and work, that is, between thermal and mechanical energy. The operational definition of heat in terms of energy provides the conversion factor between thermal and mechanical energy, the mechanical equivalent of the calorie:

$$J = 4186 \text{ J/Kcal} \qquad (2.6)$$

This conversion factor is obtained through Joule's apparatus (see Figure 2.1).

It is important to note here that the first principle applies to any kind of transformation: reversible or irreversible. The inclusion of heat between the forms of energy leads to this wide generalization of the principle of conservation of energy of classical mechanics. The subsistence of this very general law of energy conservation allows the introduction of mass/energy balances by writing a continuity equation in the form of Equation 2.1. In this equation, energy or mass can be used interchangeably to describe ecological constraints such as the carrying capacity or resource availability (nutrient or space limitation, etc.).

2.5 Information

As already stated, at first glance, the concept of information seems ambiguous. We should wait until the introduction of information theory (Shannon and Weaver, 1963) and of "far-from-equilibrium thermodynamics" (Prigogine, 1967) to have a satisfactory definition:

$$I = S_0 - S \qquad (2.7)$$

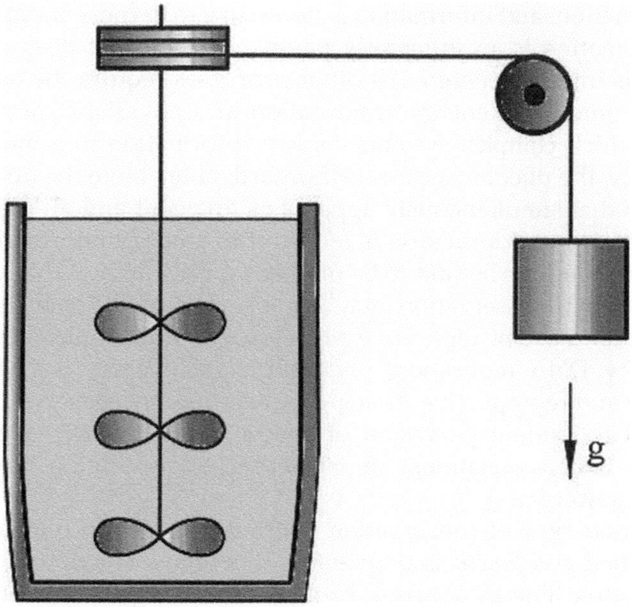

FIGURE 2.1
Joule's experiment allowing the establishment of heat and work equivalence. In this experiment, a descending weight attached to a string causes a paddle immersed in water to rotate, thus causing a small increment (of the order of tenths of degree) in water temperature.

where S is the entropy of the system in its actual nonequilibrium state, while S_0 is the entropy that the same system would have in the corresponding state of equilibrium, in which the system is indistinguishable from the environment. This latter state is called the *reference state* or *heat death* (Shieh and Fan, 1982). I is, thus, a measure of the distance from the reference state.

Information may be rearranged topologically and changed into different ecological functions with a conservative operation; however, from a physical point of view, the existence of a universal information conservation principle seems unconceivable. According to Equation 2.7, in fact, dissipative processes play a distinctive role rather than conservative processes. Such a law would imply that information can be neither created nor destroyed. Half the sea lies between saying and doing. This old popular maxim expresses a fundamental fact of information theory: Amplifying information is easier than amplifying power. For example, in materials and energy technology, researchers are accustomed to working at the boundary of fundamental limitations, and engineers are used to designing their structures based on the tolerances imposed by these limitations. In information technology, fundamental theorems are not inherently as restrictive as in energy and materials technology.

The phenomena that take part in information can be divided into four categories: storage, preparation, processing, and distribution. Energy is needed to get information, and information is necessary to harness energy. The storage of information is an inherently passive activity that does not require a continuous input of energy. All other processes require the employment of a certain amount of energy. Information and physical structure are the means by which complex systems convey information to a more efficient use of energy, the phenomenon of self-organization. Note the double role of information that simultaneously appears as an agent and an object of self-organization. Other information is required to amplify information. In this sense, information can be said to be mathematically autocatalytic.

In this context, conservation may be observed at different hierarchical levels, such as those of replicative processing. For example, a genotype is conserved by DNA replication; phenotypic biodiversity is conserved by competition and reproductive strategies; structures are conserved by growth with fractal redundant pervasion of space; and allometric relations (e.g., metabolism–biomass relations) are conserved by replication through scale invariance (power laws).

This peculiar type of conservation is often included in our descriptions using so-called *goal functions* or *orientors*. These are descriptors of the system; for example, Energy, Entropy, Exergy, and Emergy may be used to construct a continuity equation by also including some evolutive tendency,* for instance, maximizing.

Interestingly, the widespread definition of sustainable development is a good example of conservation in this broadest sense:

> The development that can satisfy the needs of the present generation without compromising the ability of future generations to meet their own needs. ([Our Common Future: Brundtland Report, 1997 UNCED] McChesney and Lincoln University, Centre for Resource, 1991)

The orientor here is "the ability of future generations to meet their own needs," which should not decrease (be compromised) in order to consider our environment conserved.

* Sometimes, these goal functions may also describe the structural dynamics of the system (see, e.g., structural dynamic modeling in Chapter 19).

3

Energy Dissipation

3.1 Introduction

Dissipation in physics is strictly related to the concept of dynamics. Indeed, friction or turbulence causes a degradation of energy over time. This energy is typically converted into heat, which raises the temperature of the system. Such systems are called *dissipative systems*. Formally, dissipating forces are those that cannot be described by a Hamiltonian. This includes all those forces that result in the conversion of coherent or directed energy flow into an undirected or more isotropic distribution of energy (e.g., the conversion of light into heat).

As discussed in Chapters 1 and 2, this is exactly what happens in ecosystems. However, at first glance, this seems to be in contradiction with the observation that living objects embody order and information. In the 1940s of the twentieth century, in his famous book *What is Life?* (Schrödinger, 1944), Schrödinger threw new light on the relationship between order, disorder, entropy, and living organisms. The instance "order from order" introduced by Schrödinger was a startling premonition of what would be the results of the research that led Watson and Crick in 1954 to the discovery of DNA. The other fundamental idea expressed in Schrödinger's book, the instance "order from disorder" was a first attempt to apply the fundamental theorems of thermodynamics to biology. Schrödinger recognized that living systems, being embedded in a network of flows of matter and energy, are far from equilibrium systems. He realized that the study of such systems from the perspective of nonequilibrium thermodynamics would enable the reconciliation of biological self-organization and thermodynamics.

Every biological process, event, or phenomenon is inherently irreversible, leading to an increase of the entropy of that part of the universe in which the event occurs. So, every living organism produces positive entropy, thus coming dangerously close to the state of maximum entropy, the heat death. It manages to stay away from this state only through continuous negative entropy or Helmholtz free energy absorption from the surrounding environment in the form of matter and energy. This follows immediately from the natural tendency of organisms to maintain the steady state (homeostasis). In this case, the condition of stationarity implies that the flow of entropy is null:

$$0 = \frac{dS}{dt} = \frac{dS_{in}}{dt} + \frac{dS_{out}}{dt}$$

and, hence,

$$\frac{dS_{in}}{dt} = -\frac{dS_{out}}{dt}$$

Along with Schrödinger, we can say that "… essential for metabolism to work is that the body is able to get rid of all the entropy that produces in its lifetime." Thus, the organization in living systems is maintained by absorbing order from the surrounding environment and emitting disorder in the form of thermal radiation in the infrared. In this context, Schrödinger, without mentioning them, explicitly makes reference to the biological necessity of dissipative processes. He states, "… the highest body temperature of warm-blooded animals includes the advantage of making them able to get rid of its entropy at a faster rate so that they can afford a more intense process of life."

3.2 The Second Principle

From everyday experience, we know that there are processes that meet the first principle but which, nevertheless, never occur spontaneously. The transformations that occur spontaneously in nature follow predefined orientations. In order to account for these experimental facts, a general criterion is needed that determines the transformations which can occur spontaneously. This is the subject of the second law of thermodynamics.

The second principle or the law of entropy states that for every physical system, a new thermodynamic state function can be defined, the entropy* S, which should satisfy the following set of rules:

1. The entropy S is a state function.
2. S is an extensive quantity: as U or V, if we divide the system into two subsystems α and β that are not interacting with each other, the total entropy is given by $S_{tot} = S_\alpha + S_\beta$.
3. If the system changes its state, the entropy S varies. This variation can be decomposed into two contributions $dS = d_iS + d_eS$, where d_eS represents the entropy change due to interactions with the external

* From Greek εντροπειον, evolution.

environment, whereas d_iS represents that due to processes that take place exclusively within the system.

4. A quantitative definition is given: $d_eS = \frac{\partial q}{T}$, where ∂q indicates the infinitesimal* amount of heat absorbed by the considered system in the transformation, and T is the temperature of the system at which this exchange takes place.
5. For d_iS, an operative definition is not given, and one can only say that for spontaneous processes, $d_iS \geq 0$, where the equality holds if the transformation is reversible or quasistatic.

If we consider a transformation between two states a and b, for the entropy change, we can write

$$\Delta S_{a,b} = \int_a^b dS = \int_a^b d_iS + \int_a^b d_eS \tag{3.1}$$

In general, the integrals in the last member depend on the trajectory followed by the transformation between states a and b (i.e., d_iS and d_eS, and are not state functions), except in the event that the transformations along which we measure ΔS are reversible or quasistatic. In this case, we have

$$\Delta S_{a,b} = \int_a^b dS = \int_a^b d_eS = \int_a^b \frac{\partial q}{T} \tag{3.2}$$

Note that the last integral can be written only when it makes sense: We know that if the transformation is not reversible, intermediate states are not equilibrium ones, and then, in general, the state variables are not well defined. From point 5, it follows immediately that for irreversible transformations between a and b, it is always

$$\int_a^b \left(\frac{\partial q}{T}\right)_{\text{irrev.}} > \int_a^b \left(\frac{\partial q}{T}\right) = \Delta S_{a,b} \tag{3.3}$$

The integral of the heat transfer in a reversible transformation from a to b is always less than what it would be by making an irreversible transformation between these two states. In the former case, the integral coincides by definition with the change in entropy; in the latter case, the entropy is not defined. From point 4, it is evident that for *adiabatic* transformations, that is, transformations in which $\partial q = 0$, is $d_eS = 0$, hence, for a generic adiabatic transformation between states a and b, $\int_a^b \frac{\partial q}{T} \geq 0$. We can, therefore, formulate the second principle with the following sentence: For an adiabatic system,

* It is not denoted by dq to emphasize the fact that this is not an exact differential.

entropy can only increase, or, at most, in the case of reversible transformations, remain constant.

Illustration 3.2.1: Statistical Significance of Entropy

From the definition given in the previous paragraph, it follows that entropy is a physical measurable quantity as, for instance, the length or the temperature. For example, a cube of ice that melts increases its entropy of a quantity, which is given by the heat of fusion divided by the temperature of the melting point. Boltzmann (1896) was the first who showed that the entropy is a measure of disorder as well. This fact can be demonstrated by applying the microscopic perspective to macroscopic thermodynamic systems. Consider a system consisting of n particles in equilibrium at temperature T, which can be found at various energy levels. The basic assumption is that the system is in a condition of complete molecular chaos. This assumption is equivalent to an assumption that all microscopic states are equally probable. Statistical mechanics for equilibrium systems tells us that at a given value of energy level, an equilibrium distribution, called the *microcanonical ensemble*, exists. The probability P for the system to be in a state whose energy is E is given by

$$P(E) \propto e^{-\frac{E}{k_B T}}$$

where $k_B = 1,3803 \times 10^{-23}$ J/K is the Boltzmann's constant. By normalization of this function, the energy distribution function $f(E)$ is obtained:

$$f(E) = e^{-\frac{(E-F)}{k_B T}}$$

with F Helmholtz free energy and

$$\sum_E e^{-\frac{E}{k_B T}} = e^{-\frac{F}{k_B T}}$$

From the conservation of the Hamiltonian flow, it follows that all the trajectories of the system states in the phase space are constrained to lie on constant energy surfaces. These are indeed states that are not distinguishable from the point of view of the energy or the macroscopic point of view. A measurement of this indistinguishability, for a system in equilibrium at temperature T, is provided by the area of the surface of constant energy, or the number of possible microscopic states or complexions. If we denote this number by Π, from the energy distribution function, we get

$$\Pi = \int_{E=E_0} e^{-\frac{(E-F)}{k_B T}} dE$$

Boltzmann's intuition was that the entropy S of the system increases with the number of possible microscopic states Π. In other words, the entropy is a measure of the macroscopic indistinguishability. This insight led to the demonstration of the renowned relation, valid for isolated systems, named *Boltzmann's order principle*:

$$S = k_B \log \Pi$$

By using the energy distribution function $f(E)$, the expression for the kinetic energy, $E_c = \frac{1}{2}mv^2$, Maxwell–Boltzmann distribution is obtained. This function describes the distribution of the particle velocity moduli for n moles of gas at temperature T:

$$f(v) = \frac{n}{(2\pi m k_B T)^{3/2}} e^{-(1/k_B T)(mv^2/2)}$$

which is represented in Figure 3.1 for three different values of the temperature. It is readily understood here how the existence of a more probable velocity, corresponding to the maximum of the curve of Maxwell–Boltzmann, is associated with the natural tendency of the system to reach the most probable state. The more likely macro states are those that are achievable with a greater number of microstates Π; that is, they are more indistinguishable from a macroscopic point of view and, therefore, more disordered.

An isolated system increases its entropy and more or less rapidly reaches the inert state of maximum entropy, which is the equilibrium. Thermodynamic systems tend to evolve toward states of greater disorder. The tendency to depart from an ordered state to reach a less ordered one, to move from a less probable to a more probable one, to an entropy increase are all different manifestations of the same natural law expressed by the second principle. For an isolated system, the state of equilibrium is the most likely and it is characterized by a maximum of the entropy.

Boltzmann's order principle can be immediately extended to non-isolated systems. In fact, if the system is closed, that is, can exchange energy but not matter, and it is at a constant temperature T, its behavior is described by Helmholtz free energy:

$$F = U - TS$$

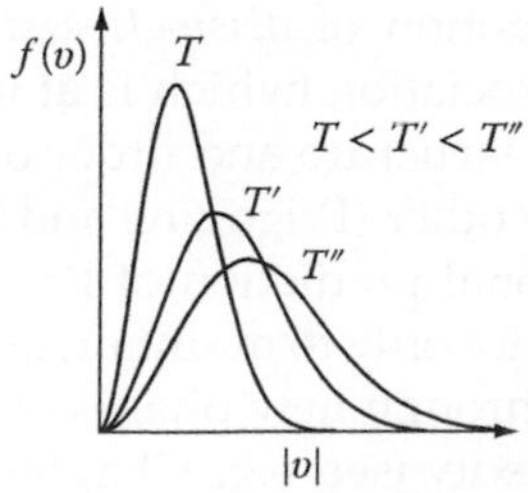

FIGURE 3.1
Maxwell–Boltzmann distribution of particle velocity moduli for n moles of a gas at three different temperatures.

The second principle says that the state of thermodynamic equilibrium of the system, which corresponds to that of maximum entropy, is that in which F is a minimum. The structure of this latter definition reflects a competition between the internal energy U and the entropy S. At low temperatures, the second term is negligible, and the minimum of F corresponds to a minimum of U. In these conditions, the realization of low entropy structures such as crystals and solids is experimentally observed. The higher the temperature, the system evolves progressively toward high entropy states such as gaseous states. Boltzmann's order principle does not apply to dissipative structures.

3.3 Dissipative Structures

The first and perhaps the most authoritative in-depth description of self-organizing systems was the theory of dissipative structures that had been developed during the 1960s of the twentieth century by Ilya Prigogine and his collaborators of the School of Brussels (Prigogine, 1967; Glansdorff and Prigogine, 1971; Nicolis and Prigogine, 1977). Prigogine's crucial insight was that far-from-equilibrium systems should be described by nonlinear equations. Prigogine did not start his study on living systems; instead, he focused on much easier phenomena as *Bénard instability* (Bénard, 1901) or Belousov–Zhabotinsky autocatalytic reactions* (Nicolis and Prigogine, 1977). These are spectacular examples of spontaneous self-organization phenomena. In one case, the state of nonequilibrium is maintained by the continuous flow of heat through the edges of the system; whereas in the other case, it is maintained by the presence of catalytic compounds that are capable of sustaining nonequilibrium oscillating chemical reactions.

Prigogine's work resulted in the development of a nonlinear thermodynamics that is apt at describing the phenomenon of self-organization in far-from-equilibrium open systems. Classical thermodynamics provides the concept of "equilibrium structure," such as crystals. Bénard cells are also examples of structures, but of a very different nature. For this reason, Prigogine introduced the notion of *dissipative structures* to emphasize by the name itself the close association (which is at first glance truly paradoxical) that may exist between structure and order on one side, and losses and wastage (dissipation) on the other (Prigogine and Strangers, 1984).

Indeed, the most sensational prediction of Prigogine's theory is that dissipative structures are able not only to maintain far-from-equilibrium steady states, but can also evolve through new phases of instability, becoming new structures of greater complexity (see, e.g., Chapter 19).

* The so-called chemical watches.

Classical thermodynamics has solved the problem of competition between chance and organization for equilibrium situations. If the temperature drops, the contribution of the energy E to Helmholtz free energy $F = E-TS$ becomes dominant. Increasingly complex structures can appear, which correspond to increasingly lower values of entropy. A phase transition as a liquid → solid one is, in fact, characterized by a definite loss of entropy (or increase in organization). Actually, steady states are also characterized by lower entropy. Therefore, dissipative processes might lead to an increase in organization (Prigogine, 1967). This increase in organization is usually continuous. Prigogine wonders whether "discontinuous changes in structure are possible due to dissipative processes? Such situations would be the same for nonequilibrium systems of phase transitions."

The detailed analysis carried out by Prigogine shows that dissipative structures receive their energy from the outside, whereas instabilities and jumps to new forms of organization are the result of microfluctuations that are amplified by positive feedback (Glansdorff and Prigogine, 1971).

3.4 Irreversible Processes

Classical thermodynamics displays severe limitations, because it is based on concepts such as state of equilibrium and reversibility. Biological systems instead prosper in thermodynamic states that are far from static equilibrium and are characterized by the presence of irreversible processes. It is, therefore, desirable to extend the results of classical thermodynamics to this class of processes as well. Irreversible thermodynamic processes can be divided into two subclasses: linear not-very-far-from equilibrium processes and nonlinear ones.

In its more general formulation, the second principle is equally applicable to situations of equilibrium as well as to those of nonequilibrium (Nicolis and Prigogine, 1977). However, the most significant results of classical thermodynamics, developed since the nineteenth century, belong to the domain of the equilibrium phenomena. Just think, for example, of the law of mass action, to the rule of Gibbs or to the equation of state of a classical ideal gas. In classical thermodynamics, the state of nonequilibrium is seen as a disturbance that temporarily prevents the system from reaching equilibrium. The implicit assumption is that the natural condition, or rather the only describable one, is that of equilibrium. The concept of dissipation enters the theory only as a disturbing element that is significantly linked to the inability to transform all the heat energy absorbed in useful work.*

The situation changed dramatically with the discovery of Onsager's reciprocity relations (Onsager, 1931). This fact led to an extension of the domain

* The concept of dissipation is then bound to the second principle in the impossibility to build thermal machines with a 100% efficiency.

of classical thermodynamics with the introduction of the thermodynamics of linear nonequilibrium states. These methods are applicable in situations where the flows or the reaction rates of irreversible processes are linear functions of the generalized thermodynamic forces, such as temperature or concentration gradients.

It is experimentally observed that linear nonequilibrium systems can evolve into states with lower entropy.* However, we cannot talk in this context of the emergence of new structures. These are simply structures, so to speak, that are inherited from the regime of equilibrium, modified and maintained in nonequilibrium conditions by external constraints. In the 1960s, the School of Brussels extended the description in terms of macroscopic thermodynamic variables also to far-from-equilibrium conditions. If we consider a system in equilibrium, we know that, depending on external constraints, the state will be stationary for an appropriate thermodynamic potential. For example, the state will be that of maximum entropy S for an adiabatic system, or, that of minimum free energy F for a closed system at constant temperature. The solution of the Hamiltonian system is unique. Prigogine calls this solution the *thermodynamic branch* (Nicolis and Prigogine, 1977). If the external constraints are changed in order to force the system to gradually further from equilibrium, the velocity of the processes or the intensity of the flows begins to have a nonlinear dependence on the generalized forces. As already mentioned in the Bénard experiment, reaching a critical value of forcing the system evolves by creating new structures (convective cells), exhibiting scale invariance and correlations at macroscopic dimensions. Prigogine and collaborators established a criterion for the stability of the thermodynamics branch. If the criterion is not verified, the thermodynamics branch may become unstable and bifurcations may appear.

When dealing with systems that are not in thermodynamic equilibrium, some remarks are needed with regard to the definition of thermodynamic potentials. Indeed, it was said that under nonequilibrium conditions, thermodynamic quantities that characterize every system that is under nonequilibrium conditions are not well defined. In order to allow the definition of potentials, thermodynamics of irreversible processes assumes the validity of the hypothesis of local thermodynamic equilibrium (LTE):

Every thermodynamic system can be decomposed into a certain number of macroscopic subsystems within which the state variables are substantially constant ($\Delta x/x \ll 1$), and are locally linked by the same equations of state $f(V, p, T) = 0$, which are extensively valid for equilibrium systems.

Statistical mechanics explains why this assumption is reasonable. In fact, it is shown that all the relaxation times, which are characteristic for the achievement of the equilibrium in microscopic systems, are proportional to $\sqrt{N}$, where N is the number of elementary constituents. In other words if a

* For example, by applying a temperature gradient to a box containing a mixture of two gases, we observe an increase in the concentration of one of the two gases in the vicinity of the warmest wall and a decrease in the vicinity of the coldest one (Nicolis and Prigogine, 1977).

large system is considered, this moves quickly to the equilibrium on small scale while remaining in nonequilibrium on a larger scale. Under conditions of LTE, Gibbs fundamental equation of thermodynamics holds

$$TdS = dU + pdV - \sum_{j=1}^{k} \mu_j dn_j \tag{3.4}$$

with S entropy, T absolute temperature, V volume, U total energy, p pressure, μ_j chemical potential of the jth element, and n_j molar concentration of the jth element.

This equation is initially obtained for equilibrium states, and it is subsequently considered valid locally also for nonequilibrium systems. Note that this assumption has an important consequence: The function $S(U,V,n_i)$ that appears in Equation 3.4 is point by point the same S which has been defined for equilibrium systems in the previous paragraphs. Finally, we note that S is an extensive state function, and, thus, the total entropy of the system is the sum of the entropies of the macro subsystems.

3.5 Entropy Production

From Equations 2.3 and 3.4 and recalling that if we consider only mechanical work is $w = pdV$, we obtain for the entropy change

$$dS = \frac{dq}{T} - \frac{1}{T}\sum_{j=1}^{k} \mu_j dn_j$$

In general, any extensive quantity can vary for two broad categories of causation: variations due to internal processes and changes due to external ones. For example, we have seen in Equation 3.1 that the entropy change can be written as $dS = d_iS + d_eS$. Similarly, the k variables n_j (number of moles of the jth element) appearing in Equation 3.4 are extensive quantities whose variations can be separated into two contributions:

$$dn_j = d_i n_j + d_e n_j; \qquad 1 \le j \le k$$

$d_e n_j$ represents the change due to flows of matter through the surface of the system, whereas $d_i n_j$ represents the variation of matter due to internal processes such as chemical reactions or phase changes.

As we did for dn_j, the variation of heat dq can also be decomposed into two contributions:

$$dq = d_i q + d_e q$$

The variation in time of the entropy is

$$\frac{dS}{dt} = \frac{1}{T}\frac{dq}{dt} - \frac{1}{T}\sum_{j=1}^{k}\mu_j \frac{dn_j}{dt}$$

and separating internal and external contributions, we may write

$$\frac{dS}{dt} = \frac{d_iS}{dt} + \frac{d_eS}{dt} = \frac{1}{T}\frac{d_iq}{dt} - \frac{1}{T}\sum_{j=1}^{k}\mu_j \frac{d_in_j}{dt} + \frac{1}{T}\frac{d_eq}{dt} - \frac{1}{T}\sum_{j=1}^{k}\mu_j \frac{d_en_j}{dt}$$

The entropy production p is defined as

$$p = \frac{d_iS}{dt} = \frac{1}{T}\frac{d_iq}{dt} - \frac{1}{T}\sum_{j=1}^{k}\mu_j \frac{d_in_j}{dt} \tag{3.5}$$

According to what has been stated in point 5 of Section 3.2 and under LTE hypothesis, the total entropy production is given by the sum of the entropy production of the single irreversible processes. Definition 3.5 can be generalized. In fact, we observe that each of the terms appearing in Equation 3.5 is formally the product of a state function and a flow. In a very general way, we can, therefore, redefine the entropy production of a thermodynamic system with a form

$$p = \frac{d_iS}{dt} = \sum_k X_k J_k \tag{3.6}$$

where X_k and J_k denote, respectively, the generalized forces and generalized flows related to the considered irreversible processes. It is shown that the value of p does not depend on the choice of X_k and J_k, and ultimately, it depends on the choice of the potentials. From the second principle of thermodynamics, it immediately descends that for irreversible processes,

$$p = \frac{d_iS}{dt} = \sum_k X_k J_k > 0 \tag{3.7}$$

Relation 3.7 says that the entropy production p of an open thermodynamic system in which irreversible transformations take place is always positive. For a system at equilibrium instead, $d_iS = 0$ and, consequently, $p = 0$. This result is a direct consequence of the second law and is of general validity.

Equation 3.7 has an important consequence: There should be some causal relation, a coupling, between generalized forces and flows of the type

$$J_k = J_k(X_1, X_2, \ldots, X_n) \tag{3.8}$$

This is clear when we consider a thermodynamic system in which a single irreversible process occurs. Equation 3.7 for such a system becomes

$$p = X \cdot J > 0 \tag{3.9}$$

This relationship obviously imposes a constraint on the signs of X and J. The second law of thermodynamics says that forces and fluxes should have equal signs. For example, if we think of a system consisting of two compartments that can only interact with each other by exchanging heat, we can write

$$X = \left(\frac{1}{T_1} - \frac{1}{T_2}\right); \quad J = \frac{\mathrm{d}_i q_1}{\mathrm{d}t}$$

From Equation 3.9, we will have only two possibilities:

$$T_1 < T_2, J > 0 \qquad \text{or} \qquad T_1 > T_2, J < 0.$$

Heat always flows from the body at a higher temperature to the coldest one. As long as a difference or gradient in temperature exists, there will be a flux $J \neq 0$. Temperature difference is progressively reduced, and the system will move to a state of equilibrium with $X = 0$ and $J = 0$. At equilibrium, entropy production is null, and we conclude that the thermodynamic equilibrium is a state of absolute minimum for p. The existence of a gradient is necessary for the flow to persist. Is henceforth crucial the action on the system from outside by imposing external constraints or forcings.

We may define a *steady state* as a condition in which some state variables are time independent. An equilibrium state is trivially a steady state. The existence of external constraints is a necessary condition for the existence of nonequilibrium stationary states. In the condition of the nonequilibrium steady state, the following equations hold:

$$p = \frac{\mathrm{d}_i S}{\mathrm{d}t} > 0, \qquad \frac{\mathrm{d}_e S}{\mathrm{d}t} + \frac{\mathrm{d}_i S}{\mathrm{d}t} = 0$$

From these, it necessarily follows that

$$\frac{\mathrm{d}_e S}{\mathrm{d}t} < 0$$

This condition may be interpreted by saying that for nonequilibrium stationary states, the entropy of matter/energy entering the system should be minor of that released by the system to the outside. Along with Prigogine, we can say that, from the thermodynamic point of view, open systems degrade the material that comes in and it is this degradation that maintains the steady state (Prigogine, 1967).

For a system that is subject to external constraints from Equation 3.7 and the assumption of linearity of the relation expressed by Equation 3.8, it can be shown that

$$\frac{\mathrm{d}p}{\mathrm{d}t} \leq 0 \tag{3.10}$$

This inequality follows directly from the second law of thermodynamics. It says that the entropy production of a system, under the linearity condition, decreases over time. The linearity condition is demonstrated by Onsager (1931). From Equation 3.7, for steady-state systems that are subject to external constraints, as long as constraints persist, $p > 0$. This fact as well as Equation 3.10 leads to the conclusion that for constrained systems, the stationary states are local minima of the entropy production p. In summary, we will have $\mathrm{d}p/\mathrm{d}t < 0$ far from the steady state, and $\mathrm{d}p/\mathrm{d}t = 0$ at the steady state. The following theorem can be proved: A thermodynamic system under linear regime and subject to outside constraints tends to evolve toward states of minimum entropy production. The previous statement is a first example of evolutional criterion.

The entropy production p, in the regime of irreversible linear processes, plays a similar role as thermodynamic potentials in the theory of equilibrium states. However, this decrease in entropy does not reflect the emergence of macroscopic order, as is observed while continuously increasing the constraints that drive the system away from equilibrium. The stability of the not-too-far-from-equilibrium states implies that, in a system which obeys linear laws, the spontaneous emergence of superior ordered structures is impossible.

All the results presented so far are based on the assumption of validity of the fundamental Gibbs Equation 3.4. This was first demonstrated for equilibrium states. The physical interpretation of this assumption is that under nonequilibrium conditions also, entropy depends on the same independent variables as in equilibrium states. This can happen in far-from-equilibrium situations as well. The assumption of linearity is rather more restrictive. Even in the range of validity of Gibbs equation, the relationship between generalized flows and forces may be nonlinear. Since the theorem of minimum entropy production is valid only for the linear regime, it is not very interesting for biological systems. In 1971, Prigogine and Glansdorff proposed a method to extend the results of the thermodynamics of stationary states to nonlinear situations (Glansdorff and Prigogine, 1971). Indeed, it can be shown that there is always a part of the variation of the entropy production in time that keeps a definite sign:

$$\frac{\mathrm{d}_X p}{\mathrm{d}t} \leq 0 \quad \text{or} \quad \frac{\mathrm{d}_X p}{\mathrm{d}t} \geq 0 \tag{3.11}$$

This part identifies a local extreme for that part of p that has been defined as in Equation 3.6 by some external constraints.

An in-depth analysis of this theory would lead us very far. However, the remarkable point is that in these developments, the theoretical basis for the definition of *evolutional criteria* might be found.

4

A Perspective from Information and Systems Theory

4.1 Introduction

In 1948, C. E. Shannon and W. Weaver (1963) initiated a new research trail that became successively known as *information theory*. The interest of these scientists essentially was in signals communication. However, research in this new field soon found application in many other fields of science. From the cybernetics of Shannon's precursors Wiener and von Neumann (Wiener, 1999), in which the concept of feedback is central, to the theory of Eigen's hyper cycles (Eigen, 1971), to the theory of *autopoiesis*,* or the "Santiago theory" of Maturana and Varela (1980), concepts such as information, network, cyclical processes, cognitive processes, feedback, autopoiesis, and self-organization entered with full rights into the heritage of technical terms shared by many areas of scientific research.

The key concept, as will be discussed later, is that of information, which, in its more appropriate definition (see Shannon and Weaver [1963] and Brillouin [1962]), possesses the same characteristics as thermodynamic entropy. The advantages of this approach are given by the generality of the definition of information, which, once fixed a scaling factor that determines the unit of measure, allows a comparison on the same basis of different types of energies, otherwise energies with different qualities. The transition from the quantitative point of view to the qualitative one is a significant epistemological change. Along with Bateson (Bateson and Bateson, 1987), one may say that "the question is quantitative while the answer is qualitative"; in other words, a certain amount of energy can be informed during the cognitive process and then gain a qualitative value.

All the definitions of information (as well as that of entropy in statistical mechanics) lead to a logarithmic function of the probability. Actually, this is fairly obvious once it is recognized that, in order to have an information flow, significant differences should exist. The term *significant* here refers to the scale.

* Literally meaning self-production.

It is a well-known fact that organs of sense display a logarithmic response, enabling them to perform their functions on a large range of intensity of the input stimuli. Probability is, therefore, an exponential function of the information. The exponential is strictly related to the properties of scale invariance that appear whenever there is any conserved observable quantity. As seen in Chapter 2, at the level of the biosphere, in addition to the usual conservation laws of physics, it may be convenient to introduce another law: process conservation or conservation of the activity (rate) of information exchange.

Parallel to the development of information theory, Ludwig von Bertalanffy, an Austrian biologist, was among the founders of a new interdisciplinary topic: the *general systems theory* (Bertalanffy, 1968). This theory provides descriptions of open systems with interacting components and is applicable to biology, cybernetics, and many other fields. Systems theory offers a conceptual bridge as well as a mathematical toolbox that is used for applying information theory to living systems. As stated in Chapter 1, biological systems are better described with a systemic approach or in a holistic perspective. Dealing with these systems is crucial to assign a specific hierarchical level to each relationship among parts. The identification of the hierarchy in the nested levels enables a correct establishment of the units, making it possible to describe the flows that are associated to relations in terms of information. It is straightforward, based on the theoretical aids discussed in Chapter 3, to derive evolutional criteria from these descriptions.

In systems theory, formally, three different equivalent and interchangeable mathematical models of a dynamic system are available: the input-state-output model (ISO), which displays the internal state of the system, the perturbations or forcing functions acting on it, that is, the inputs, and finally the output; the input–output model (auto regressive moving average [ARMA] model), which binds directly to the inputs (and its derivatives) the outputs (and its derivatives), hiding the state variables; and, finally for linear time-invariant systems (LTI), transfer function models obtained in the domain of Laplace, Fourier, or the Z transform. The ISO model, through the state, depicts more information and properties of the system and is directly obtained from the input–output system, expressing the state variables that, in general, may not be unique, being the target of the analysis that often dictates their selection. This is the choice when our description is to be deterministic, the so-called *white box* model. In non-fully deterministic cases instead, *gray or black box* models are used; the input–output ARMA model is directly obtained as a differential or integro-differential equation from the "balance equations" of the physical system in question (mechanical, thermodynamic, or biologic). For example, Bertalanffy mathematical model of organism growth, which was published in 1934 and is still in use these days (Bertalanffy, 1950), is of this type. In general, n linear differential equations of the first order are obtained from a linear input–output differential model of order n (LTI), and these may be expressed in a compact manner using the matrix formalism.

These two complementary views, "white" and "black," of the same system illustrate the following general principle: Systems are structured hierarchically. The widest abstraction pertains to the higher hierarchical levels, those encompassing a view of the whole in a holistic perspective, regardless of the details of the components or parts. Instead, at the lower hierarchical levels, a variety of interacting parts are observed, but without much hint on how they are organized to form a whole. As stated in Chapter 1, according to reductionism, the laws governing the parts determine or cause the behavior of the whole. This is referred to as *bottom up causation*: from the lowest level to the higher levels. In emergent systems, however, the laws governing the whole system also constrain or cause the behavior of the parts. This is called *top down causation*. Although each hierarchical level has its own laws, these laws are often similar. The same type of organization can be found in subsystems that belong to different hierarchical levels. For example, all open systems necessarily have a boundary, an input, an output, and a throughput function. The cells in our body as well as the body as a whole need food and energy. The material is different, but the function is the same, that is, allowing the cell or organism to grow and repair itself. Similar functions can be seen at the level of society, which also needs an input of energy and materials that is used for self-maintenance and growth. The basic assumption of systems theory is that there are universal organizational principles, be they physical, chemical, biological, or social, which hold for all systems. The reductionist worldview seeks universality by reducing everything to its material constituents. On the contrary, the systemic holistic worldview seeks universality by ignoring the concrete material out of which systems are made, so that their abstract organization comes into focus.

4.2 Information: Definition and Measure

Information theory has been fully developed since the forties of the twentieth century to provide answers to practical problems. Basic questions have originated from communications engineering problems: How is it possible to define the amount of information contained in a message? How can the amount of information transmitted by a system of telegraphic signals be measured? How can both these quantities be compared and then determine the efficiency of the coding mechanisms?

A new theory with a sound practical interest and a well-defined mathematical basis emerged from these discussions. Through this theory, which is based on probabilistic considerations, a close analogy between information and entropy is expressed. Thermodynamic entropy measures the lack of information with regard to a given physical system. In 1929, this fact was noticed by L. Szilard (Szilard, 1929) in an article that was the forerunner of information theory. During a laboratory experiment, the inevitable price to be paid is an increase in the entropy of the laboratory and, from the second principle, it follows that this increase should always be greater than the information gained.

The second principle appears, therefore, in this context as a new restriction on the experimental procedure of the type of the Heisenberg's uncertainty principle, but of a completely different nature.

In recognizing the equivalence of the expressions "distinguishable from the environment" and "not in equilibrium," Shannon established a connection between his definition of information and that of thermodynamic information (negative entropy):

$$I = S_0 - S \tag{4.1}$$

where S_0 is the entropy of the system at thermodynamic equilibrium, in which the system is diluted and indistinguishable from the surrounding environment (the reference state or heat death [Shieh and Fan, 1982]). When I is defined in this manner, it is, thus, a measure of the distance from the reference state.

Science begins when the meaning of terms is clearly delimited. The first requirement for an information theory is a precise definition of the term *information*. This definition is obtained from statistical considerations.

Knowledge can be represented as a probability distribution. For example, if we formulate a question, complete knowledge occurs in a situation when it is possible to assign probabilities $P = 0$ for all answers but one for which is $P = 1$. In 1948 (Shannon and Weaver, 1963), Shannon defined *information* as that which causes a change in the probability distribution. Consider, for example, a situation in which π_0 different events can take place, with the condition that all of these π_0 events are equally probable a priori. This is the initial situation, when no information is available on the particular system under consideration. If some information is acquired, we may be able to determine that only one of the π_0 events will be effectively realized. The greater the initial uncertainty, the greater will be the amount of information required to operate this selection. In summary, in the initial situation, $I_0 = 0$ with π_0 equiprobable events; in the final situation, $I_1 \neq 0$ with $\pi_1 = 1$; that is, there is only one possibility left.

The symbol I indicates the information and is defined as (Brillouin, 1962)

$$I_1 = K \ln \pi_0 \tag{4.2}$$

where K is a constant and ln is the natural logarithm with base e. In the case of equiprobable events, the information coincides with the uncertainty. This is not true anymore for nonequiprobable events, for which the probability distribution can vary over time. In this case, the information is indicated by the difference in uncertainty before and after the measurement. The use of the logarithm in Equation 4.2 is grounded on the fact that information should be additive. Let us, in fact, consider two independent problems with π_0^{I} and π_0^{II} solutions that are equally probable a priori.

Each solution of the first problem can be associated to each solution of the second one. The total number of initial cases is given by

$$\pi_0 = \pi_0^{\mathrm{I}} \cdot \pi_0^{\mathrm{II}}$$

from which

$$I_1 = K \ln\left(\pi_0^{\mathrm{I}} \cdot \pi_0^{\mathrm{II}}\right) = I_1^{\mathrm{I}} + I_1^{\mathrm{II}} \tag{4.3}$$

with

$$I_1^{\mathrm{I}} = K \ln \pi_0^{\mathrm{I}} \text{ and } I_1^{\mathrm{II}} = K \ln \pi_0^{\mathrm{II}}$$

Therefore, the quantity of information needed to solve both problems is simply given by the sum of two separate contributions: I_1^{I} and I_1^{II}.

Let us consider a generic signal that is constructed using an alphabet of M different symbols. It is easily understandable that in a meaningful sentence, different symbols are present with different probabilities. Think, for example, of an English sentence in which the letter "z" appears much more seldom than the letter "e."

Therefore, we assume that the symbols have different a priori probabilities to appear: $P_1, P_2, \ldots, P_M$. The information content of a signal that consists of a sentence built with G symbols is calculated using the formula introduced by Shannon in 1948 (Shannon and Weaver, 1963):

$$I = G \cdot \zeta = -G\ K \sum_{j=1}^{M} P_j \ln P_j \tag{4.4}$$

where ζ is the average information per symbol and

$$\sum_{j=1}^{M} P_j = 1$$

We can immediately observe that each constraint or additional condition decreases the information content of the signal. In fact, when conditions that restrict the variability of symbols (of the probabilities P_j) are imposed, these conditions will eliminate the possibilities that were earlier acceptable. The average information per symbol is minimal when $P_j = 1$ and, is in this case, 0. Another way of explaining this general result is by considering that the constraints on the symbols represent a quantity of information in excess I_c. Then,

$$I' = I - I_c$$

is the quantity of useful information contained in the signal. This way of reasoning can be generalized. Given a question Q, if we denote the value of knowledge by X, Shannon's information is then defined as the difference between the uncertainty a priori and a posteriori of the cognitive process (Tribus and McIrvine, 1971):

$$I = S(Q|X) - S(Q|X') \tag{4.5}$$

where $S(Q|X)$ is the uncertainty on the question Q having the knowledge X and is given by

$$S(Q|X) = -K \sum_j P_j \ln P_j \tag{4.6}$$

This equation defines *Shannon's entropy*. Shannon's measure does not depend on the mechanisms by which information is generated. In communication theory, it is demonstrated that the merit of a channel does not depend on how well the current signal is transmitted, but on how well all the other possible messages may be transmitted. For example, an ohmmeter that indicates 1 kΩ when connected to a 1 kΩ resistance certainly does not provide any information if it continues to indicate the same value even when connected to different resistances.

Information has been defined as the result of a choice, however, totally ignoring its value. No distinction has been stressed between useful and useless information. In order to make a later choice, a certain amount of information could, in fact, serve as the basis for a prediction. In other words, information and knowledge are distinct, and a numerical measure is not provided for the latter. The definition presented here is based only on scarcity. If a possibility is remote, when this occurs, the information transmitted is large. The idea of "value" refers to the possible use of information by the living system. These considerations go well beyond the scope of information theory and represent a limit of the mathematical definition. Shannon's information is an absolute quantity that has the same numerical value regardless of the observer. From the definition presented, Shannon's uncertainty is always a positive quantity. On the other hand, the value of information is, and sometimes should be, negative. This value can be abstractly identified with the variation of knowledge of the observer that is caused by the information exchanged. For example, in the following sequence of statements:

1. My name is Erwin.
2. The previous sentence is false.

(1) It provides information with a positive value and (2), destroying the information conveyed by (1), has an information content with a negative value.

In Equation 4.5, the knowledge X about the question Q has been used to define the entropy S or the uncertainty about the answer. Similarly, we could use S to define X by saying that every X which maximizes $S(Q|X)$ is a state of maximum ignorance of the answers to Q. In this way, Shannon's formula can be used to quantitatively describe X; otherwise, X is a qualitative concept. From Equation 4.5, we understand that the information defined by Shannon is a measure of the variation of the uncertainty of the observer with regard to the question Q. A new message produces a new X, which, in turn, induces a change in the probability distribution P_j and then a new value of S.

4.3 Thermodynamic Entropy and Shannon's Information

It has been shown (Brillouin, 1962) that the only function that has the basic requirements to obtain a useful measure* of what is passing through communication channels is a function of the form 4.6. Shannon's entropy is a fundamental concept in information theory, and it is, therefore, a useful starting point for a discussion of information processes in general.

At first glance, there seems to be nothing in the definitions given, indicating that Shannon's entropy 4.6 and thermodynamic entropy (Section 3.2), which were introduced by Clausius in 1864, are actually the same function. Was Shannon itself that, as suggested by von Neumann, called his function entropy? The entropy of Clausius, although it has important practical implications, is a concept whose definition in physical terms turns out to be quite obscure and surrounded by an aura of mystery. The appearance of the definition of Shannon evinced considerable interest due to its formal simplicity and due to the fact that, different from thermodynamic entropy, it is relatively easy to calculate. Later, Leon Brillouin (1962), through the statistical definition provided by Boltzmann (1896), gave a first proof of the equivalence of the two functions and used them as if they were the same in his work.

Shannon based his definition of *entropy* on the simple logical criteria that should be met by such a function. In retrospect, it could be shown that based on the criteria used by Shannon, the only measure of information that can be obtained is coincident with thermodynamic entropy (Tribus and McIrvine, 1971). However, the formulation of Shannon is somewhat more general, as it is based entirely on a mathematical theory. Instead, *thermodynamic entropy* is defined with reference to real systems that consist of atoms, molecules, and energy. There is no conflict between Shannon's abstract information and thermodynamic entropy. The apparent difference between these two quantities is nonessential and lies in the fact that thermodynamic entropy, being associated with real systems, is numerically much more relevant than the information that humans can manage with the currently available technology.

* That is, a mathematical measure that satisfies the following properties: positivity, countable additivity, and null on the empty set.

The constant K that appears in Shannon's Equation 4.6 determines the unit of measurement of information. Usually, information is considered a dimensionless quantity, and K is a pure number. For practical applications, it is convenient to base the measurement system on binary digits (abbreviated, bits). Let us consider a problem with n independent choices, each of which corresponds to the binary selection of 0 or 1; the total number of possibilities is given by

$$P = 2^n$$

If I is identified with the number n of binary digits that are necessary for the determination of a choice, from Equation 4.2, we get

$$I = n = K \ln 2^n$$

and, thus,

$$K = \frac{1}{\ln 2} = \log_2 e = 1\text{bit} \tag{4.7}$$

It is possible to establish a relationship between bits and Joules. Consider, in fact, a thermodynamic system of N molecules, the natural choice, following the reasoning of Boltzmann (1896), it appears to be to put $K = N\,k_B$, where J/°K is Boltzmann constant. With this definition, the information is measured in J/°K. The conversion factor with bits is given by the relation

$$[1\text{bit}] = \frac{1}{Nk_B \ln 2}\left[\frac{\text{J}}{°\text{K}}\right] \tag{4.8}$$

Now, let us consider one mole of a given element in which two different isotopes are present. By complete mixing, a uniform blend is produced; the entropy change will be given by $N_a k_b \ln 2$, with $N_a = 6.02 \times 10^{23}$ Avogadro's number. This variation is numerically ~ 6 J/°K, or ~ 6×10^{23} bits, a value which corresponds to the number of choices that should be performed to separate the two isotopes by extracting them one by one from a uniform mixture.

4.4 Solar Energy and Information

The solar radiation on our planet provides a source of energy that keeps the ecosystems in states that are far from thermodynamic equilibrium. As implied by Equation 4.1, in this way, the sun appears to be a source of information and, thus, of useful energy. In considering the use that living systems make of this source of energy and information, the energy balance at the Earth's surface should be considered. As seen in Section 1.6, the solar energy incident on the Earth amounts to about 10^{22} kcal/year. Most of this energy is reradiated as blackbody radiation. The Earth's energy budget is, thus, approximately null. Living systems make use of the incident energy through internal transactions, by thermally degrading it at the average temperature of the planet's surface. Assuming the commonly accepted values for the temperature of the sun (5800°K) and of the earth (286°K), the energy reradiated to outer space ~ 10^{22} kcal/year, by using Equation 2.6, is easy to demonstrate as it carries with it the capability of decreasing the entropy of the planet with a negative entropy flow of -10^{23} J/°K per year. By virtue of Equation 4.8, this value corresponds to 10^{46} bits per year or more than 10^{38} bits per second (Tribus and McIrvine, 1971). At the level of the biosphere, energy and negative entropy (order) are accumulated in various subsystems (consider e.g., fossil fuels, clouds, organisms, and so on). These accumulations are mandatory in order to be sustained far-from-equilibrium stationary systems such as living ones. However, for practical purposes, these quantities of energy and negative entropy are negligible compared with the flows that originate from the sun. Considering, therefore, that on the Earth's surface the maximum velocity at which information can be processed is ~ 10^{38} bits per second, Tribus and McIrvine concluded that this is not a very strong limitation; such an information flow is largely able to sustain all biological activities, including human cognition.

From experimental observations comes the suggestion that energy flows are the principal factors that are responsible for generating hierarchies of structures that are typical of complex self-organizing processes (see Section 1.5). In the 1980s, H. T. Odum introduced a methodology for energetic analysis that is applicable to all complex systems, when properly generalized in terms of networks, energy, and information. He introduced a symbolic formalism for modeling living systems by using symbols similar to those used in electronics, in order to focus attention of energy flows. Observations on natural self-organizing systems show that the energy should be related with the systemic hierarchy that it competes with as well as with the information associated to it (Odum, 1983b). Odum introduces a change in perspective by striking that the quality of the energy involved in the game is an essential aspect. The Joules of energy that originate from processes that occur at different hierarchical levels are not directly comparable.

The *transformity* (T^R), a factor by which scaling the energy, is introduced. The transformity can be defined quite generally as one type of energy that is required to produce a unit of a different type of energy. In the case of processes that occur at the level of the biosphere, with solar energy being the primary source, *solar transformity* is defined as the quantity of solar energy that is required for the construction of 1 *Joule* equivalent of organic matter. Solar transformity varies from one for the incident solar radiation to 10^{13} for the categories of the information that are shared by the human community (Odum, 1988). In order to compare energy contributions that are provided by various sources on the same basis, all energies are expressed in terms of a new quantity, which is the embodied energy or *emergy* (Em). The solar emergy of a given energy flow or storage is given by the solar energy that is required to generate it. The emergy is measured in solar-em-Joules (seJ). The transformity is the emergy per unit energy and is expressed in seJ/J. Let E_i be the solar energy required to obtain a quantity of *i*th type energy; the emergy can be written as

$$\mathrm{Em}_i = T_i^{\mathrm{R}} \cdot \mathrm{E}_i \quad [\mathrm{seJ}] \tag{4.9}$$

and

$$T_i^{\mathrm{R}} = \frac{\mathrm{Em}}{\mathrm{E}_i} \left[\frac{\mathrm{seJ}}{\mathrm{J}}\right]$$

Sunlight is transformed into organic matter of plants by primary production, which, in turn, is converted into herbivore matter, and, finally, into carnivore matter. At each stage, energy is degraded as a necessary part of the process of transformation of low-quality energy into higher-quality energy but in lower quantities (Figure 4.1).

The transformity represents the quality factor for the energy. Note that this is not a dimensionless coefficient; rather, it plays a similar role as *K* in Equations 4.6 and 4.7. Therefore, T^{R} may be interpreted as a measure of information that accounts for the different energy qualities. See Table 4.1 for a list of transformities for different environmental products and flows.

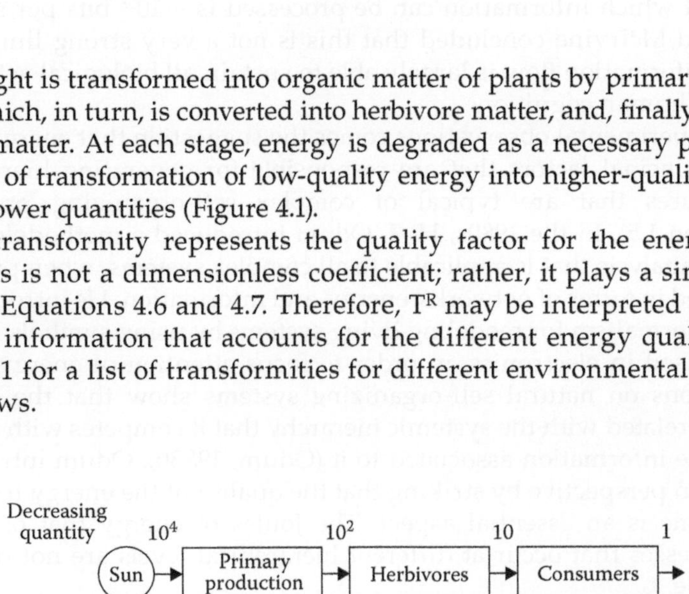

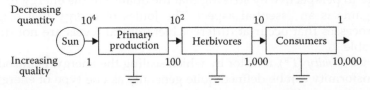

FIGURE 4.1
Solar energy transformation hierarchy. Transformities represent increasing quality. Further, solar energy is needed to obtain a smaller amount.

TABLE 4.1

Transformities Estimated by Odum (1996) for Environmental Accounting and Emergy Calculations

Transformities	sej/J (× 10^4)
Fuels	
Rainforest logs	3.2
Rainforest wood, transported and chipped	4.4
Liquid motor fuel	6.6
Ethanol	6.0
Crude oil	5.4
Natural gas	4.8
Coal	4.0
Peat	1.9
Lignite	3.7
Plantation pine	0.7
Charcoal	10.6
Electric power	
Coal power plant	16
World stream geopotential	9.4
Hydroelectric power	8–16
Wood power plant	6–20
Oil power plant	20
Lignite power plant	15–20
Global flows	
Global solar insolation	0.0001
Surface wind	0.00015
Convective earth heat	0.0006
Physical energy, rain on land	10.5
Chemical energy, rain on land	18.2
Tidal energy absorbed	16.8
Waves absorbed on shores	30.5
Continental earth cycle, heat flow	34.4
Oceanic upwelling, inorganic carbon	78
Oceanic upwelling, phosphate	3800
Plant and animal products	
Corn	8.3
Cotton	86
Butter	130
Caterpillar pupae	200
Mutton	200
Silk	340
Veal	400
Wool	440
Aquaculture shrimp	1300
Upper consumers	3000

Note: See Odum (1996) for the details on the estimations.

In order to generalize Equation 4.9 to a product whose creation and maintenance requires multiple energy flows J_i^E with diverse transformities, we may write of the emergy flow or *empower*:

$$\mathrm{Em} = \sum_i T_i^R \cdot J_i^E \quad \left[\frac{\mathrm{seJ}}{\mathrm{Time}}\right] \tag{4.10}$$

Illustration 4.4.1: Energy and Emergy Balance Calculation

Odum applied an ARMA systemic modeling approach to obtain a black box model by balancing energy flows. Consider, for example, a system in which two components, I and II, are sustained by two different energy sources s_1 and s_2 with diverse transformities (Figure 4.2).

In order to calculate the emergy as defined in Equation 4.10, the following rules should be applied:

1. Different coproducts of a given process possess the same emergy (the total emergy required for maintaining the father process).
2. If a flow of emergy is split, the emergy assigned to each branch is proportional to the quantity of energy flowing through it.
3. The only flows that should be considered in emergy calculus for a compartment are those that have originated by independents inputs (≠ feedbacks).

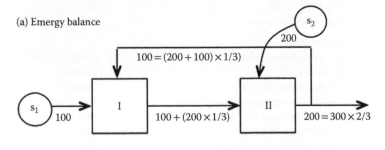

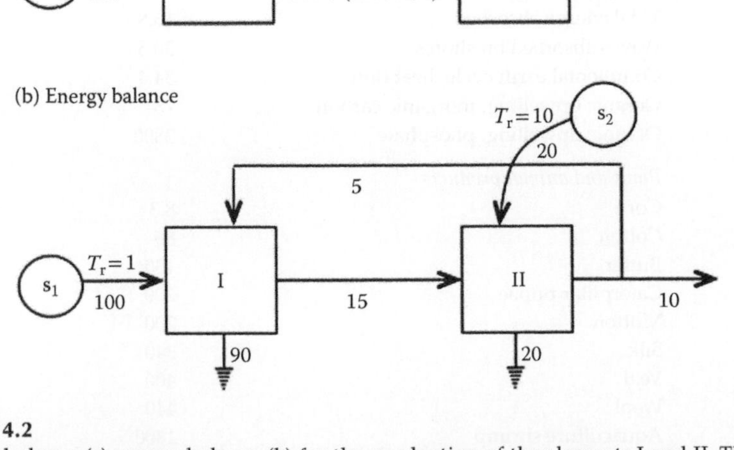

FIGURE 4.2
Emergy balance (a) energy balance (b) for the production of the elements I and II. The numbers near the arrows represent the flows; transformities of the sources s_1 and s_2 are also indicated.

By applying these rules to the example depicted earlier, we obtain

$$\begin{aligned} \mathrm{Em}_\mathrm{I} &= 100 + 200 \cdot \frac{1}{3} \approx 166 \\ E_\mathrm{I} &= 15 \\ Tr_\mathrm{I} &= 11 \end{aligned}$$

and

$$\begin{aligned} \mathrm{Em}_\mathrm{II} &= 15 \cdot \mathrm{Tr}_\mathrm{I} + 20 \cdot \mathrm{Tr}_{S2} \approx 366 \\ E_\mathrm{II} &= 15 \\ Tr_\mathrm{II} &\approx 24 \end{aligned}$$

The definition of *empower* (Equation 4.10) has striking analogies with that of *entropy production* (Equation 3.6). We may, therefore, apply Glansdorff–Prigogine criterion (Equation 3.11) and write

$$\eta = \frac{\mathrm{d}_{T^R} \mathrm{Em}}{\mathrm{d}t} \geq 0 \tag{4.11}$$

The efficiency in transformity η, other than that part of the variation in empower production per unit time due to energy transformations, keeps a definite sign.

Odum generalized Lotka's maximum power principle (Lotka, 1922) in terms of emergy. Along with Lotka, he observes that a characteristic of the functional structure of ecosystems is to self-maximize the energy flow (and, hence, energy dissipation or entropy production). However, when studying systems in which more than one type of energy is circulating, energy cannot be simply used as a measure of work. Higher-quality energy (in a lower quantity) should enter into a feedback mechanism with a control function, reinforcing (amplifying) the production process. According to Odum, the essence of self-organization is "the automatic reinforcement of available choices." Odum's hypothesis is that ecological systems are characterized by mathematically defined structures whose design is aimed at maximizing the empower (i.e., the rate of transformation of the available energy into higher-quality energy) for each possible hierarchical level. In formulating the principle of maximum empower, Odum hypothesized that the development of an ecosystem is made in a compromise between the maximization of the rate of energy dissipation (transformation of energy into work) and the maximization of efficiency in transformity (variation in empower per unit time). The result of this compromise is the maximization of the empower (Jørgensen, 1997). Figure 4.3 illustrates the maximum empower principle.

This principle, which may be formally expressed by Equation 4.11, represents an evolutional criterion according to which ecosystems tend toward local maxima of Em and p.

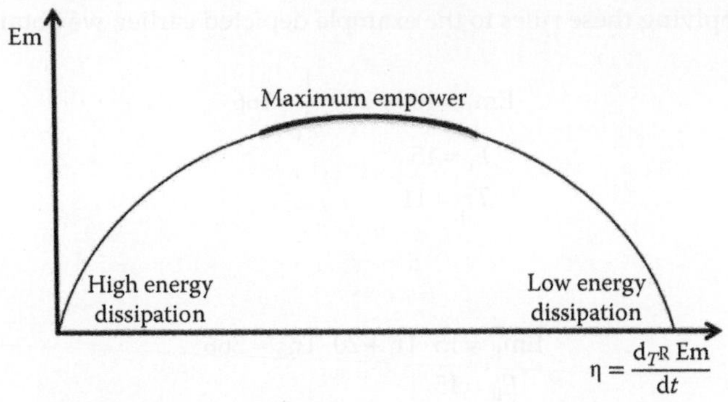

FIGURE 4.3
Maximum empower principle. The maximum emergy flow is realized when an optimal efficiency in transformity η rules energy dissipation.

4.5 Ecological Network Analysis

As seen in Chapter 3, Prigogine introduced a formalism for describing the interactions between the elements of nonequilibrium systems in terms of generalized forces and flows. However, in ecological systems, generalized forces (the causes) are not easily identifiable; whereas it is possible or relatively easy to obtain quantitative estimates of flows and storages. Thermodynamics provides a phenomenological description without inferring the causes. Hence, in this context, forces are of secondary importance. In the ecological perspective, while is difficult to describe the force that pushes an insect into the bird's beak, on the other end it is more immediate the description of flows of matter and energy. Although the observer is not allowed to know all the details of the processes involved in a complex system, some clues on the entity of flows are often helpful in assessing how well the system is working. Gross domestic product (GDP) is an example of a flow indicator for the system "nation" as a whole, and the heartbeat or body temperature is indicative of the biochemical flows in organisms.

Ecological network analysis is an LTI modeling approach in which ecosystems are described as networks of flows (arrows describing interactions) and storages (nodes or boxes referring to populations of taxa) (Ulanowicz, 2004). Once the network is drawn and the currency circulating in it (often referred to as the *medium,* typically energy or matter, e.g., C, N, P, Si, etc.) is defined, the observer measures storages and at least some predominant flows. The remaining unknown flows may then be estimated by balancing the network (Allesina and Bondavalli, 2003). The network of flows is a very general model that is applicable to all the transformation phenomena occurring in nature. Consider, for example, an ecological network that consists of three compartments (taxon or groups of taxa). Let J_{ik} be the flow originating from i and entering k; an example of such a network is depicted in Figure 4.4.

Once the network is balanced, it may be considered as being in a time-independent or stationary configuration. Therefore, some quantities that are useful for network analysis may be immediately defined:

$J = \sum_{i,k} J_{ik}$	(4.12)	*Total system throughput* or the sum of all that is flowing in the network gives an idea of how much medium is processed by the system and is, hence, related to the size.
$P_{i,k} = \frac{J_{ik}}{J}$	(4.13)	*Unconditional joint probability* or the probability of concomitant happening of one unit medium emission from i and entering k.
$P_{i\vert k} = \frac{J_{ik}}{\sum_q J_{iq}}$	(4.14)	*Conditional probability* or the probability that the emission is coming from i. This quantity estimates the probability that a unit of medium enters compartment k knowing a priori that originated in i.
$P_k^* = \frac{\sum_i J_{ik}}{J}$	(4.15)	*A priori probability* that one unit enters k.

The network model conceptualized with graphs such as the one depicted in Figure 4.4 can also be studied by using a set of various n × n matrices, where n is the number of compartments. The matrixes of flows J_{ik} as well as those of joint probabilities P_{ik} belong to this set of adjacency matrixes.* Indeed, using the linear algebra approach, two significant advantages are gained: (1) systems of arbitrary dimension n may be treated; (2) through the analysis tools provided by the matrix theory, a quite detailed picture of the network is obtained, for example, indirect effects, relative importance, and length of different pathways.

Since the birth of information theory, some ecologists have begun speculating on the applicability of this theory in ecology. MacArthur (1955), for example, proposed to apply the definition of Shannon (4.6) to obtain a qualitative test of Odum's flow optimization hypothesis:

$$H = -K \sum_{i,j} P_{i,j} \ln P_{i,j} \tag{4.16}$$

where H is an information measure of flows in the network, and is referred to as *diversity of flows*. Ulanowicz (1986) provided an inspiring description of ecological networks by translating matter/energy flows into flows of

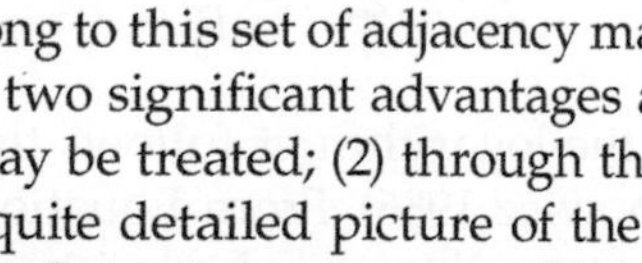

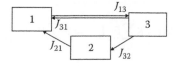

FIGURE 4.4
General conceptualization for a network of flows.

* A matrix whose elements are 1 or 0 depending on the existence or not of a relationship (flow) between the two related compartments. Such a matrix describes the topology of the network.

information. In general, one may say that information is a quantity that is carried by changes in real and observable physical quantities (e.g., matter or energy). These observables are the media through which information can be stored and propagated. Regardless of the medium, Equation 4.5 defines *information* as the difference between the a priori uncertainty on the demand $S(Q)$ and that subsequent to an observation $S'(Q)$. Equation 4.6 relates the uncertainty with a probability distribution. If we denote the a priori and a posteriori probability distributions of the outcomes of an observation by $\{P^*_k\}$ and $\{P_k\}$, respectively, based on Equation 4.6, we may derive the a posteriori and a priori average uncertainties as (see in Shannon and Weaver [1963])

$$S'(Q \mid X) = -K\sum_j P_j \ln P_j \text{ and}$$
$$S(Q \mid X) = -K\sum_j P_j \ln P_j^*$$

Then, from Equation 4.5, we obtain the average information gained after the observation:

$$I = K\sum_j P_j \ln \frac{P_j}{P_j^*} \tag{4.17}$$

From the concavity of the logarithm, it follows that the sum in Equation 4.17 is nonnegative (Ulanowicz 1986). From Equations 4.14 and 4.15, having the certitude that one unit medium has been issued by i, the variation of uncertainty associated with the flow $i \to j$, that is, the information exchanged through the medium, is given by

$$I_{i\to j} = S_{i|j} - S_j = K \ln P_{i|j} - K \ln P_j^* = K \ln \frac{P_{i|j}}{P_j^*}$$

The quantities $I_{i\to j}$ are not always positive for any possible choice of the pair i,j. However, if we sum over the indices by weighing the information flows exchanged with the probability that emission and input occur concurrently (joint probability, $P_{i,j}$) a nonnegative quantity is obtained, which is the average mutual information:

$$\text{AMI} = K\sum_{i,j} P_{i,j} I_{i\to j} = K\sum_{i,j} P_{i,j} \ln \frac{P_{i|j}}{P_j^*} \tag{4.18}$$

AMI, being a measure of the synergy between the structure and function of the flow network, is an indicator of the level of organization. It is easily demonstrated that H of Equation 4.16 can be rewritten as (Ulanowicz, 2004)

$$H = \text{AMI} + H_c \tag{4.19}$$

where

$$H_c = -K\sum_{i,j} P_{i,j} \ln \frac{P_{i|j} P_{i,j}}{P_j^*} \tag{4.20}$$

is the conditional or residual diversity, a measure of freedom (disorder) that remains available to the machinery constituting the network.

Total system throughput J of a system characterizes the size of the system, without saying anything about the structure of the network, that is, the level of organization. On the other hand, the AMI quantifies the organization of the network of flows, but it is not significantly sensitive to the size of the system. Constant K in Equations 4.16 through 4.19, as seen in Section 4.3 by convention, defines the units of information (the scale). For example, setting 2 as the base of the logarithm, a unit of K corresponds to a bit. In information theory, the units of information are often defined by specifying the base of the logarithm and assuming $K = 1$. Tribus and McIrvine (1971) suggest that the role of K is to give physical meaning to the quantity of which is the scale factor. Since total system throughput is a feature of the size (scale) of the network, Ulanowicz (1986) follows this suggestion, stating that it is appropriate to assume $K = J$.

$$K = J = \sum_{i,j} J_{i,j}$$

With such scaling by total system throughput, Equation 4.19 is rewritten as

$$C = A + \phi \tag{4.21}$$

with

$$C = -J \sum_{i,j} P_{i,j} \ln P_{i,j} \quad \text{Network capacity}$$

$$A = J\sum_{i,j} P_{i,j} \ln \frac{P_{i|j}}{P_j^*} \quad \text{The ascendency}$$

$$\phi = -J\sum_{i,j} P_{i,j} \ln \frac{P_{i|j} P_{i,j}}{P_j^*} \quad \text{The overhead}$$

Ulanowicz (2004) proposes the following evolutional criterion: *Ecosystems show a propensity to increase in ascendency*. According to Ulanowicz, this new quantity includes both aspects: size and organization of the network. The dimension (extensive) is represented by the term J and the number n of compartments, whereas the network structure or the degree of organization (intensive) is represented by the summation. Organization is measured by AMI, whereas the factor J calibrates the level of activity of the global system. An increase in system activity J can be interpreted by saying that the system is growing in the economic sense of the term. An increase in AMI means that the system is developing new constraints that channel the flow in more specific and efficient pathways. Thus, an increase in the quantity A is a sign that the system in analysis is undergoing growth and development (Ulanowicz and Abarca-Arenas, 1997). In the past, many of the ecological processes that contribute to an increase in the ascendency had been identified as favorable factors for evolution. For example, a greater number of species (compartments), a higher retention of resources, and the trend toward specialization of the food chain are among the primary signs detected by Odum (1983), showing that the ecosystem is developing.

The approach that is properly synthetized by Equation 3.11 may be applied to ascendency as well. The propensity to maximization of the ascendency and then an increase in organization is a consequence of the second principle applied to far-from-equilibrium nonlinear systems. Thanks to network conceptualization, a potential, the ascendency, is found, and its maximization as caused by some perturbations or variations is formally endured by Equation 3.11.

It is instructive to briefly analyze the domains of C, A, and ϕ. For clarity, let us assume that $J = 1$. Being C, A, ϕ nonnegative quantities from Equation 4.21, both $0 \leq A \leq C$, and $0 \leq \phi \leq C$. For a network of N compartments in which everything is connected with everything else by equal flows, $A = 0$ and $C = \phi = 2\log_2 N$ (point b in Figure 4.5).

- Systems in "o" are purely theoretical: The compartments are available, but no flows connect them. In this case, $A = \phi = C = 0$.
- If all compartments are connected in a closed chain by equal flows, $A = \phi = C/2 = \log_2 N$ (point "a").

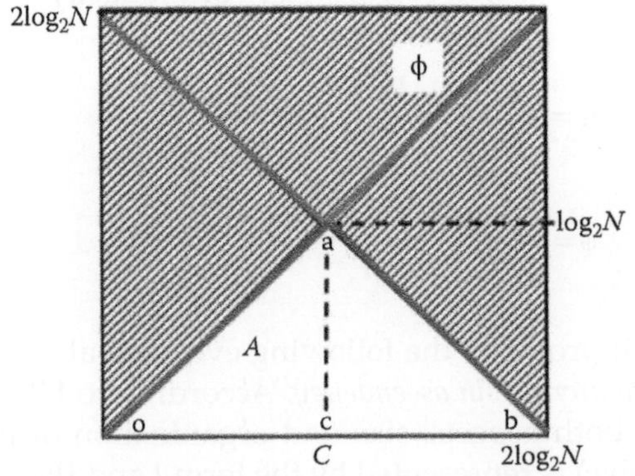

FIGURE 4.5
(See color insert.) The domain of AMI. A and ϕ are plotted against C. (o) no connection, (a) linear closed loop, (b) everything connects with everything, and (c) star network where one single compartment feeds all the other compartments.

- If all compartments are connected in a closed chain by arbitrary flows, $C < \log_2 N$, $A < \log_2 N$, $\phi < 2\log_2 N$ (proximity of segment oa).
- If only one compartment feeds all the others by equal flows, $A = 0$, $\phi = C = \log_2 N$ (point c).

Possible networks are those for which the point (*C*,*A*) falls in the area of the white triangle. Actually, based on Equations 4.12 through 4.15 and Equation 4.21, it is possible to demonstrate that

$$0 \leq A \leq \frac{C}{2} \leq \phi \leq C \leq 2\log_2 N \tag{4.22}$$

Networks falling in the neighborhoods of point a have an ascendency that is comparable to the overhead. According to Ulanowicz (2004), these are overdetermined systems that have reduced freedom to evolve, harmonize, or change their structure. If *C* grows further (going toward hyper-connected systems), ϕ starts increasing again, whereas *A* begins decreasing. In practice, for real ecosystems in which $N \gtrsim 20$ and the flow matrices are sparse, *C* are small and the overheads are quite high compared with ascendency values. By and large, living systems appear in the lower left part of the domain of ascendency. It seems as if Ulanowicz' evolutional criterion of growing ascendency is in reality the result of a tradeoff between the maximization of ascendency and the minimization of overhead.

4.6 Exergy

Similar to the analysis carried out in Section 4.4, another way of taking energy qualities into account is by using *exergy*, which is a thermodynamic concept. The notion of exergy has attracted increasing interest in the general framework of theoretical ecology. In particular, this quantity seems to be a good candidate as a goal function (orientor) that is useful to define an evolutional criterion for ecological models. Exergy is a thermodynamic function that is classically derived from the concept of Helmoltz free energy. It is a measure of the useful energy that can be supplied or of the work that may be performed by a far-from-equilibrium system during the process that brings the system into equilibrium with its surroundings. In the context of ecosystems theory, this function is employed as an estimator of the distance from the equilibrium and is often interpreted as a measure of the complexity of the system. At equilibrium, the exergy is zero.

Note that, in spite of being an extensive quantity, exergy is, however, not properly a state function of the system; in fact, its very definition hangs on external conditions. Even if exergy of an entire ecosystem cannot be calculated, some authors purpose to compute an exergy index that can also serve as a relative indicator. Apart from the difficulties related to the high complexity of ecosystems, which entangle numerical evaluation, the classical definition of the thermodynamic quantity *exergy* is indeed physically rigorous. In the context of ecosystems theory, the thermodynamic approach has frequently been applied in order to study the processes of growth and development. This approach has led to the definition of several thermodynamic functions, acting as indicators, orientors, or goal functions for models of ecosystems. The physical interpretation of these thermodynamic goal functions is based on the classical thermodynamics evolutionary principle (Schneider and Kay, 1994), that is, the second principle or the law of entropy as described in Chapter 3.

Several attempts have been made in order to frame the role of thermodynamic indicators in the context of a coherent ecosystem theory. Jørgensen and Fath (2004) propose the construction of a consistent theoretical framework that relies on eight physically based observational principles. The authors show how several ecological observations can be explained based on these principles. A similar systematization effort is presented in Bendoricchio and Palmeri (2005), where a cost/benefit approach is used to characterize ecosystems indicators from both the thermodynamics and network perspectives.

Of the indicators proposed in the literature, emergy (introduced in Section 4.4) and exergy are those having probably the most reliable thermodynamic theoretical definition. The latter was first applied to ecosystem studies by Jørgensen and Meyer (1977). Exergy is a generalization of the classical function free energy, and it is obtained by including an arbitrary reference state in its definition. Both these indicators are often assumed to measure different aspects of the complexity related to ecosystems. However, system evolutionary principles derived from classical thermodynamics (e.g., the exergy-storage hypothesis of Jørgensen and Meyer, 1977) are firmly established only for a system's states that are not too far from equilibrium. In the far-from-equilibrium domain (that in which all living systems operate), the classical law of entropy does not hold (Fath et al., 2001). This fact, along with the huge complexity of ecosystems and the enormous number of elements involved, makes problematic the practical application of these indicators. Scientists are compelled to accept very rough estimations, in addition, based on non-fully verifiable assumptions. However, one can always turn to the resorts provided in Section 3.5 for a generalization of the law of entropy, thus allowing the definition of some evolutional criterion.

Evans (1969) introduces a new definition of exergy based on thermodynamic information (4.1):

$$\mathrm{Ex} = T_0 \cdot I \tag{4.23}$$

where T_0 is the temperature of the environment, and I is the thermodynamic information defined by Evans as

$$I = S_0 - S = \frac{U + p_0 V - T_0 S - \sum_i \mu_i^0 n_i}{T_0} \tag{4.24}$$

with U as the system's internal energy, p_0 as the pressure of the environment, V as the system's volume, μ_i^0 as the chemical potential of ith element in the surroundings, and n_i as the number of moles of the ith element in the system. Evans proves that the exergy defined by Equations 4.23 and 4.24 is fully consistent with that of thermodynamics.

Thermodynamic information as defined by Equation 4.24 may be interpreted as a signal-noise ratio for a system in an environment at temperature T_0. In this view, Ex represents the signal intensity (energy) that is needed to sustain the transmission of information I against the environmental temperature T_0. At higher temperatures, more energy is required to exchange the same I. Moreover, at a fixed temperature, different energy quantities are employed to sustain various communication channels. In this way, exergy is seen as the cost in energy of the process of information exchange (see Table 4.2).

The exergy introduced by Evans (Equation 4.23) turns out to be a convenient tool in ecological energy analysis. Energy is neither created nor destroyed, and it is indeed conserved. Consequently, energy cycles, being almost balanced, do not give many clues about a system's dynamics. On one hand, exergy changes enable a comparison of different flows of energy and

TABLE 4.2

Cost in Energy of the Information Exchange Processes

	Energy (J)	Information (bits)	Energy/ Information
Characters			
Typed page	30,000	21,000	1.4
Facsimile	20,000	21,000	1
Reading a page (lighting)	5400	21,000	0.3
Photocopy	1500	21,000	0.07
Digital			
Transmission of data (3000 characters)	14,000	21,000	0.7
Reading a page of computer output	13,000	50,400	0.3
Sorting data from a binary file (3000 characters)	2000	31,000	0.06
Printing a page of computer output (60 lines × 120 characters)	1500	50,400	0.03
Audio			
Phone conversation (1 minute)	2400	288,000	0.008
hifi sound reproduction (1 minute)	3000	2,400,000	0.001
AM radio broadcast (1 minute)	600	1,200,000	0.0005
Video			
Projection of a frame 35 mm (1 minute)	30,000	2,000,000	0.02
TV transmission of a frame (1/30 second)	6	300,000	0.00002

Source: Tribus, M. and McIrvine, E.C., *Sci. Am.*, 225, 179–188, 1971.

information, highlighting advantages and inefficiencies. The manipulation of information requires that a certain amount of energy, the exergy, be used to support the constraints sustaining information flows. On the other hand, the use of energy that is anyhow limited by thermodynamic availability requires an exchange (flow) of information.

It should be noted that exergy is not conserved except in the case of processes in which energy is transferred without entropy production (reversible processes). In natural irreversible processes, exergy is dissipated, and there is a net production of entropy. Exergy seems to be more readily usable than the entropy that is used to describe the irreversibility of natural processes. It is difficult to associate the definition of entropy with our descriptions of nature, whereas exergy, being an energetic measure, can be more easily estimated for real systems. By applying the concept of exergy to ecosystems, it might be argued that if a system is removed from the state of thermodynamic equilibrium through the application of a flow of exergy, it will use all available means to build dissipative structures (storage of exergy, i.e., information) in order to reduce the effects of the imposed gradient (Jørgensen, 1997). This is clearly another example of an evolutional criterion. This last criterion is established by applying an ARMA modeling approach (balancing energy flows) and introducing the exergy (Equation 4.23). Glandsdorff–Prigogine criterion (Equation 3.11), thus, enables the application of the generalized law of entropy to the exergy. Some part of variation of the exergy production maintains a definite sign, so that maximization of exergy production is attained.

Table 4.2 contains a catalog of information content and energy that are required by some information transmission activities. From these estimates, it is possible to compare the flows of energy and information confronting the variations of exergy as energy per unit information. For example, the minor quantities of energy per unit information are those of broadcast TV. However, the information transmitted in this case corresponds to an exergy variation of only 2×10^{-5} J/bit. So, in terms of energy consumption, the aspects related to information processing are negligible. Most of the energy is dissipated into an increase in temperature of the apparatus. Likewise, looking at the other activities listed leads to the conclusion that these activities are actually mainly of energetic type, that is, in which only small quantities of information are involved. At the level of the biosphere, however, the flow of energy of ~ 4.2×10^{25} J/year is accompanied by an information flow that corresponds to a negative change in exergy of about ~ 4.2×10^{-23} J/bit. This information is used to harness energy in complex biological structures in the biosphere.

Applying Equations 4.23 and 4.24 to an ecosystem formed by N components (taxa or bio-geochemical species) with an input of inorganic matter and an output of dead organic matter (detritus), Jørgensen and Meyer (1977) demonstrate that the exergy density may be written as

$$\mathrm{Ex} = R \cdot T \sum_{i=0}^{N} \left[c_i \cdot \ln\left(\frac{c_i}{c_i^{\mathrm{eq}}}\right) - (c_i - c_i^{\mathrm{eq}}) \right] \qquad (4.25)$$

where $R = 8.31$ J/mole·°K is a gas constant; c_i is the concentration of the ith element in the ecosystem; c_i^{eq} is the concentration of the same element at chemical equilibrium (the state of heat death); and quantities with $i = 0$ and $i = 1$, respectively, refer to inorganic matter and detritus.

Further, noting from Equation 3.4, we may write the internal energy of the system as

$$U = TS - pV + \sum_i \mu_i n_i$$

and substituting this in Equation 4.24, we obtain the following expression for the exergy:

$$\mathrm{Ex} = T_0 \cdot I = (p_0 - p)V - (T_0 - T)S + \sum_i (\mu_i - \mu_i^0) n_i \tag{4.26}$$

Jørgensen (2006) notes that if T and p of the system, as is likely in normal conditions, do not differ substantially from those of the surrounding environment, Equation 4.26 becomes

$$\mathrm{Ex} \approx \sum_i (\mu_i - \mu_i^0) n_i$$

and, hence,

$$\frac{\mathrm{dEx}}{dt} \approx \sum_i (\mu_i - \mu_i^0) \frac{dn_i}{dt} = \sum_i X_i J_i \tag{4.27}$$

The exergy production is formally identical to Equation 3.6. Glansdorff–Prigogine criterion says that exergy should display steady behavior.

Illustration 4.6.1: Exergy as a Measure of Useful Work—Example Calculations

Exergy is defined as the amount of work (= entropy-free energy) that a system can perform when it is brought into thermodynamic equilibrium with its environment (Jørgensen et al., 1999). Figure 4.6 illustrates the definition of *exergy* as it is used in technology; for instance, to find the work capacity of a power plant. The considered system is characterized by the extensive state variables S, U, V, $N1$, $N2$, $N3$......., where S is the entropy; U is the energy; V is the volume; and $N1$, $N2$, $N3$..... are moles of various chemical compounds, and by the intensive state variables, T, p, μ_{c1}, μ_{c2}, μ_{c3}.... The system is coupled to a reservoir, a reference state, by a shaft. The system and the reservoir form a closed system. The reservoir (the environment) is characterized by the intensive state variables T_o, p_o, μ_{c1o}, μc_{2o}, μ_{c3o}.... and since the system is small compared with the reservoir, the intensive state variables of the reservoir will not be changed by interactions between the system and the reservoir. The system develops toward thermodynamic equilibrium with the reservoir

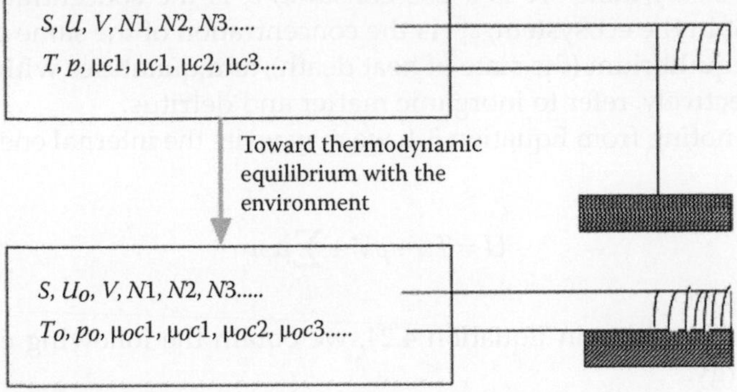

FIGURE 4.6
Definition of technological exergy is shown. The experimental setting corresponds to the conceptual reversal of the setting of Joule's experiment (Figure 2.1) in which mechanical work is converted into heat.

and is simultaneously able to release entropy-free energy to the reservoir. During this process, the volume of the system is constant, as the entropy-free energy should be transferred through the shaft only.

The entropy is also constant, as the process is an entropy-free energy transfer from the system to the reservoir, but the intensive state variables of the system become equal to the values for the reservoir. The total transfer of entropy-free energy in this case is the exergy of the system. From this definition, it is observed that exergy is dependent on the state of the total system (= system + reservoir) and not entirely dependent on the state of the system. Exergy, therefore, is not a state variable. In accordance with the first law of thermodynamics, the increase of energy in the reservoir, ΔU, = $U - U_0$, where U_0 is the energy content of the system after the transfer of work to the reservoir has taken place. According to the definition of exergy, Ex, we have

$$\text{Ex} = \Delta U = U - U_0$$

Since

$$U = TS - pV + \sum{}_c \mu_C N_C \tag{4.28}$$

(when we only consider heat, spatial energy/work, and chemical energy; see any textbook in thermodynamics and Chapter 2), and correspondingly for U_0,

$$U_0 = T_0 S - p_0 V + \sum{}_c \mu_{CO} N_C \tag{4.29}$$

we get the following expression for exergy, excluding, of course, in this case kinetic energy, potential energy, electrical energy, radiation energy, and magnetic energy:

$$\text{Ex} = S(T - T_0) - V(p - p_0) + \sum{}_c (\mu_C - \mu_{CO}) N_C \tag{4.30}$$

These energy forms could, however, be easily included. The equation just cited also emphasizes that exergy is dependent on the state of the environment (the reservoir = the reference state), as the exergy of the system is dependent on the intensive state variables of the reservoir. Exergy is conserved only if entropy-free energy is transferred, which would imply that the transfer is reversible. In reality, all processes are, however, irreversible, which means that exergy is lost and entropy is produced (see the discussion in Chapter 3).

Energy is, of course, conserved by all processes according to the first law of thermodynamics. It is, therefore, wrong to discuss an energy efficiency of an energy transfer, because it will always be 100%; whereas the exergy efficiency is of interest, because it will express the ratio of useful energy (it means work energy) to total energy, which is always less than 100% for real processes. All transfers of energy imply that exergy is lost, because energy is transformed into heat at the temperature of the environment.

It is, therefore, of interest to set up an exergy balance, in addition to an energy balance, for all environmental systems. Our concern is loss of exergy, because that means loss of work capacity or "first class energy" which can do work is lost as "second class energy" (heat at the temperature of the environment) that cannot do work. So, the particular properties of heat, including that temperature, serve as a measure of the movement of molecules, and give limitations in our possibilities to utilize this energy form to do work. The total amount of heat energy is entropy times the absolute temperature, but the amount of heat energy that we can utilize to do work is only the high temperature that we may have provided, for instance, in steam generators minus the temperature of the environment times the available entropy = $S\ (T_g - T_{env})$. Due to these limitations, we have to distinguish between exergy that can do work and anergy, which cannot do work, and all real processes imply inevitably a loss of exergy as anergy (see also Chapter 3 for details with regard to the second law of thermodynamics).

Exergy seems more useful to apply than entropy to describe the irreversibility of real processes, as it has the same unit as energy and is an energy form; whereas it is more difficult to relate the definition of *entropy* to concepts that are associated to our usual description of reality. In addition, entropy is not clearly defined for "far from thermodynamic equilibrium systems," particularly for living systems; see, for instance, Tiezzi (2003). Moreover, it should be mentioned that the self-organizing abilities of systems are strongly dependent on the temperature, as discussed in Jørgensen et al. (1999). Exergy takes the temperature into consideration as shown in the definition; whereas entropy does not do so. This implies that exergy at 0 K is 0 and at minimum. Negative entropy is sometimes used, but it does not express the ability of the system to do work (or we may call it the *creativity* of the system, as creativity requires work). Exergy becomes a good measure of the creativity, which increases proportional with the temperature. In classical thermodynamics, entropy cannot be negative. Furthermore, exergy facilitates the understanding of the difference between low entropy energy and high entropy energy, as exergy is entropy-free energy.

Finally, information content mirrors exergy = energy that can do work. Boltzmann (1905) showed that the free energy of the information that we actually possess (in contrast to the information that we need to describe the system) is $k \cdot T \cdot \ln I$, where I is the information we have about the state of the system; for instance, that the configuration is 1 out of W possible (i.e., that $W = I$) and k is Boltzmann's constant = 1.3803×10^{-23} (J/molecules × deg).

It implies that one bit of information has exergy that is equal to $k\,T \ln 2$. The transformation of information from one system to another is often almost an entropy-free energy transfer. If the two systems have different temperatures, the entropy lost by one system is not equal to the entropy gained by the other system; whereas the exergy lost by the first system is equal to the exergy transferred and equal to the exergy gained by the other system, provided that the transformation is not accompanied by any loss of exergy, which is always the case in reality. In this case, it is obviously more convenient to apply exergy than entropy.

Exergy of the system measures the contrast—it is the difference in free energy if there is no difference in pressure and temperature, as may be assumed for an ecosystem or an environmental system and its environment—against the surrounding environment. If the system is in equilibrium with the surrounding environment, the technological exergy is, of course, zero. The only way to move systems away from equilibrium is to perform work on them. Therefore, it is reasonable to use the available work, that is, the exergy, as a measure of the distance from thermodynamic equilibrium.

It is not possible to measure exergy directly, but it is possible to compute it. As illustrated in Figure 4.7, we assume a reference environment that represents the same system (ecosystem) but at thermodynamic

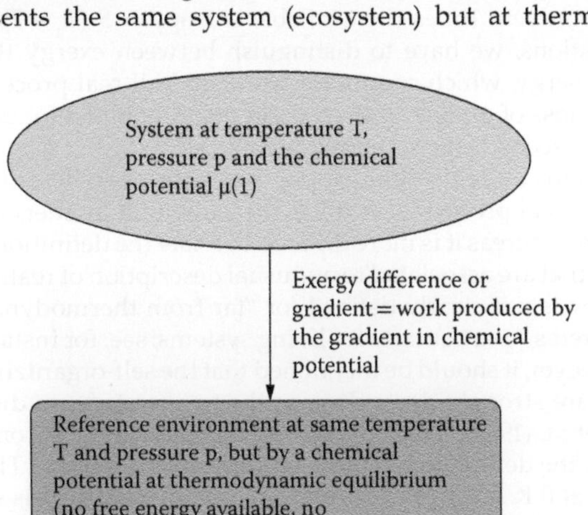

FIGURE 4.7
The exergy content of the system is calculated in the text for the system relatively to a reference environment of the same system at the same temperature and pressure, but as an inorganic soup with no life, biological structure, information, or organic molecules. (From Shieh, J.H. and Fan, L.T., *Energy Sources*, 1–46, 1982.)

equilibrium, which means that all the components are inorganic at the highest possible oxidation state and homogeneously distributed in the system (i.e., no gradients, see in Shieh and Fan [1982]).

Since the chemical energy embodied in the organic components and the biological structure contributes far most to the exergy content of the system, there seems to be no reason to assume a (minor) temperature and pressure difference between the system and the reference environment. Under these circumstances, we can calculate the exergy, which we will name *eco-exergy* to distinguish it from the technological exergy defined earlier, as coming entirely from the chemical energy:

$$\sum_{c}(\mu_C - \mu_{CO})N_C \tag{4.31}$$

This represents the nonflow chemical exergy. It is determined by the difference in chemical potential ($\mu_c - \mu_{co}$) between the ecosystem and the same system at thermodynamic equilibrium. This difference is determined by the concentrations of the considered components in the system and in the reference state (thermodynamic equilibrium), as is the case for all chemical processes. We can measure the concentrations in the ecosystem, but the concentrations in the reference state (thermodynamic equilibrium) can be based on the customary use of chemical equilibrium constants. If we have the process

$$\text{Component A} \Leftrightarrow \text{Inorganic decomposition products} \tag{4.32}$$

It has a chemical equilibrium constant, K

$$K = \frac{[\text{Inorganic decomposition products}]}{[\text{Component A}]} \tag{4.33}$$

It is difficult to find the concentration of component A at thermodynamic equilibrium, but we can find the concentration of component A at thermodynamic equilibrium from the probability of forming A from the inorganic components, as will be presented next.

Eco-exergy is a concept that is close to Gibb's free energy but contrary to Gibb's free energy, eco-exergy has a different reference state from case to case (from ecosystem to ecosystem) and it can be used far from thermodynamic equilibrium; whereas Gibb's free energy in accordance to its exact thermodynamic definition only is a state function that is close to thermodynamic equilibrium.

Through these calculations, we find the eco-exergy of the system by comparison with the same system at the same temperature and pressure but in the form of an inorganic soup without any life, biological structure, information, or organic molecules. Since ($\mu_c - \mu_{co}$) can be found from the definition of chemical potential by replacing activities with concentrations, we get the following expressions for the eco-exergy (see also Equation 4.25):

$$\text{Ex} = \text{RT}\sum_{i=0}^{n} C_i \ln\left(\frac{C_i}{C_{i,o}}\right) \tag{4.34},$$

where R is the gas constant (8.317 J/K moles = 0.08207 liter × atm/K moles); T is the temperature of the system; and Ci is the concentration of the ith component expressed in a suitable unit, for instance, phytoplankton in a lake Ci could be expressed either as mg/l or as mg/l of a focal nutrient. Ci,o is the concentration of the ith component at thermodynamic equilibrium, and n is the number of components. Ci,o is, of course, a very small concentration (except for $i = 0$, which is considered to cover the inorganic compounds), and it corresponds to a very low probability of forming complex organic compounds spontaneously in an inorganic soup at thermodynamic equilibrium. Ci,o is even lower for the various organisms, because the probability of forming the organisms is very low with their embodied information, which implies that the genetic code should be appropriate.

By using this particular exergy based on the same system at thermodynamic equilibrium as reference, the eco-exergy becomes dependent only on the chemical potential of the numerous biochemical components that are characteristic for life. This is consistent with Boltzmann's statement, that life is a struggle for free energy.

The total eco-exergy of an ecosystem cannot be accurately calculated, as we cannot measure the concentrations of all the components or determine all possible contributions to the eco-exergy of an ecosystem. If we calculate the eco-exergy of a fox, for instance, the calculations provided earlier will only give the contributions coming from the biomass and the information embodied in the genes, but what is the contribution from the blood pressure, the sexual hormones, and so on? These properties are at least partially covered by the genes but is that the entire story? We can calculate the contributions from the dominant components, for instance, by the use of a model or measurements, that cover the most essential components for a focal problem. The difference in eco-exergy by a comparison of two different possible structures (species composition) is decisive here. Moreover, eco-exergy computations always give only relative values, as the eco-exergy is calculated relatively to the reference system.

The definition of *eco-exergy* is very close to that of *free energy*. Eco-exergy is, however, a difference in free energy between the system and the same system at thermodynamic equilibrium. The reference system used is different for every ecosystem according to the definition of *eco-exergy*. In addition, free exergy is not a state function that is far from thermodynamic equilibrium. Consider, for instance, the immediate loss of free energy (or let us use the term *eco-exergy*, as already proposed, to make the use of the concepts more clear) when an organism dies. A microsecond before the death, the information can be used and after the death, the information is worthless and should, therefore, not be included in the calculation of eco-exergy. Therefore, eco-exergy cannot be differentiated.

It is possible to distinguish between the contribution to the exergy of information and of biomass (Svirezhev, 1998). Let us introduce p_i, defined as c_i/A, where

$$A = \sum_{i=1}^{n} C_i$$

is the total amount of matter in the system, as a new variable in Equation 4.34:

$$Ex = A\,RT\sum_{i=1}^{n} p_i \ln \frac{p_i}{p_{io}} + A \ln \frac{A}{A_o} \tag{4.35}$$

Since $A \approx A_o$, exergy becomes a product of the total biomass A (multiplied by RT) and Kullback* measure:

$$K = \sum_{i=1}^{n} p_i \ln\left(\frac{p_i}{p_{io}}\right) \tag{4.36}$$

where p_i and p_{io} are probability distributions, a posteriori and a priori, respectively, to an observation of the molecular detail of the system. It means that K expresses the amount of information that is gained as a result of the observations. If we observe a system that consists of two connected chambers, we expect the molecules to be equally distributed in both the chambers; that is, $p_1 = p_2$ is equal to 1/2. On the other hand, if we observe that all the molecules are in one chamber, we obtain $p_1 = 1$ and $p_2 = 0$.

Table 4.3 gives an overview of the exergy of various organisms expressed by weighting factor

$$\beta = RTK \tag{4.37}$$

which is introduced in order to cover the exergy for various organisms in the unit detritus equivalent or chemical exergy equivalent per unit of volume or unit of area (or the eco-exergy density):

$$\text{Ex total density} = \sum_{i=1}^{n} \beta_i c_i \text{ (as detritus equivalents g at the temperature } T = 300\,\text{K)} \tag{4.38}$$

The β-value embodied in the biological/genetic information is found from Equations 4.36 and 4.37 based on how much information the various organisms are carrhing in their genome. In accordance with Equation 4.37, detritus has the β-value = 1.0. By multiplication of the result obtained with Equation 4.38 by the average energy content of 1 g of detritus, that is 18.7 kJ, the exergy can be expressed in kJ per cubic-meter or square meter, given that Equation 4.38 is giving g detritus equivalents.

The eco-exergy is found (1) from the knowledge to the entire genome for the simplest organisms and (2) for a number of other organisms by the use of a correlation between various complexity measures and the information content of the genome (see Jørgensen et al., 2005). The values of β for various organisms have been discussed in the journal *Ecological Modeling* by several papers, but the latest published values in

* Kullback measure is a generalization of Shannon's measure of information, Equation 4.4, and is widely used in statistical mechanics.

Jørgensen et al. (2005) (see Table 4.3) probably come closest to eventually truth β-values. This interpretation of eco-exergy (the work capacity of living organisms) is completely in accordance with Boltzmann (1905), who provided the following relationship for the work, W, that is embodied in the thermodynamic information (see also Illustration 3.2.1):

$$W = RT \ln M \qquad (ML^2T^{-2}) \tag{4.39}$$

TABLE 4.3

β-Values Represent the Exergy Content Relatively to the Exergy of Detritus, or the Eco-Exergy Content Relatively to the Eco-Exergy of Detritus

Early Organisms	Plants	β-Values	Animals
Detritus		1.00	
Viroids		1.0004	
Virus		1.01	
Minimal cell		5	
Bacteria		8.5	
Archaea		13.8	
Protists	Algae	20	
	Yeast	17.8	
		33	Mesozoa, Placozoa
		39	Protozoa, amoebe
		43	Phasmida (stick insects)
	Fungi, molds	61	
		76	Nemertina
		91	Cnidaria (corals, sea anemones, and jellyfish)
	Rhodophyta	92	
		97	Gastrotricha
		98	Porifera
		109	Brachiopoda
		120	Plathyhelminthes
		133	Nematoda
		133	Hirudinea
		143	Gnathostomulida
	Mustard weed	143	
		165	Kinorhyncha
	Seedless Angiosperms	158	
		163	Rotifera
		164	Entoprocta
	Moss	174	
		167	Insecta
		191	Coleodiea (sea squirt)

TABLE 4.3 (*Continued*)

Early Organisms	Plants	β-Values	Animals
		221	Lepidoptera
		232	Crustacea
		246	Chordata
	Rice	275	
	Gymosperms	314	
		310	Mollusca
		322	Mosquito
	Angiosperms	393	
		499	Fish
		688	Amphibia
		833	Reptilia
		980	Aves
		2127	Mammalia
		2138	Monkeys
		2145	Anthropoid apes
		2173	*Homo sapiens*

Source: Jørgensen, S.E. et al., *Ecol. Model.*, 185, 165–176, 2005.

where M is the number of possible states, among which the information has been selected. M is, as seen for species, the inverse of the probability to spontaneously obtain the amino acid sequence that is valid for the enzymes controlling the life processes of the considered organism.

Kullback measure of information covers the gain in information, when the distribution is changed from p_{io} to p_i. It should be further noted that K is a specific measure (per unit of matter). Viruses DNA code for about 2500 amino acids. The β-value is, therefore, only 1.01. The smallest known agents of infectious disease are short strands of RNA. They can cause several plant diseases and are possibly implicated in enigmatic diseases of human beings and other animals. Viroids cannot encode enzymes. Their replication, therefore, relies entirely on enzyme systems of the host. A viroid has typically a nucleotid sequence of 360, which means that it would be able to code for 90 amino acids, although they are not translated. The β-value can, therefore, be calculated to be 1.0004. It should be discussed whether viroids should be considered living materials.

Eco-exergy is, of course, a relative measure. It is, therefore, proposed that the exergy found by these calculations should always be considered a relative minimum eco-exergy index to indicate that there are other contributions to the total exergy of an ecosystem, although they may be of minor importance. In most cases, however, a relative index is sufficient to understand and compare the reactions of ecosystems, because the absolute exergy content is irrelevant for the reactions and cannot be determined due to the extremely high complexity. In most cases, the change in eco-exergy is of importance in order to understand the ecological reactions.

The weighting factors presented in Table 4.3 have been successfully applied in several structurally dynamic models, and furthermore, in many illustrations of the maximum exergy principle; see Chapter 19. The relatively good results in application of the weighting factors, in spite of the uncertainty of their more exact values, seems only to be explicable by the robustness of the application of the factors in modeling and other quantifications. The differences between the factors of the microorganisms, the vertebrates, and the invertebrates are so clear that it seems not to be important whether the uncertainty of the factors is very high—the results are robust.

5

Ecological Processes: An Overview

Ecological models are developed with the aim of overviewing and understanding complex ecological systems and of describing responses to changing impacts on the ecosystems. They consist of five elements:

1. State variables that describe the most important components of the ecosystem.
2. Forcing functions that describe the impacts on the ecosystem. They encompass all exchanges between the ecosystem and the environment.
3. Equations of the ecological processes that describe how state variables are changed, including transfers of matter, energy, and information between two or more state variables.
4. Parameters that describe the properties of state variables.
5. Physical and chemical constants.

Ecological processes are, therefore, indispensable elements in all ecological models. They link two or more state variables and describe how the ecological components are functions of time. In principle, process description can be expressed as a sentence with no quantifications, but in far most cases, ecological processes are described by the use of a mathematical equation.

Figure 5.1 depicts an ecological model, and all the processes involve a transfer from one state variable to another, except the processes denoted by 10, 13, 14, and 15 that are actually forcing functions, as they cover an exchange between the ecosystem and the environment.

Ecological processes can be classified in a number of ways. As the processes are transfer processes, one possible classification is based on what the process transfers: information, matter, energy, a number of organisms, for instance, or two or more of these four possibilities. The processes could further be classified in accordance with the applied mathematical description.

Many processes could, however, be described by more than one equation, and it may be of great importance for the results of the final models that the right equations are selected and applied for the case under consideration. This book presents the possible mathematical formulations of most ecological processes, but a short overview of the most applied mathematical equations is presented in this chapter. There are a number of slight modifications to the overview of equations that is presented next. Nevertheless, more than

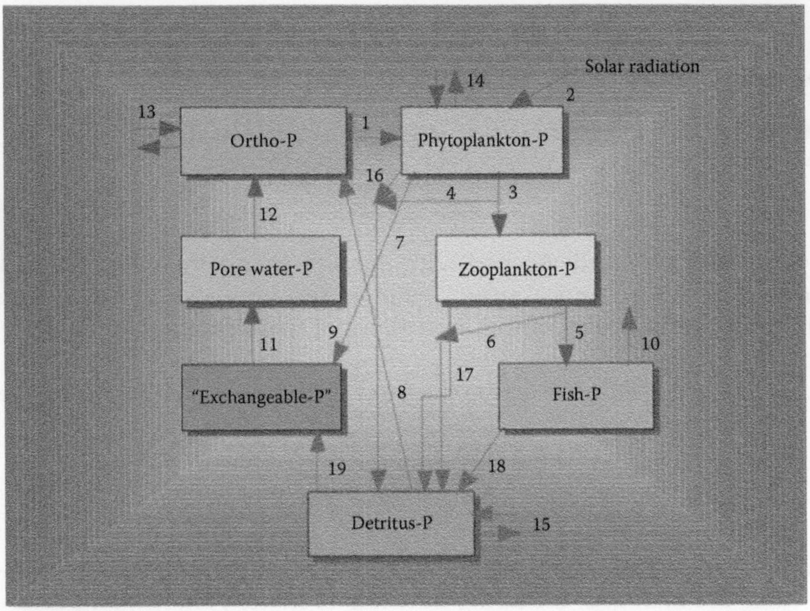

FIGURE 5.1
The model presents a phosphorus cycle in a lake by using a simple food chain nutrient–phytoplankton–zooplankton–fish to describe the processes of the cycle. The cycles are closed by mineralization of detritus and by the release of nutrients by the sediment (Jørgensen and Bendoricchio, 2001). The processes are indicated by numbers from 1 to 19, where 1 is the uptake of phosphorus by phytoplankton; 2 is the growth of phytoplankton by photosynthesis; 3 is the grazing of zooplankton on phytoplankton; 4 is the feces resulting from the grazing; 5 is fish predation on zooplankton; 6 is the feces resulting from the predation; 7 and 9 are the settling of phytoplankton; 8 is the microbiological decomposition of the detritus forming inorganic phosphorus; 10 is the removal of fish from the lake by fisheries; 11 is the decomposition of the settled exchangeable phosphorus in the sediment, forming soluble phosphorus in the pore water of the sediment; 12 is the diffusion of phosphorus from the pore water to the lake water; 13 covers the inflow and outflow of phosphorus by tributaries; 14 is the inflow and outflow of phytoplankton by tributaries; 15 is the inflow and outflow of phosphorus in detritus by the tributaries; 16 is the mortality of phytoplankton; 17 and 18 are the mortalities of zooplankton and fish, respectively; and 19 is the settling of detritus phosphorus. Notice that all the processes involve transfer from one state variable to another except 10, 13, 14, and 15, which, however, can be considered forcing functions, because they involve the exchange between the ecosystem and the environment.

95% of all ecologically relevant processes can be mathematically formulated (sometimes, minor modifications are required) by one of the following six equations:

1. A constant flow rate—also denoted as zero-order expression: rate = $dC/dt = K$
2. A first-order rate expression, where the rate is proportional to a variable such as a concentration of a state variable: rate = $dC/dt = k_1 \cdot C$.

This expression corresponds to exponential growth. The following expression is obtained by integration $C(t) = C_0 \cdot e^{k_1 * t}$

First-order decay has the rate $= dC/dt = -k \cdot t$ and $C(t) = C_0 \cdot e^{-k_1 * t}$ The expression is often used to cover not only population growth (see Chapter 22) but also decomposition processes (Section 17.4), and radioactive decay can be approximated as first-order reactions.

3. A second-order rate expression, where the rate is proportional to two state variables simultaneously; for instance, rate $= dC/dt = k_2 \cdot C_1 \cdot C_2$. This expression is often needed to describe chemical processes, as will be discussed in Section III.
4. A Michaelis–Menten expression or Monod kinetics known from kinetics of enzymatic processes (Section 9.3) in biochemistry: rate $= dC/dt = k_3 \cdot C/(C + k_m)$.

   At small substrate concentrations, the process rate is proportional to the substrate concentration; whereas at high substrate concentrations, the process rate is maximum and constant and the enzymes are fully utilized. The same expression is used when the growth rate of plants is determined by a limiting nutrient according to Liebig's minimum law. The so-called "Michaelis–Menten's constant," k_m, or the half-saturation constant corresponds to the concentration that gives half the maximum rate. At small substrate or nutrient concentrations, the rate is very close to a first-order rate expression; whereas at high concentrations, it is close to a zero-order rate expression. The rate is gradually regulated from a first-order to a zero-order expression as the concentration increases.
5. A first-order rate expression with a regulation due to limitation by an additional factor, for instance the space or the nesting areas. C could be limited by the Monod kinetics or by a supply process, for instance if the growth rate k_{max} is limited by the amount of available food. The limitation by the additional factor may be expressed by the introduction of a carrying capacity. The general first-order expression (or the Monod kinetics) is regulated by the following factor: (1–concentration/carrying capacity). The expression is rate $= dC/dt = k_{max} \cdot C(1 - C/K)$, where K is the carrying capacity. When the concentration reaches the carrying capacity, the factor becomes zero and growth stops. This process rate expression is denoted as logistic growth, and it is illustrated in greater detail in Chapter 22, including the graphic differences between exponential growth by a first-order reaction and logistic growth. Both these growth expressions are extensively applied in population dynamic models.

6. A rate governed by diffusion often uses a concentration gradient dC/dx to determine the rate, as it is expressed in Fick's law: rate $= k_5 \cdot (dC/dx)$ (Fick's First Law, see Equation 7.2).

There are many modifications of these six expressions. For instance, a threshold concentration, tr, is often used in the Michaelis–Menten expression. The concentration (state variable) is replaced by the concentration, tr. The concentration should, therefore, exceed tr to generate any rate. For grazing and predation processes, the Michaelis–Menten expression is often multiplied by (1 – concentration/carrying capacity), which is similar to what is used in the logistic growth expression (Chapter 14). This implies that when food is abundant (concentration is high), another factor such as space or nesting area determines and limits growth. These modifications will be discussed in Section IV.

The modifications of the six equations and the more rarely applied equations are presented in this book to give a more complete overview of ecological processes and the equations that are applied to cover the wide range of these processes. We present these processes by using the following classifications:

1. **Physical processes**, where the mathematical formulation is rooted in physics. Section II gives an overview of the physical processes.
2. **Chemical processes**, which use an approach that is based on general chemistry and physical chemistry. Indeed, many ecologically important processes are chemical, and as many of the chemical processes take place simultaneously, we have found it necessary to include how it is relatively easily to calculate the subsequent chemical composition resulting from several simultaneous processes by the use of log–log presentations. The Appendix at the end of this book presents these chemical calculation methods. Section III presents the chemical processes of importance for ecosystems, including partition processes between the ecological phases (the four ecospheres) and the ecotoxicological implications.
3. **Biological processes**, which include the processes at all the levels of the ecological hierarchy: the ecosystem, the population, the organism, and the cell level. The biological effects of ecotoxicological processes are discussed here.
4. **Landscape processes**, where the perspective is enlarged to that of landscape ecology, with its large-area and long-term focus. Here, some of the processes presented in the previous chapters are appropriately contextualized and discussed for aquatic, terrestrial, urban, and aerial ecosystems.

Section II

Physical Processes

6

Space and Time

6.1 Introduction

Space and time are the basic concepts that encompass our mathematical descriptions of ecosystems. Usually, space is interpreted as being three dimensional; whereas time, playing the role of a fourth dimension, is of a different type from the spatial dimension. This is the common Euclidean paradigm adopted for the description of a continuous space. In nonrelativistic contexts, time is considered a universal parameter that is independent of the state of motion of an observer.

Actually, according to the characteristics of the system under analysis and of the description that has to be implemented, a zero-, one-, two-, or three-dimensional space representation is adopted. The zero-dimensional description is adopted when space discrimination is not vital; one-dimensional description may be useful in describing rivers; and two- or three-dimensional descriptions may be important, depending on vertical variability, for processes taking place, for example, in lakes or coastal zones (Figure 6.1). As soon as changes, flows, or processes are considered, time parameterization comes into play.

The various factors governing ecological processes pertain to different layers in space and occur over different time spans. Hence, in order to make operative the measurement procedure, the units of space, time, and the other characteristic quantities should be established. In practice, this ends up in defining the scale, as done in Section 4.2, for information.

6.2 Scale and Hierarchy

State variables that describe ecological processes are usually expressed as some combinations of the basic SI units. Of these latter, those that are of interest for ecosystems are the meter for distance, the kilogram for mass,

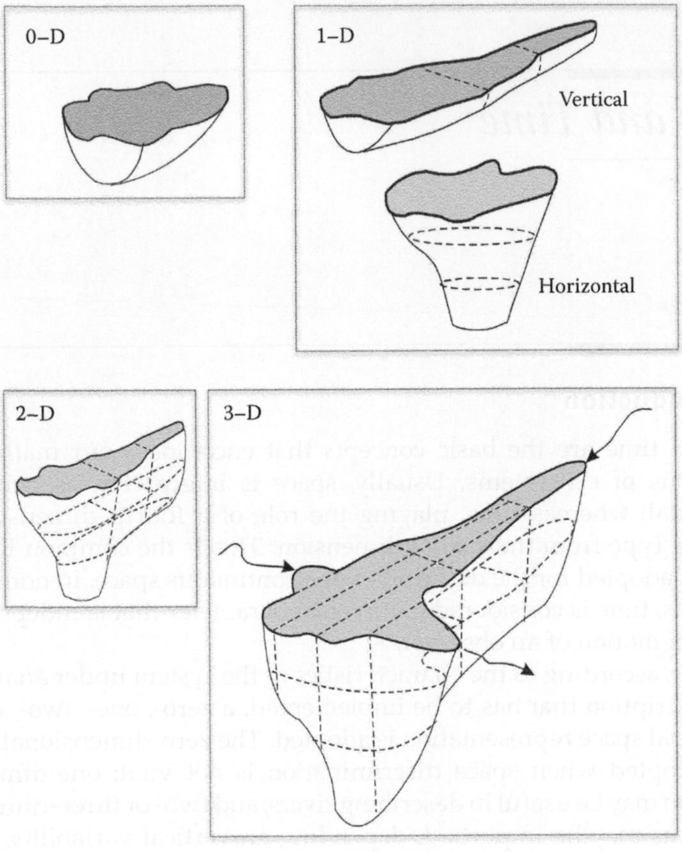

FIGURE 6.1
Spatial representation in different dimensions. (From Jørgensen, S.E. and Bendoricchio, G., *Fundamentals of ecological modeling* (3rd ed.), Elsevier, Amsterdam, 628 pp, 2001.)

the second for time, the Kelvin for temperature, the mole for amount of substance, and the candela for intensity of light. Often, standard units are not convenient for all ecological processes, and, therefore, rescaling these units may simplify the description. When introducing the conceptual model on which the description of the system under analysis is implemented, the earliest step is identifying the correct hierarchical level. In doing this, a tradeoff between elegant simplicity and realistic detail is pursued. For example, if the focus is on eutrophication processes, the appropriate scale lies somewhere between nutrient generation processes and phytoplankton growth (see Figure 6.2). At lower scales, phytoplankton dynamics may be considered relevant for detailed descriptions. At higher scales, nutrient generation processes (e.g., in the catchment) turn out to be constraints or forcing functions for eutrophication. Therefore, every description of this

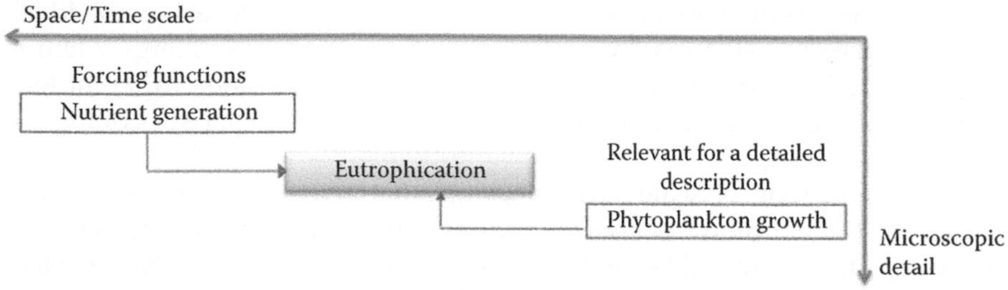

FIGURE 6.2
(See color insert.) Hierarchical levels linking scales in time and space and the description's detail. For example, a satisfactory understanding of eutrophication requires the inclusion of forcing loads as well as other relevant microbial processes.

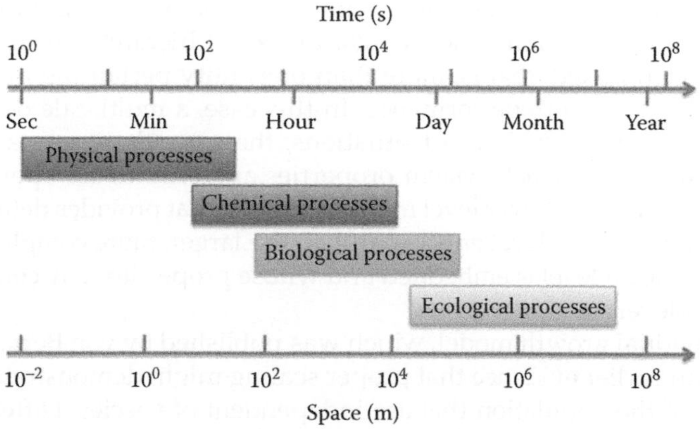

FIGURE 6.3
(See color insert.) Typical space and time scale ranges for physical, chemical, biological, and ecological processes. (From Jørgensen, S.E. and Bendoricchio, G., *Fundamentals of ecological modeling* (3rd ed.), Elsevier, Amsterdam, 628 pp, 2001.)

phenomenon cannot do without some partial backing from next hierarchical levels.

In order to select the appropriate scales, the hierarchy of physical, chemical, biological, and ecological processes taking place in the system should be properly specified (Figure 6.3).

Physical processes need a fine spatial mesh and a short time step for proper description. By ascending the hierarchy, the coarse spatial mesh and larger time steps become appropriate.

Generally, when dealing with ecological processes, two types of hierarchy patterns may be recognized: nested and non-nested. Nested hierarchies arise when subsystems are combined together to form larger systems. The properties of the focus level are determined by the characteristics of lower-scale systems as well as by those of the context in which the focal level is embedded, that is, upper-scale systems. Instead, non-nested hierarchies emerge when it is possible to identify chains of functional connections among different levels. Responses coming from lower levels regulate the characteristics of higher ones (Maurer, 2008).

The identification of scales and hierarchy patterns is important, in particular, for the definition of control entities. These are peculiar elements in the system that may guide the behavior of the whole system. Non-nested arrangements often enable deciding whether the control is top down or bottom up. In the first case, entities (organisms and/or processes) pertaining to higher levels control the overall activity. In the latter case, those at the lower levels are the key elements of the system. Of course, the scale of control elements should be included as focal for the description. This is usually the case when dealing with food webs or trophic networks (see Section 19.5). In these descriptions, the temporal dynamics of predators and prey is usually resolved over similar time extents. In nested hierarchy arrangements, control is often mixed; that is, more than one entity pertaining to different levels guides a system's performance. In this case, a multiscale description is appropriate. Actually, in most situations, there are three relevant scales: (1) the focal scale at which system properties and entities are operationally measured; (2) the next lower level in the hierarchy that provides details about the properties of focal-level entities; and (3) the larger, more complex system in which the focal level is embedded and whose properties may constrain or direct focal-level entities.

The individual growth model, which was published by von Bertalanffy in 1934, was an earlier evidence that proper scaling might demonstrate general properties of the population that are independent of species. Different species of organisms perform life processes with various spatial meshes and time spans. Therefore, the characteristic scale for population dynamics is directly linked to body size. By rescaling metabolism to mass-independent units, allometric principles spanning over several scales are found. Indeed, scaling appears in living systems when similar constraints on processes that occur at different scales are operating. As will be further discussed in Chapter 14, allometric principles may prove general properties of living systems.

6.3 Space Delimitation: Boundaries

Hierarchical arrangement in ecosystems and ensuing spatial scaling is the origin of topological space delimitation. A boundary that represents an interface through different scales may be identified between different hierarchical levels. A boundary among objects at the same hierarchical level may be defined based on the rates of processes within entities. Boundaries are recognized by the existence of sharp gradients or discontinuities; namely, when a boundary is crossed, rates in space and/or time of some processes show a marked shift. As a consequence to what is argued in Section 2.2, boundaries imply limitation; ecological competition for resources (space, food, etc.) arises as soon as boundaries at higher hierarchical scales are established. At lower scales, however, boundaries are intimately linked to two very important features of living systems: the differentiation of functions and the ability to maintain stable internal conditions.

Typically, at lower scales, biological systems establish boundaries by building membranes. As sketched in Figure 6.4, membranes appeared during life evolution when special organic molecules started assembling in structures that circumscribed portions of space and matter. These structures are the basic components on which cell membranes are built. The cell membrane is an assembly of organic molecules that separates the interior of all cells from the outside environment (see Section 2.2). The cell membrane is a very complex interface that is selectively permeable to ions and organic molecules and which has the ability to control the movement of substances in and out of the cell. Besides protecting the cell from accidental perturbations coming from the outside, the cell membrane represents the very machinery through which all information flows are processed (Lipton, 2005). Cell membranes, even if all consist of a phospholipid bilayer that incorporates embedded proteins with various configurations and functions, are indeed very specific for each typology in the variety of cellular assemblages. During morphogenesis, totipotent stem cells become the various pluripotent cell lines of the embryo, which, in

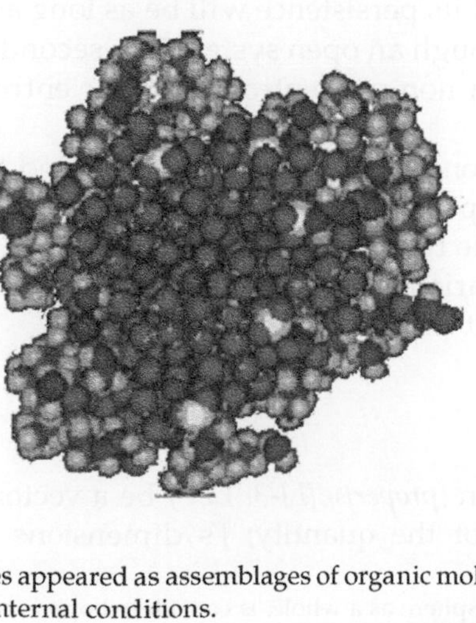

FIGURE 6.4
(See color insert.) Membranes appeared as assemblages of organic molecules (lipids, proteins), enabling the onset of stable internal conditions.

turn, become fully differentiated cells. Each cell constructs a representation of the outside world through its proper membrane. This interface is also involved in gene expression control, which ultimately comes from the surrounding environment. At larger scales, boundaries are established by tissues (cell collections). This differentiation process gives rise to the incredible variety of hierarchies and functions that are recognized in living systems.

6.4 Flows through Open Systems

As discussed in Chapter 1, ecological systems are sustained by flows of energy and matter. Therefore, boundaries are necessarily permeable to a certain extent. For example, at lower scales, cell membranes are traversed by photons and chemical elements that carry energy and information. At larger scales, feeding and migration serve as examples of cross-boundary flows. Hence, by definition, living systems are open to material and energy flows.* The existence of boundaries produces a conceptual separation between inside and outside environments and is, thus, the first move toward individualization. Homeostasis is a property of living systems, and it is the ability of maintaining stable internal conditions. Under the LTE hypothesis presented in Section 3.4, it is possible to describe internal properties by introducing state variables (e.g., pressure, temperature, or chemical concentrations) and potentials. As soon as a step difference (a gradient) between the different parts of a system is established, it is always associated to a flow. The entity of the flow depends on the extent of the gradient, and its persistence will be as long as the gradient subsists. Existing a flow through an open system, the second law of thermodynamics translates into a non-null entropy flow or entropy production (see in Section 3.5).

As stated in Sections 2.3 and 2.4, the idea behind the continuity equation is the flow of some property, such as mass, energy, or information, through boundaries from one region of space to another. Let this property be represented by a scalar variable, x, and let the volume density of this property (the amount of x per unit volume V) be

$$\varphi = \frac{dx}{dV}$$

φ has the dimension *[property][L]-3*. Let J be a vector field representing the flow or transport of the quantity; J's dimensions are *[property][T]-1[L]-2*.

* Actually, Gaia, the biosphere as a whole, is considered open to energy but closed to matter, that is, a closed nonisolated system.

Then, the differential form for a general continuity equation is

$$\frac{\partial \varphi}{\partial t} + \nabla \cdot J = \sigma \tag{6.1}$$

where ∇ is the vector operator divergence, and σ is the production of x per unit volume per unit time. Terms that generate ($\sigma > 0$) or remove ($\sigma < 0$) x are referred to as a sources and sinks, respectively. The divergence of a field at a given point tells how many field lines are born (or die) at that point, hence describing the degree to which a small volume around the point is a source or a sink for the vector flow. In other words, this term accounts for the advective transport of the quantity x. The divergence theorem states that the outward flow of a vector field through a closed surface Σ (the boundary) is equal to the volume integral of the divergence of the region of volume V inside the surface:

$$\oint_{\Sigma} J \cdot \mathrm{d}\Sigma = \int_{V} \nabla \cdot J \, \mathrm{d}V$$

Intuitively, the sum of all sources minus the sum of all sinks gives the net flow out of a region. The general Equation 6.1 may be used to derive any continuity equation, ranging from the volume continuity equation to the Navier–Stokes equations (6.1), and also generalizes the advection equation that will be discussed in Chapter 7. If x is a conserved quantity, that is, which cannot be created or destroyed (such as energy), the continuity equation becomes

$$\frac{\partial \varphi}{\partial t} + \nabla \cdot J = 0 \tag{6.2}$$

Flows formalize the idea of change in time and space and are, therefore, ubiquitous in science. In a world of changes, this concept is involved in the definition of several interesting quantities with regard to ecological processes. For example, the production of entropy, information, biomass, emergy, and exergy are special instances of flow. Power as well is energy flow per unit time. Efficiency may be defined by a flow that describes how time or action is well used for the planned scope. A trivial definition for this quantity may be

$$\eta = \frac{p}{c}$$

where η is the efficiency, p represents the amount of production of some valuable resource (output or benefit), and c indicates the amount of consumed valuable resources (input or cost). If the same units are used for products and consumables and if the transformation process is conservative, efficiency is expressed as a percentage. In the case that time is the only valuable consumable resource employed, efficiency is formally identical to the production of something (see, e.g., Equation 4.11).

6.5 Equilibrium, Stability, and Steady State

In the general context of ecological thermodynamics, it is helpful to make a categorization of system states in terms of stability properties, which also include far-from-equilibrium nonlinear processes. Usually, there is common agreement on the following definitions (Ulgiati and Bianciardi, 1997):

- *Thermodynamic equilibrium* is a state in which the system is indistinguishable from the surrounding environment, the disorder is maximal (maximum entropy), and the total absence of structures and organization is evident. Sometimes, this state is referred to as the *state of thermal death* or *reference state* (Shieh and Fan, 1982).
- *Static equilibrium* is a state in which the system is distinguishable from the surrounding environment (e.g., is bounded), state variables are well defined, no entropy flows take place, and, hence, entropy production is null, $p = 0$ (see Section 3.5). For these states, variational principles are applied, and conservation laws hold, allowing the definition of classical thermodynamic potentials. The second law of thermodynamics for such systems says that maximization of entropy is the evolutional criterion.
- *Dynamic equilibrium* or *steady state* is, in the classical sense, a nonequilibrium state in which the system is distinguishable (bounded, confined), and some state variables or other relevant descriptive quantities (or, depending on time scales, their temporal averages) nearly do not depend on time. Let x be one of such quantities, then

$$\frac{dx}{dt} \approx 0$$

In these conditions, under the only assumption of LTE (see Section 3.4), the entropy production may be written as in Equation 3.6 and $p = \sum_k X_k J_k > 0$; that is, the system is forced to entropy exchanges. Potentials for such states cannot be defined in the usual way. In far-from-equilibrium states, several quantities are not preserved, and then, the system is not allowed to forget "initial conditions." However, the search for conserved quantities is always the initial step in process description. According to the intensity of external forcing, processes can occur in linear or nonlinear regimes. In the first case, the evolutional criterion is that of minimum entropy production. In the latter case, variational principles cannot be directly applied. Indeed, some additional assumptions on constraints and process control are needed in order to use Glansdorff–Prigogine criterion (Equation 3.11). At the scale of living systems, the steady state of an organism is called *homeostasis*.

- *Evolutionary* or *progressive dynamic equilibrium* is that characterizing complex, self-organizing systems that are capable of operating continuous or discontinuous evolutionary transitions between more-or-less stable states. It is essentially a series of steady states observed on a sufficiently long time scale and includes events such as drifts, bifurcations, or evolutionary jumps (see, e.g., Illustration 19.2.1). If the evolution is continuous, it is said that the system has the ability to maintain the *homeoresis*.

A useful and operational definition of *stability* is obtained by measuring the effects of perturbations. A stable system, if disturbed, reacts in such a way as to cancel the perturbation; the efficiency in absorbing its effects depends on the intrinsic characteristics of the system. There are indeed different degrees of stability. For example, if a perturbation δx acts so as to produce fluctuations in the value of some descriptive quantity x of the system's state,

$$\tilde{x} = x + \delta x$$

and the system responds by eliminating the variation in a time δt, it is useful to introduce the following two quantities:

$$\begin{aligned} &\textit{Resistance} \quad R = \frac{x}{\delta x} \\ &\textit{Resilience} \quad r = \frac{1}{\delta t} \end{aligned} \tag{6.3}$$

Resistance and resilience are measures of the degree of stability of the steady or progressive state. In general, it is observed that more developed ecosystems of higher hierarchical levels, which are more specialized and with comparatively low input flows per unit biomass, are more resistant (small δx). Less developed systems, corresponding to minor scales or lower levels of ecosystem hierarchy and characterized by higher input flows, are, on the contrary, more resilient. That is, recovery after perturbations is attained quickly (small δt). Dynamically, resistance is analogous to elastic stability, whereas resilience corresponds to elastic stability. In Figure 6.5, homeostasis of steady state and homeoresis of progressive state continuous evolutions are illustrated along with the concepts of resistance and resilience to perturbations.

In ecology, the process of growth and development of ecosystems is called *ecological succession*, and the term adopted for the stable and mature steady state is *climax*. However, from experimental observations, it is evident that the stability of this state is not indefinitely ensured. Circulating matter may be irrecoverably lost; abiotic components such as temperature or air humidity may drift over time because of global change, geological or astronomical events. As a consequence, there might be large-scale

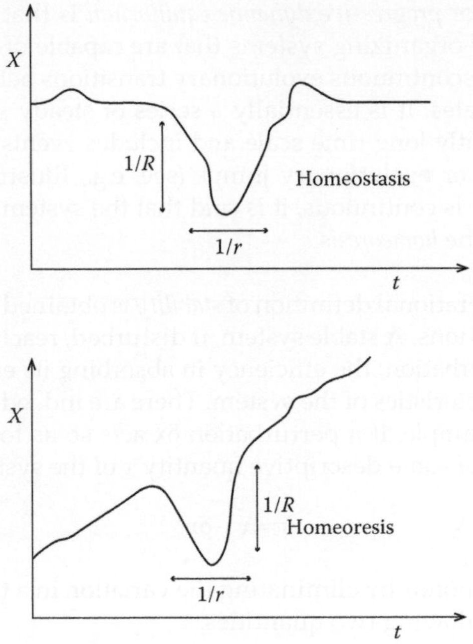

FIGURE 6.5
Homeostasis of steady state and homeoresis of progressive continuous evolution.

trans-boundary alterations that lead to radical reorganization of the entire ecosystem.*

Recent studies in ecology suggest that the state of climax is rather pulsing in time, following a criterion of performance maximization. According to the control scheme, pulses may not only be generated by long-term external forcing (higher hierarchies) but also be induced by elements of the food web. For example, pulses may be originated at higher levels (consumers) as well as at lower ones (producers). In the first case, the control is said to be top down; whereas in the latter case, it is said to be bottom up. Occasionally, these fluctuations may induce a discontinuous transition from one climax state to another (see Figure 6.6). For example, this is the case of mass extinctions or alien species invasion.

The concept of resistance as defined by Equation 6.3 is a special instance of the more general concept of buffer capacity:

$$\beta = \frac{\delta F/F}{\delta x/x} \tag{6.4}$$

* Analogously to what has been anticipated in Section 3.3: Energy dissipation by life processes entrains discontinuous jumps.

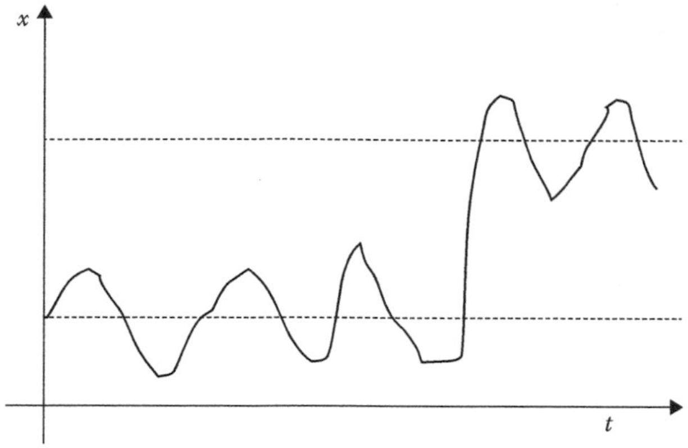

FIGURE 6.6
Discontinuous jumps between pulsing climax states are often observed during ecosystem growth and development.

which is conventionally used in systems modeling in order to study model sensitivity to the variations of control functions F (*sensitivity analysis*). High resistance to perturbations means high buffer capacities to the fluctuations of control functions. Resilience has the drawback that it is difficult to quantify, because recovery times are subject to some unknown control factors; whereas resistance (or buffer capacities) may be conveniently estimated by a direct observation of measurable system indicators such as mass, energy, and chemical potentials. In addition, several possible buffer capacities may be employed that, in a multidisciplinary perspective, enable an assessment of the stability of different system elements (the xs), which confront various forcing controls (the Fs).

7

Mass Transport

7.1 Introduction

Chapters 1 and 6 pointed out that an ecosystem is influenced by matter and energy exchanges with the surrounding environment, that is, transboundary fluxes that link the ecosystem to its neighboring systems. An ecosystem is also affected by the relationships between its abiotic and biotic components, which can be represented as flows of energy, matter, and information. For this reason, transport phenomena represent fundamental ecological processes underlying the functioning of any kind of ecosystem. In this chapter, the physical processes that determine the transport of mass found in fluids toward, from, or within ecosystems are described; the transport of energy and the role of energetic factors in physical processes will be illustrated in Chapter 8.

Here, *mass* means the amount of any kind of matter whose physical transport in fluids (air and water) is relevant for the functioning or for a particular characteristic of an ecological system. Typical examples are not only chemical elements and substances, such as food and resources (e.g., biomass, carbon, and nutrients), but also toxins and pollutants. Even living beings can be transported, along with the substances they carry in their own body tissues, such as nutrients and bioaccumulated contaminants, as well as with the ecological processes that they carry out (predation, competition, etc.; see Chapters 16 and 19). For instance, plankton are, by definition, those drifting organisms that are transported by water currents. Why should mass transport in fluids be focused on? Fluids are a major constituent of most ecosystems and typically represent their most dynamic media; therefore, mass transport can strongly influence the availability or concentration of many key substances in space and time. Indeed, mass transport plays an important role in the biogeochemical cycles of key ecological elements, which are detailed in Chapter 12. Transport between two different fluids (water and air), or between a fluid and another medium (e.g., water–sediment fluxes), is described in this chapter.

Much of the description of mass transport processes provided here relies on a basic property of many kind of media in ecosystems, that is, mass conservation (Chapter 2), and the mass budgets that can be derived from it. In several cases, the concepts and models presented are generally valid for any kind of fluids, that is, for both air and water, with only one (key) difference represented by changes in fluid properties such as density and viscosity. Moreover, the substances whose transport is described in ecological applications are typically not the same in air and water.

7.2 Advection and Diffusion

Advection is probably the simplest form of physical mass transport in fluids that one can think of. It is the transport of a substance, such as a solute or a solid, contained in a fluid, and is caused by the motion of the fluid's bulk in one direction. A substance is said to be advected if it moves steadily along with the fluid in such a direction, without changing concentration, as if it were a part of the fluid. Examples include the transport of a pollutant in a water pipe or silt–clay sediments that are carried by the current in a river, when the flow is relatively slow and streamlined (or, more precisely, laminar) and no significant diffusive transport takes place with regard to the dominant advective water motion. In oceanography, advection commonly indicates transport along horizontal directions only; whereas vertical mass transport, which generally includes both advective and diffusive processes, is usually called *vertical mixing*.

Diffusion, another basic transport process, is the spontaneous spreading of a substance in a fluid (Figure 7.1a). Very dissimilar processes acting at different spatiotemporal scales can cause diffusion: At microscopic scales, molecular diffusion takes place, that is, the substance spreads due to the random thermal agitation (Brownian motion) of particles; at larger scales, mixing by turbulent eddies (of varying sizes) causes turbulent (or eddy) diffusion. Even if the causes of these diffusive processes are distinct and act at different scales, the practical outcome is always the same from a qualitative point of view: Diffusion causes a net diffusive flux of matter from areas at high to low concentration, along the direction of the gradient of concentration. Thus, diffusion tends to reduce concentration gradients and results in a spatially homogeneous concentration, and the concentration gradient can be considered as the potential driving the diffusive flux. The only difference between diffusion at a microscopic scale and that at a macroscopic one, for example, molecular versus eddy diffusion, is that the order of magnitude of mass fluxes can be quite different.

Although diffusion is always present in nature, advection can be considered the predominant process that affects mass transport, or it can happen that both processes are important (Figure 7.1b) or that diffusion preponderates, depending on both the spatiotemporal scales of the phenomenon

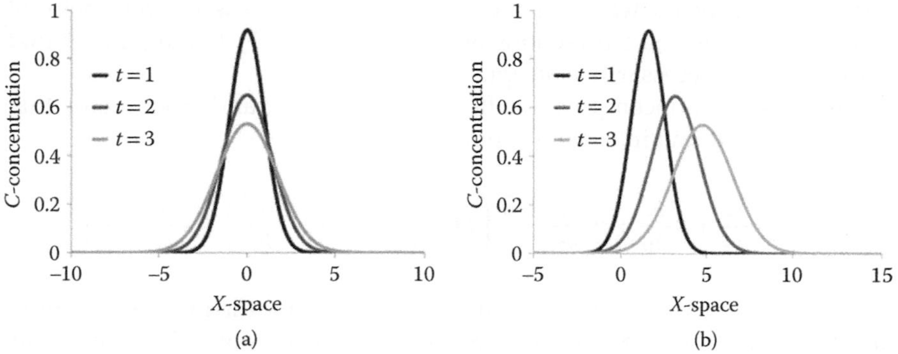

FIGURE 7.1
Change in concentration over time following an impulse of mass injected in the position $x = 0$ at time $t = 0$ and resulting from (a) pure diffusion and (b) the sum of advection and diffusion, with the same diffusion coefficient as in (a). Results are plotted for three different times. Units are arbitrary in both cases.

analyzed and the goal of the study. For example, in the case of an accidental spill of oil from a ship in a large estuary, one should focus on advection in order to be able to predict where the pollutant travels in the short term. However, as time passes, the pollutant concentration will be more and more influenced by turbulent eddies, which will tend to dilute and spread it, and diffusion will become more important.

Based on these definitions, we can write with regard to the monodimensional case, the flow of mass J_A (whose currency is $[M]\ [L]^{-2}\ [T]^{-1}$) caused by advection for a given substance, in a fluid with current velocity u in the x-direction and containing a concentration of mass C:

$$J_A = uC \tag{7.1}$$

Instead, the net diffusive flux of mass in the x-direction is proportional to the spatial gradient of the concentration through the diffusivity coefficient D_x:

$$J_D = -D_x \frac{\partial C}{\partial x} \tag{7.2}$$

The minus sign indicates that the net flow is from areas with high to low concentration. In the case of molecular diffusion, this is known as *Fick's first law*.

Equations 7.1 and 7.2 can be extended to three spatial dimensions and combined with Equation 6.2 to write a mass balance for the transported substance in a control volume of fluid:

$$\frac{\partial C}{\partial t} = -\nabla \cdot (C\mathrm{v}) + \nabla \cdot (D\nabla C) \tag{7.3}$$

where v is the vector whose elements u, v, and w represent the velocity of water in the x-, y-, and z-directions of space; $D = (D_x, D_y, D_z)$ highlights that diffusion is not necessarily isotropic.

In the case of an incompressible fluid, such as water and other media that are commonly found in ecology, Equation 7.3 becomes

$$\frac{\partial C}{\partial t} + u\frac{\partial C}{\partial x} + v\frac{\partial C}{\partial y} + w\frac{\partial C}{\partial z} = \frac{\partial}{\partial x}\left(D_x\frac{\partial C}{\partial x}\right) + \frac{\partial}{\partial y}\left(D_y\frac{\partial C}{\partial y}\right) + \frac{\partial}{\partial z}\left(D_z\frac{\partial C}{\partial z}\right) \quad (7.4)$$

This advection–diffusion equation can be easily modified to describe pure advection (by choosing all diffusivities = 0) or pure diffusion (all velocities = 0), and it can be simplified to study mono- or bidimensional problems by deleting the appropriate terms. In the case of molecular diffusion, which is an isotropic process, $D_x = D_y = D_z = D_m$, and the value of D_m depends on factors such as the substance that is being transported, the fluid it is found within, and temperature. For example, polar substances diffuse more quickly in water, which is itself polar and, thus, interacts with them, than apolar molecules such as oil or proteins (for which D_m is of the order of 10^{-10} m^2 s^{-1} in water, whereas for salts, D_m is about 10^{-9} m^2 s^{-1} or higher); see Figure 7.2.

Instead, turbulent diffusion is an anisotropic process, and diffusivity coefficients can vary depending on the spatial direction along which transport is assessed, due to the shape of the eddies that cause diffusion. Unlike molecular diffusion, the value of eddy diffusivity depends on the flow regime and, moreover, on the characteristic spatial scale of the problem: As a substance diffuses in a water body, it covers a wider area, and, thus, larger eddies are involved, which should be considered (by modifying the diffusivity coefficient) as they can influence transport. For example, the horizontal diffusivity in large and deep water bodies (sea, lakes, etc.) increases with the length scale

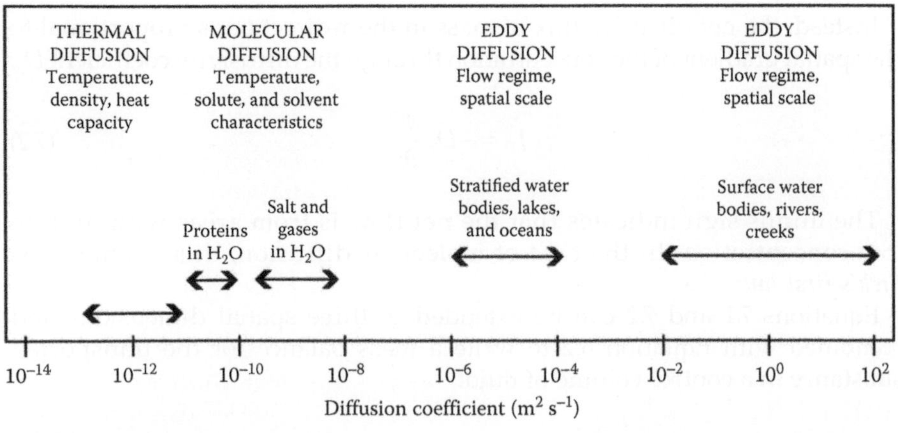

FIGURE 7.2
Diffusion coefficients. Orders of magnitude depend on flow regime.

L as $D_{hor} \propto L^{4/3}$. This observation is important, because, for example, as a spilled pollutant spreads in a lake, the length scale increases because a larger area is covered, and diffusivity increases as well, accelerating the whole process. Turbulent diffusivity coefficients range from $10^{-6}-10^{-4}$ m^2 s^{-1} in deeper zones of seas and lakes to $10^{-2}-10^{2}$ m^2 s^{-1} in horizontal, surface waters (channels and rivers): Thus, molecular diffusion can be neglected when the much-stronger eddy diffusion is present (Figure 7.2).

Analytical solutions can be found for Equation 7.4 for some particular initial and boundary conditions, especially in the monodimensional case, based on the superimposition of fundamental solutions for point sources (Fischer et al., 1979). In a monodimensional spatial domain, the following relationship can be derived, which describes the concentration resulting from an instantaneous injection of a mass M at time $t = 0$ in point $x = 0$ of a domain containing a still fluid that was previously free from that substance:

$$C(x,t) = \frac{M}{\sqrt{4\pi Dt}} e^{-x^2/4Dt} \tag{7.5}$$

Thus, concentration is represented by a bell-shaped curve that enlarges over time and whose center of mass is fixed in the point $x = 0$, where the injection was made (Figure 7.1a). If the fluid is not still but displays a nonzero velocity u, Equation 7.5 is modified to account for both advection and diffusion; that is, the bell-shaped curve spreads and, simultaneously, its center of mass moves with speed u due to advection (Figure 7.1b):

$$C(x,t) = \frac{M}{\sqrt{4\pi Dt}} e^{-(x-ut)^2/4Dt} \tag{7.6}$$

When exact solutions for Equation 7.4 cannot be found, numerical solutions can be computed (Hundsdorfer and Verwer, 2003).

7.3 Dispersion

Besides advection and diffusion, another important transport process should be mentioned: dispersion, which is the spreading of a substance in a fluid along the direction of an advective motion, due to the velocity gradient or profile. For example, when water flows in a round pipe, velocity is close to zero near the walls due to shear stress and it increases as one moves toward the center of the pipe section, where velocity is highest. Therefore, the effect of different water velocities on the pipe cross section will cause the particles of a substance, uniformly distributed over such a section, to spread along the axis of the pipe. Similarly, mixing takes place in rivers and channels due to velocity gradients, as currents near the bottom and the banks are slower than those at the center of the flow cross section. Flows displaying velocity gradients are called *shear flows* and, interestingly, over sufficiently long time scales, the effect of shear flow dispersion is equivalent to a Fickian diffusive process (Fischer et al., 1979). Therefore, the same mathematical description that is used for the combined effect of advection and diffusion can be applied. For example, if one averages concentration and water velocity over the cross section of a river, the following equation is obtained, which is equivalent to the one-dimensional version of Equation 7.4:

$$\frac{\partial \bar{C}}{\partial t} + \bar{u}\frac{\partial \bar{C}}{\partial x} = \frac{\partial}{\partial x}\left(K_x \frac{\partial \bar{C}}{\partial x}\right) \tag{7.7}$$

where K_x is the longitudinal dispersion coefficient. Generally, shear flow dispersion predominates in water bodies that are characterized by strong shear stresses due to shallowness, presence of banks, and large mean flow, such as rivers, channels, and estuaries. Diffusion can be considered more important in deep and wide water bodies, for example, lakes.

7.4 Mass Balance and Reactors

Mass balance is a basic physical principle, which, despite its simplicity, represents a powerful tool that simulates the fate of substances, such as biomass, pollutants, or nutrients, in an ecological system. If a system has well-defined spatial boundaries, that is, it can be considered a reactor, we can easily write a budget for the fluxes of mass entering (J_{in}) and exiting (J_{out}) from the system:

$$Accumulation \pm Reaction = J_{\text{in}} - J_{\text{out}} \tag{7.8}$$

Thus, the difference between mass inflows and outflows yields the accumulation of matter inside the system plus a reaction term representing the generation (e.g., after a chemical reaction) or degradation (biodegradation, but also settling, etc.; see Section 7.5, Section 18.2) of matter that can take place inside the reactor. As pointed out in Chapter 6, mass balance is commonly used to model a large number of ecological processes. For instance, the transport equations shown in Sections 7.2 and 7.6 are based on such an approach. Here, the most common reactor models used in ecological applications are reviewed. The word *reactor* simply highlights the fact that these mathematical tools are common in chemical engineering applications; for our purposes, a reactor could be an ecosystem, for example, a lake or a wetland.

The continuously stirred tank reactor (CSTR) is probably the simplest and most intuitive reactor. A CSTR is based on the idealized assumption that the system is perfectly mixed, meaning that the concentration C (currency: $[M]\,[L]^{-3}$) of the substance analyzed is the same at every point of the reactor and, consequently, in the outflow from the CSTR (Figure 7.3a). A CSTR formulation is often used as a useful first approximation to study the effects of discharge of pollutants or nutrient loads in lakes, or to model one or more stages in wastewater treatment plants.

Equation 7.8 can be adapted to such a system, yielding a differential equation:

$$V\frac{\partial C}{\partial t} = L(t) - QC(t) - kVC(t) - vA_sC(t) \tag{7.9}$$

where V is the reactor volume $[L]^3$, which is assumed to be constant over time; t is time; $L(t)$ is the mass load entering the reactor $[M]\,[T]^{-1}$; Q is the fluid flow (e.g., water) exiting from the reactor $[M]^3\,[T]^{-1}$ and assumed to be constant over time; the third term on the right-hand side of Equation 7.9 represents a first-order reaction, as the rate of mass change $V\,\mathrm{d}C/\mathrm{d}t$ is proportional to the mass $VC(t)$ found in the system through the rate coefficient k, and the minus sign indicates a degradation process; and the fourth term represents the net loss of mass from the system through settling, which is again simulated as a first-order kinetics that is dependent on the apparent settling velocity $v\ [L]\,[T]^{-1}$ and the area $A_s\ [L]^2$ of the sediment–water interface. Even if Equation 7.9 includes only one source of mass load and one sink of mass leaving the system, there is no loss of generality because the CSTR is perfectly mixed: Therefore, the sum of multiple point sources and even distributed sources can be lumped into the term $L(t)$, as instantaneous mixing will cancel the effects of the different localizations of such sources in space.

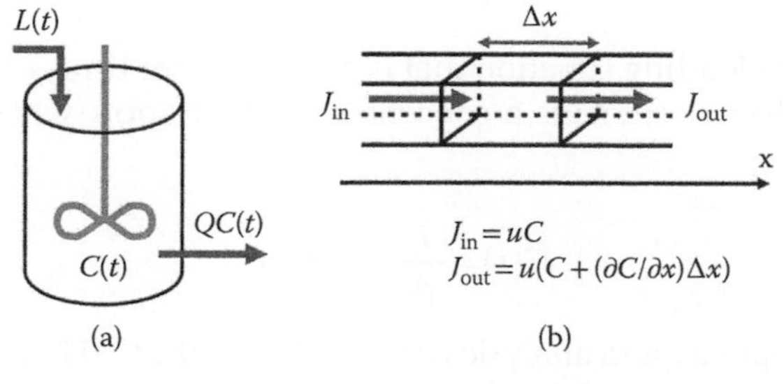

FIGURE 7.3
Mass budgets in two commonly used reactors: (a) a continuously stirred tank reactor, which is completely mixed, and (b) a plug flow reactor, where advection dominates and the balance is made over an infinitesimal element, and then integrated over the longitudinal x-axis. Thick arrows represent mass fluxes (a) or surface-specific mass fluxes (b). In both cases, reactions such as degradation and settling can take place inside the reactor.

On further assuming that the water inflow is constant over time and equal to the outflow, we can write

$$V\frac{\partial C}{\partial t} = QC_{in}(t) - QC(t) - kVC(t) - vA_sC(t) \quad (7.10)$$

Thus, the concentration of a substance inside the system will depend on a combination of its geometric and hydraulic characteristics (Q, V, A_s); the concentration C_{in} in the inflow, that is, the mass input; and rates of degradation and settling. Under steady-state conditions, from Equation 7.10, one can directly calculate the concentration as

$$C = C_{in}/(1 + k\tau + vA_s/Q) \quad (7.11)$$

where $\tau = V/Q$ is the hydraulic residence time in the reactor. This formulation shows that a higher concentration in the inflow will increase the concentration in the CSTR; whereas such a concentration will decrease due to longer residence time, lower water throughflow, and higher degradation/settling rates. If steady-state conditions are not met, Equation 7.9 can be solved to predict $C(t)$ in a numerical manner, or even analytically, for particular forms of the loading function $L(t)$ (Chapra, 1997), exploiting the fact that the general solution of Equation 7.9 is the sum of the solution of the homogeneous equation that is associated with it and of a particular solution depending on $L(t)$:

$$C(t) = C(0)e^{-\gamma t} + C_{part} \quad (7.12)$$

where $\gamma = Q/V + k + vA_s/V$. For example, in a CSTR with zero initial concentration, a Dirac-delta impulse input of mass M at time $t = 0$; that is, $L(t) = \delta(0)M$ yields a decaying exponential concentration:

$$C(t) = \frac{M}{V}e^{-\gamma t} \quad (7.13)$$

Instead, a step-loading function that is constant over time, $L(t) = L$ for $t \geq 0$ and $L(t) = 0$ otherwise, yields a concentration that approaches a saturation level over time:

$$C(t) = \frac{L}{\gamma V}(1 - e^{-\gamma t}) \quad (7.14)$$

Given the simplicity and analytical tractability of the CSTR model, this kind of a reactor is also used to simulate nonideal systems that are incompletely mixed, for example, because of spatially heterogeneous concentration distributions or complex spatial structures, such as a wastewater treatment plant with multiple stages, or a wetland consisting of several basins. In order to do this, the whole system is represented as a number of interconnected CSTR subsystems, all of which are completely mixed and characterized by their own volume, reaction rate, and so on. The output of each CSTR can be fed into other reactors, so that they are coupled by mass fluxes. Depending on the topology of the fluxes among reactors, the CSTR system can have a simple linear structure, as in the case of a series (or cascade) of CSTRs, or it can be more complex, as in the case of a network of CSTRs that involve feedbacks. The latter topology permits the simulation of the recycling of flows, which is a typical feature of ecological systems (including the ones made by man); for example food webs or wastewater treatment plants.

In the case of an elongated system in which advection along the main longitudinal direction dominates, the assumption of complete mixing can be considered inappropriate for the whole system, but can still hold for its cross section. Examples include open systems, such as rivers and channels, and constrained systems, such as pipes and blood vessels. In all these reactors, the water flow can be assumed to be predominant over all other mixing processes along the longitudinal direction, and mixing can be considered perfect in the radial direction, that is, in the cross section, but not along the axial direction. In this case, the plug flow reactor (PFR) (Chapra, 1997), which is another ideal reactor, can be used to model the change in the concentration of a substance along the longitudinal axis. This is done by building a mass balance for an infinitely thin slice that represents one cross section of the PFR (Figure 7.3b), yielding a differential equation whose integration over the length of the whole reactor links the concentration and position x along the longitudinal axis of the PFR. In the case of a first-order degradation reaction with rate k taking place within the reactor, such mass balance yields

$$\frac{\partial C}{\partial t} + u\frac{\partial C}{\partial x} = -kC \tag{7.15}$$

where u is the water velocity in the longitudinal direction. This relationship is a monodimensional advection equation (Section 7.2) where a first-order reaction term has been added.

If dispersion in an elongated system cannot be neglected, PFR is not the right model choice and the mass balance should include a mass flux which is caused by turbulent diffusion, that is, proportional to the spatial gradient of concentration, in addition to the advective mass transport. In such a case, the reactor is termed mixed flow reactor (MFR), and the transport equation becomes

$$\frac{\partial C}{\partial t} + u\frac{\partial C}{\partial x} = DA\frac{\partial^2 C}{\partial x^2} - kC \tag{7.16}$$

where D is the turbulent diffusivity coefficient, and A is the surface of the reactor cross section. Chapra (1997) describes an analytical solution for this equation, which resembles the advection–diffusion Equation 7.4 in one dimension, with the addition of a degradation term.

7.5 Settling and Resuspension

Settling is the vertical transport of particulate substances from the water column (or pelagic compartment) where they are suspended toward the sediments (or benthic compartment) of a water body. Many substances are prone to settling in aquatic ecosystems, making settling a key ecological process for both the pelagic and benthic compartments, although due to different reasons. In the latter compartment, settling represents a source of mortality for primary producers. When sinking phytoplankton cells quit from the well-lit part of the water column, known as the *euphotic zone*, their growth is no longer possible (Chapters 14 and 15). On the other hand, settling matter, including but not limited to microalgae, can represent the major source of energy and nutrients for organisms living at the bottom of the aquatic ecosystem, where photosynthetic primary production is either scarce or cannot take place due to light limitation. Besides phytoplankton cells, settling can involve other planktonic organisms and, in general, particulate, agglomerated, or flocculated matter, including resuspended sediments. One important example of settling matter in marine ecosystems is that of marine snow, which is (mainly organic) matter that reaches the seafloor of deep-sea ecosystems through the water column; it originates from settling plankton, exudates of microalgae and bacteria, fecal pellets, excretion of zooplankton, pelagic fishes, and invertebrates, carcasses of dead organisms, and so on. The settling of marine snow can represent the main source of energy for these marine ecosystems that are not reached by sunlight. Settling is also a significant process in many environmental engineering applications, for example, in removing solids from wastewater in treatment plants.

When the matter settling in a water body reaches the bottom, becoming a part of the sediments (a process known as *sedimentation*), there are several possibilities: It can be resuspended back into the water column (resuspension) by tidal and wave energy, which is transmitted to the bottom by eddies; it can be covered by a large amount of other sediments (burial) and become unavailable in the ecosystem; and it can disappear because of physical–chemical–biological reactions (biodegradation, adsorption, etc.) that are caused by the conditions found at the bottom, such as the presence of a solid phase or anoxic conditions. Here, settling and resuspension are treated as physical processes; however, it is important to remember that both settling and resuspension can sometimes be affected by biological processes. For example, settling rates can be influenced by the physiological state of phytoplanktonic cells; biofilms at the bottom can trap and bind sediment particles; and resuspension can be enhanced by the bioturbation that is caused by benthic organisms mixing and ingesting sediments, digging burrows, and so on.

In this section, we will focus on settling and resuspension in water; however, equivalent processes can be observed in other fluids, for example, in air, the deposition of particles upon the soil or the erosion caused by wind.

Indeed, some of the most common models for settling in water can be easily adapted to other fluids by changing the proper parameters (fluid density and viscosity). This is the case with Stokes' law, which represents a particular instance of a simple but very general model of settling, which is now described, based on Newton's second law. The study of this model is instructive, because it highlights the main factors that influence settling in water. The second law of motion for a particle with mass m, mass density ρ_p, and volume V, settling in a fluid with settling velocity v, is (Figure 7.4a)

$$ma = \rho_p V \frac{\partial v}{\partial t} = F_g - F_b - F_d = \rho_p V g - \rho_f V g - C_d A \rho_f v^2/2 \tag{7.17}$$

That is, vertical acceleration a times mass is given by the gravitational force F_g subtracted from the buoyancy force F_b and the frictional (or drag) force F_d due to fluid resistance. The letter g stands for gravitational acceleration, ρ_f is the fluid density, C_d is a drag coefficient depending on both the flow regime through the Reynolds number Re and the shape of the particle, and A is the particle surface as projected on the plane which is orthogonal to the direction of motion, that is, the area opposing to settling. Since, usually, one studies the settling of multiple particles in water, a hidden assumption in this balance is that these particles are noninterfering, meaning that the settling of one particle is not influenced by the presence of other particles; for example, particles do not flocculate or form aggregates. This condition can be observed if the solution is dilute.

If particles are small, a constant settling velocity is rapidly attained, that is, steady-state conditions, and $a = 0$. Therefore, Equation 7.17 can be solved for v:

$$v = \sqrt{\frac{2Vg\left(\rho_p - \rho_f\right)}{C_d A \rho_f}} \tag{7.18}$$

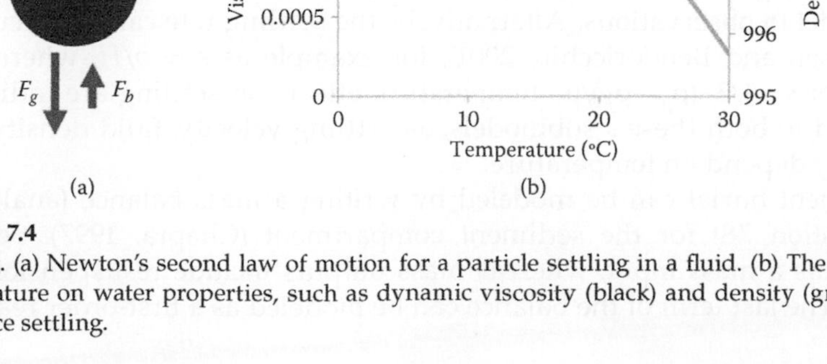

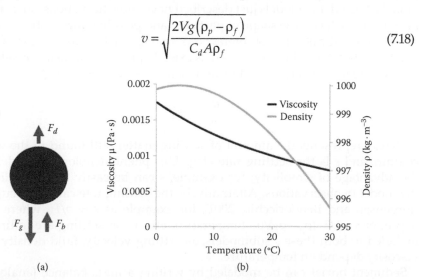

FIGURE 7.4
Settling. (a) Newton's second law of motion for a particle settling in a fluid. (b) The effect of temperature on water properties, such as dynamic viscosity (black) and density (gray), that influence settling.

Then, if we assume a spherical particle and laminar flow conditions, which correspond to $C_d = 24/\mathrm{Re}$ ($\mathrm{Re} = (\rho_f v d)/\mu$; d is the particle diameter, and μ is the dynamic viscosity of the fluid), Stokes' law can be derived:

$$v = \frac{g\left(\rho_p - \rho_f\right)d^2}{18\mu} \tag{7.19}$$

Thus, the difference of density between the particle and the fluid is proportional to settling velocity. Equations 7.17 through 7.19 highlight that, in general, the process of settling depends on factors such as particle shape and size, characteristics of the motion (flow regime; moreover, friction depends on settling velocity), and fluid properties (density, viscosity). Temperature influences settling by acting on fluid density and viscosity (Figure 7.4b), and settling velocity tends to increase in warmer waters. For other flow regimes (e.g., turbulent flow) and nonspherical particles, alternative expressions for C_d can be derived, and peculiar particle shapes, such as those of microalgae, can be modeled by modifying Equation 7.19 based on the concept of equivalent radius and correction factors for shape (Jørgensen and Bendoricchio, 2001).

We have just reviewed the main factors that influence settling; however, many additional elements, such as water column stratification, presence of currents and waves, and interactions among particles, play a role in this process. Therefore, a physically based approach to simulate settling can be inappropriate, leading to incomplete or, alternatively, unnecessarily complicated models. Indeed, the models just described have some limitations, such as the need to characterize the shape of particles, and possibly unrealistic assumptions, such as the absence of interactions among particles in the case of flocculation. Therefore, alternative models exist. In practice, for instance, settling can be modeled through a first-order reaction (see also Section 7.4):

$$\frac{\partial \xi}{\partial t} = -s\xi \tag{7.20}$$

Here, ξ represents the amount of settling matter still found in the water column, and s is the settling rate [T^{-1}]. This phenomenological model has the advantage of simplicity; for example, s can be easily found by fitting the model to observations. Alternatively, the settling rate can be calculated (Jørgensen and Bendoricchio, 2001), for example as $s = v/H$, where H is depth, or $s = 1/3 \cdot (\rho_p - \rho_f)/\mu$. Temperature effects on settling are indirectly included in both these s submodels, as settling velocity, fluid density, and viscosity depend on temperature.

Sediment burial can be modeled by writing a mass balance (analogous to Equation 7.8) for the sediment compartment (Chapra, 1997). Settling represents a mass input, whereas mass outputs include resuspension and burial. The last term of the balance can be modeled as a first-order reaction.

Alternatively, the fate and burial of a contaminant that is dissolved in the sediments can be modeled through an advection–diffusion equation (Section 7.2) which describes its vertical transport within the sediments. The diffusion term describes how the contaminant diffuses within the sediment pores, whereas the advection term accounts for burial by representing the change in the vertical position of the sediment–water interface, due to the continuous accumulation of settling material at the bottom.

Resuspension of sediments from the bottom of a water body has an opposite effect to the effects of settling and sedimentation. Resuspension can be important for aquatic ecosystems, as it brings back nutrients and organic matter to the water column, preventing their burial in the sediments due to additional settling material and their consequent disappearance from the system. Resuspended matter can drastically change the physical properties of the pelagic compartment by increasing water turbidity and, thus, the extinction of light (Section 8.3) throughout the water column. Therefore, resuspension can reduce the light available to benthic primary producers such as sea grasses and macroalgae, thus influencing local food webs (Section 19.5). In exploited marine ecosystems, strong sediment resuspension can be caused by fishing gears dredging the bottom, and the resulting reduced water clarity has a negative impact on benthic productivity, which sums to the other negative effects of fishing on bottom organisms. However, as discussed, dredging can also resuspend matter that is rich in energy and nutrients and, thus, paradoxically, sustain benthic fishery resources.

Several factors can influence resuspension (Figure 7.5), among which stands wind speed U: The energy that wind transmits to the water body creates waves, whose energy is conveyed to the bottom, in the form of a shear stress τ, through the action of the eddies that waves themselves, as well as other eddies, create in the water column. Eddies are dissipative structures; thus, only a fraction of the wind energy actually reaches the sediments.

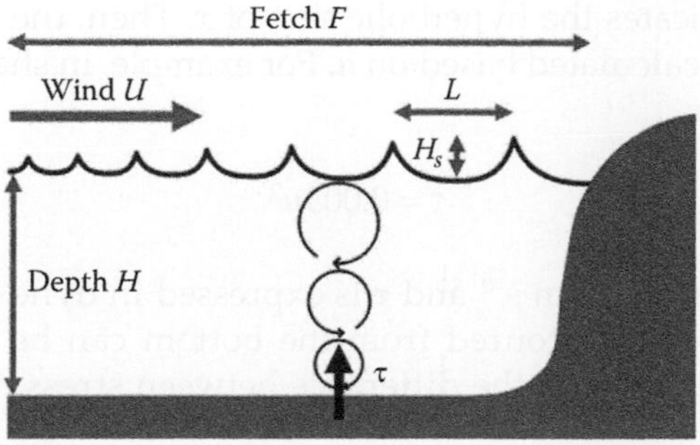

FIGURE 7.5
Factors influencing the resuspension of sediments in a water body, based on Chapra (1997) and Jørgensen and Bendoricchio (2001). The gray area represents sediments and soil. Wind creates waves, whose energy is only partially transmitted to the sediments because of dissipation in eddies, which are represented by circular arrows. Symbols are described in the text.

The geometrical characteristics of the water body, namely its depth H and fetch F, play a key role in resuspension. Intuitively, a very deep system experiences a strong dissipation of wind energy and, hence, a weak resuspension. Fetch is the length of the water body exposed to the wind, as measured along the direction over which the wind is blowing. Thus, fetch for a system can change depending on meteorological conditions. If the fetch is long, the wind can travel for a long time (or, over a long distance, which is the same) over the water body and transmit a high amount of energy to it, thus building up big waves. If the fetch is short, wind energy cannot accumulate in the water body and sediment resuspension is low. As an example, treatment wetlands should be designed in a way that minimizes the fetch along the dominant wind direction, because this choice would reduce the resuspension of settled pollutants.

Finally, resuspension is influenced by the sediment properties, which can be summarized in a single parameter, the critical stress τ_c. The critical stress depends on many factors such as the sediment grain size (which is not only an indicator of the grain weight, but is also related to the type of material: gravel, sand, clay, etc.), grain size distribution, cohesion, consolidation state (e.g., for clay sediments), presence of biofilms, plant roots, and so on. Resuspension only takes place if the shear stress τ exerted at the bottom by the eddies is higher than the opposing stress τ_c.

Complex formulas exist (e.g., Chapra, 1997; Jørgensen and Bendoricchio, 2001) that relate the effect of the energy transmitted to the water body, which is a function of U, F, and H, to wave characteristics such as significant height H_s, period T_s, and wavelength L. These wave characteristics can be related to water velocity u near the bottom (usually, at 15 cm from it), for example, as

$$u = \frac{\pi H_s}{T_s} \frac{100}{\sinh(2\pi H/L)} \tag{7.21}$$

where sinh(x) indicates the hyperbolic sine of x. Then, the stress exerted on sediments can be calculated based on u. For example, in shallow waters, this relationship holds:

$$\tau = 0.003u^2 \tag{7.22}$$

where u is expressed as cm s^{-1} and τ is expressed in dyne cm^{-2}. Finally, the amount of sediments ε scoured from the bottom can be calculated when $\tau > \tau_c$ based on the cube of the difference between stress and critical stress (Chapra, 1997):

$$\varepsilon = (0.008/49)(\tau - \tau_c)^3 \tag{7.23}$$

where ε is expressed in g m^{-2}, and if $\tau \le \tau_c$, it is zero. According to Equations 7.22 and 7.23, ε depends quite strongly on water velocity, being a function of u^6, and, thus, resuspension will be disproportionately high during extreme weather events. Equation 7.23 also shows that the determination of the critical sediment stress τ_c is central to quantifying the amount of resuspended matter. This is not a straightforward task (Jørgensen and Bendoricchio, 2001) due to the interplay of the multiple factors influencing τ_c mentioned earlier. For example, it is intuitable that gravel with larger grains should be characterized by a larger τ_c than gravel with smaller grains, which is lighter, and that τ_c should be larger in consolidated clay than in unconsolidated clay with the same grain size; however, it is not trivial to state whether τ_c is larger for gravel or clay, as it depends on the relative magnitude of the effects of grain weight and consolidation.

7.6 Transport in Porous Media

In the previous sections, we have seen that several processes determine the physical transport of matter contained in surface waters. However, fluids and the substances they carry can also flow though porous media. A porous medium is a solid matrix that contains void spaces (or pores) which are interconnected, forming very small and irregular channels. When a pressure gradient is present, fluids can flow through such microscopic pathways and, therefore, transport the mass of solutes that they contain. A typical example of a flow through a porous medium in environmental sciences is groundwater flow in natural soils and rocks. In porous media, motion is generally dominated by viscosity and is slow, due to the resistance and energy dissipation arising from the presence of the solid matrix. For this reason, flow in groundwater is typically many orders of magnitude lower than flow in a channel or river, but, given the high fluid volumes that can be transported in porous media occupying a large spatial extent such as the soil, these mass transport processes can become significant on relatively long-time scales. Here, we will focus on mass transport by water in soils, a situation that is commonly encountered in environmental fields such as ecotoxicology, for example, to study the fate of toxicants in groundwater, or in watershed and eutrophication modeling, to characterize groundwater fluxes of nutrients, previously leached from agricultural land, toward coastal water bodies. However, as in previous sections, most of the considerations provided here are wide ranging and apply to liquids flowing through solid matrices in general: Environmental examples include filters for treating wastewater, microporous structures found in the body of living organisms, and chromatography.

A basic parameter that characterizes a porous medium is porosity φ, that is, the ratio of the void volume contained in the medium to its total volume (voids plus solid matrix). In the upper layer of soils, pores contain both water and air, and this zone is termed *vadose* or *unsatured*. The vadose zone is dominated by vertical flows, through which water volumes due to precipitation and irrigation reach the aquifers in the groundwater. This soil layer is particularly important because of the interactions between human activities and groundwater, such as leaching of nutrients contained in fertilizers or contaminant spilling. In the deeper layers of soil, aquifers (the "saturated zone") are found, where pores only contain water and flow in the horizontal direction is predominant.

Before dealing with mass transport in porous media, it is useful to understand how groundwater flows. This process is usually studied by adopting a macroscopic, phenomenological approach, because the fluid motion in all the small and irregular channels among the pores is impractical to model. If such microscopic details are dropped and the flow is modeled on a spatial scale much larger than that of the pores, the soil–fluid system can be considered a continuum and studied through mass-balance approaches (Section 7.4). Darcy's law states that the velocity v $[L]$ $[T]^{-1}$ of water flowing through a monodimensional porous medium is

$$v = -k\frac{\partial h^*}{\partial x} \tag{7.24}$$

where k (m s^{-1}) is called *hydraulic conductivity*, and $h^* = h + p/\gamma$ is the hydraulic head, given by the sum of elevation h with regard to a reference level and of the ratio of pressure p in the fluid to the fluid specific weight γ (= ρg, where ρ is fluid density, and g is gravitational acceleration). The fluid volume transported per unit time, or flow, is simply obtained by multiplying v by the surface A of the cross section of the flow. It should be noticed that v and A are macroscopic variables averaged over a spatial domain that is large enough for the continuum assumption to hold. As such, v is not a real velocity but rather a volumetric flux per unit of cross-section area, and it is lower than the actual mean fluid velocity through the pores (or seepage velocity) $v_p = v/\varphi_s \approx v/\varphi$, where $\varphi_s = A_s/A$ is the surface porosity, that is, the ratio of the surface of the void A_s over the cross-section area to the whole cross-section surface.

Equation 7.24 highlights that a flow only occurs when a pressure gradient is found in the porous medium, and that stronger gradients mean larger flows from zones with high to low pressure, which is analogous to mass diffusion dynamics (Section 7.2). The key parameter k depends both on the fluid and on the soil properties. This can be highlighted by calculating the permeability k_p (m^2), which is a property of the porous medium only,

$$k_p = k\frac{\mu}{\gamma} \tag{7.25}$$

based on k, the fluid dynamic viscosity μ, and γ. According to Equation 7.24, k measures the efficiency by which a pressure gradient is converted into a flow, and Equation 7.25 shows that this efficiency is low when the fluid is viscous or k_p is low. Permeability k_p can be considered the ability of the porous medium to let fluids flow through it, and it can range over many orders of magnitude depending on the medium. Consequently, k for water in unconsolidated soils also displays a wide range of variability, varying from $10^{-2}-10^{1}$ m s^{-1} for gravel to $10^{-3}-10^{-2}$ m s^{-1} for mixtures of sand and fine gravel, $10^{-5}-10^{-2}$ m s^{-1} for fine-coarse clean sand, $10^{-8}-10^{-5}$ m s^{-1} for very fine sand and silt, and $10^{-12}-10^{-6}$ m s^{-1} for clay. These numbers show that k tends to be higher for larger grain sizes, which are associated with larger inter-grain voids and channels, enabling water flow in the porous medium. However, this does not necessarily mean that porosity is proportional to k: For example, clays have a larger porosity (0.5–0.6) than sand (0.3–0.4), but a much lower permeability. Interestingly, the values reported for k show a great variability, both within and across soil categories, that cannot be explained by mean grain size alone: Hydraulic conductivity is complexly influenced by multiple factors, such as the spatial arrangement of grains (e.g., spherical grains display $\varphi = 0.476$ for cubic packing, and $\varphi = 0.259$ for rhombohedral packing), grain surface roughness, grain size distribution, the direction of motion (k_p can be anisotropic, due to the asymmetric shape of soil grains and their orientation, to the presence of fractures in rocks, etc.), and so on. It is also fundamental to realize that k_p can strongly vary in space, both randomly and because of markedly different types of soil, soils formed by nonuniform deposition, and so on (Dagan et al., 2008). Moreover, in several ecological applications, the value of k can change over time due to physical, chemical, and biological processes. For example, a filter that purifies water can be clogged by the solid particles that it retains or by the growth of bacterial biofilms.

As discussed, k is a parameter that lumps many factors, and it can be difficult to be quantified by using theoretical considerations; therefore, it can be assessed in laboratory experiments, based on soil samples (e.g., using the constant head method), or it can be measured in situ (using tracers, pumping tests, etc.), which is a better approach because it can be performed on undisturbed soil. Empirical methods that calculate k, such as Hazen's relationship, also exist:

$$k = c d_{10\%}^{2} \frac{\mu}{\gamma} \tag{7.26}$$

where $d_{10\%}$ (effective size) represents the sieve opening that retains 90% (in weight) of the soil, and $c = 0.00151$ is a constant. Thus, according to Equation 7.26, the finest fraction of the soil has a major influence on permeability.

The water velocity quantified by Darcy's law (Equation 7.24) can be used to write a mass balance for a given control volume, obtaining an equation for a tridimensional groundwater flow:

$$S_s \frac{\partial h^*}{\partial t} = \frac{\partial}{\partial x}\left(k_x \frac{\partial h^*}{\partial x}\right) + \frac{\partial}{\partial y}\left(k_y \frac{\partial h^*}{\partial y}\right) + \frac{\partial}{\partial z}\left(k_z \frac{\partial h^*}{\partial z}\right) \tag{7.27}$$

where S_s is the specific storage $[L]^{-1}$, that is, the water volume released by an aquifer, per unit volume aquifer, per unit decline in hydraulic head. This formulation can be simplified by assuming that hydraulic conductivity is anisotropic and spatially homogeneous:

$$S_s \frac{\partial h^*}{\partial t} = k\nabla^2 h^* \quad (7.28)$$

This relationship resembles a diffusion equation (Section 7.2).

The models described so far can be used to characterize water flow in soils. In order to simulate the transport of a dissolved substance in such a flow, a mass balance (Chapter 6) can be written for the solute in water, accounting for advection, hydrodynamic dispersion (see Sections 7.2 and 7.3 for definitions) within pores, and any kind of reaction that causes the production or disappearance of the solute:

$$\frac{\partial C}{\partial t} + \mathrm{v} \cdot \nabla C = \nabla \cdot \left(\bar{D} \nabla C\right) + Reactions \quad (7.29)$$

where C is the solute concentration (solute mass/fluid volume), v is the 3D-vector containing seepage velocity components and whose magnitude is v_p, and $\bar{D}$ is the tensor of dispersion coefficients, highlighting that dispersion is an anisotropic process. The second term on the right-hand side represents the sum of chemical, biological, and geochemical reactions, causing the production, degradation, or transformation of the solute (e.g., adsorption/desorption, precipitation/dissolution). The solute-transport Equation 7.29 resembles the results obtained for surface waters in Section 7.2, and it can be analytically solved in a few simplified cases (e.g., a pulse of tracer creates a Gaussian-shaped curve of concentration, as in Section 7.2) or numerically.

Solute dispersion in porous media is caused by many factors, such as shear flows in pore channels and the branching of such channels, which creates multiple microscopic flow pathways, dispersing mass due to their different lengths (longitudinal dispersion, in the flow direction), their diverging directions (transverse dispersion, normal to the flow direction), and the different velocities observed in them. Dispersion is a particularly important process in the case of pollutants, as it causes spreading and, hence, dilution of the concentration plume. The dispersion coefficients D, representing the elements of the tensor $\bar{D}$, depend on the direction considered. In the direction of the flow, $D = aD_m + D_L$, where a (tortuosity) is a dimensionless coefficient <1, D_m is the molecular diffusivity, and D_L is the longitudinal hydrodynamic dispersion coefficient. Such coefficients can be expressed as $D_L = \alpha_L v_p$, that is, the product of the magnitude of seepage water velocity (which is, by definition, the velocity component in the longitudinal flow direction) and the longitudinal hydrodynamic dispersivity α_L. In the direction transverse to the flow, $D = aD_m + D_T$, and the transverse dispersion coefficient is $D_T = \alpha_T v_p$, where α_T is the transverse dispersivity. Dispersivities represent a property of the aquifer, although they are also linked to the spatial scale over which they are measured. In general, $\alpha_T < \alpha_L$.

A typical example for the reaction term in Equation 7.29 is the case of an equilibrium, linear adsorption/desorption isotherm that is reversible (see Section 11.2). Concentration C in water can be related to S, the ratio of the mass adsorbed onto the soil grains to the fluid volume, through the relationship $S = K_D C$, where K_D is a partition coefficient. The overall effect of adsorption is that the transport of the solute in the aquifer is delayed in time, as shown by the new form taken by Equation 7.29:

$$R\frac{\partial C}{\partial t} + \mathrm{v} \cdot \nabla C = \nabla \cdot \left(\bar{D} \nabla C\right) \tag{7.30}$$

where $R = 1 + K_D$ is called *retardation coefficient*.

7.7 Transport at Interfaces

Interfaces are surface boundaries between two substances in separate phases, such as a gas and a solid, a gas and a liquid, or even two different, immiscible liquids. Although the concept of interfaces can be extended by using several other definitions, for example, an interface could be an abrupt discontinuity in the hydrodynamic properties of a fluid found in a single phase (e.g., a fluid encountering a porous barrier and flowing through it; see Section 7.6), they will not be treated in this section. However, Section 7.10 will deal with transport over membranes, as this subject is particularly relevant for biological systems.

Mass transport across interfaces is a common topic in chemical engineering, being involved in processes such as distillation, adsorption and desorption (Chapter 11), extraction, and so on. An introduction to such an approach for both fluid–fluid and fluid–solid interfaces can be found in Bird et al. (2002), who also treat the interaction of mass transfer with heat transfer processes, and multicomponent systems. In general, the chemical engineering approach remains valid and useful for ecological systems, but here, we will focus on fewer and somehow simpler models. Besides the informative value of simple models, it is important to realize that it is not easy to find general relationships for ecological systems; indeed, one reason is that it can be difficult to collect reliable measurements to quantify satisfactorily all the ecological processes taking place in ecosystems and, because of this, general models are hard to validate, but there are other justifications. An ecosystem represents a much less controlled experiment with regard to a chemical reactor, if it can be said that an ecosystem is "controlled" at all, due to the high number of forcing factors acting and the complexity of their interactions. Each ecosystem represents a peculiar, nonrepeatable experiment of nature (see Ulanowicz [2009] for an instructive discussion); moreover, the contingent effects of its past history play an important role in its dynamics. For all these reasons, simple models are most likely to represent robust and general descriptions of ecological processes.

Here, we focus on mass transport at the interface of a gas and a liquid, a common and important process in aquatic ecosystem dynamics, that serves as the basis for processes such as reaeration (Section 7.8) and volatilization (Section 7.9). In environmental engineering, transfer at interfaces represents an important process in wastewater treatment plants; for example, water should be oxygenated to ensure that aerobic processes take place in activated sludge tanks, or it should be supplied with chlorine to kill bacteria in the disinfection stage. Ammonia stripping is commonly used to remove nitrogen from water. Two well-known models of gas transfer are described here: the two-film model, which is fitter for standing water bodies such as ponds, lagoons, and lakes, and the surface renewal model, which is used in flowing waters such as those of streams and rivers. In both models, mass transport across the surface is proportional to the value of a driving force that is simply quantified by the difference between two comparable quantities, that is, concentrations or partial pressures, similar to most of the relationships described in this chapter, which are based on gradients. As a result of this driving force, the system will tend toward equilibrium, as exemplified by a zero gradient (e.g., in concentration); that is, the system will tend to cancel the imbalance in concentration represented by the driving force, thus reminding us of Le Chatelier–Braun's principle in chemistry. The coefficient of proportionality between the mass flux and the driving force depends on both the surface of the interface and the mass-transfer velocity or mass-transfer coefficient, which, in turn, depends on the two fluid phases, the transported substance and environmental conditions (temperature, turbulence caused by wind).

Whitman's two-film theory (Whitman, 1923; Lewis and Whitman, 1924) is a commonly used model of the mass transport of a substance across the interface of a gas, which may be the atmosphere, and an underlying liquid, which may be a water body, both of which are assumed to be perfectly mixed by turbulence. This assumption is equivalent to saying that the molar concentration C_l (mol m^{-3}) of the substance in the bulk of the water body is homogenous, and the same holds true for the partial pressure p_g of such a substance in the atmosphere. Therefore, mass transport should take place, and encounter resistance, where gradients are actually found, that is, near the interface, where it is assumed that turbulence dies out and two incompletely mixed laminar boundary layers are present (Figure 7.6): one in the gas and one in the liquid. In each of the laminar films, mass transport will be driven by molecular diffusion and, as such, the magnitude of the flux will depend on the concentration gradient between the liquid bulk and the interface in the liquid film, as well as on the gradient of partial pressure between the gas bulk and the interface in the gas film. The molar flux (mol m^{-2} s^{-1}) between the liquid bulk and the liquid film, that is, across the liquid film, is, therefore,

$$J_l = K_l(C_i - C_l) \tag{7.31}$$

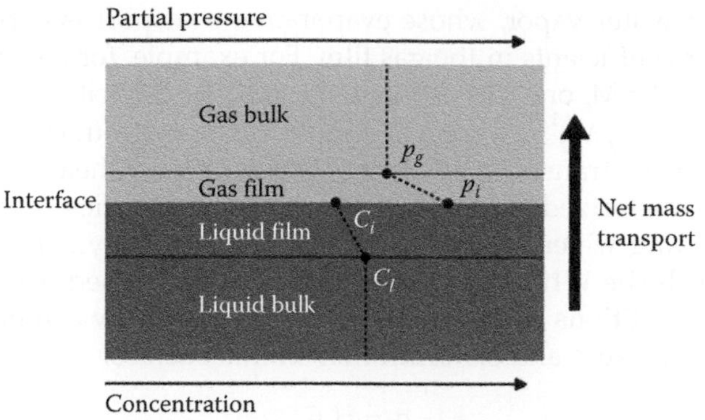

FIGURE 7.6
Whitman's two-layer model. The transport of a substance between the bulk of a gas and that of a liquid, both of which are well mixed by turbulence, is determined by the gradients of partial pressure and of concentration observed in the gaseous and liquid laminar boundary layers found at the liquid–gas interface. Partial pressure and concentration at the interface are in equilibrium and linked by Henry's law. Symbols are described in the text.

where K_l is the mass-transfer velocity in the liquid layer (m s^{-1}), and C_i is the substance concentration at the interface. The molar flux through the gas film is

$$J_g = \frac{K_g}{RT}\left(p_g - p_i\right) \tag{7.32}$$

where K_g is the mass-transfer velocity in the gas layer (m s^{-1}), p_i is the substance partial pressure at the interface, R is the universal gas constant, and T is the temperature. It should be noted that the partial pressure of a substance rescaled on RT is the number of substance moles per unit volume in the gas phase, that is, a concentration. Positive values for J_l and J_g mean that the mass flows toward the liquid. In addition, molar fluxes and concentrations are easily converted into mass fluxes and concentrations by multiplying them by the molecular weight of the substance studied, so that mass- and molar-based notations are basically interchangeable and we speak of "mass transfer."

Mass-transfer coefficients in both the liquid and the gas can be related to the respective molecular diffusivity coefficients (Section 7.2) through

$$K = \frac{D}{z} \tag{7.33}$$

where K represents the mass-transfer velocity in the liquid (or gas), D indicates the molecular diffusivity in the liquid (or gas), and z is the thickness of the liquid (or gas) layer (Chapra, 1997). Alternatively, mass-transfer coefficients are conventionally extrapolated from those of commonly studied gases (Chapra, 1997) such as oxygen (in the case of mass-transfer coefficients that describe the liquid film, as oxygen transfer is largely liquid-film controlled) or water vapor, whose evaporative transport is correlated with mass-transfer coefficients in the gas film. For example, for a substance with molecular weight M, one can estimate the transfer velocity in the liquid as being $K_l = K_{l,O_2}(32/M)^{1/4}$, where 32 g mol^{-1} is the molecular weight of oxygen and K_{l,O_2} is the transfer coefficient of oxygen. Hence, heavier compounds tend to be characterized by smaller mass-transfer velocities.

In the standing water bodies that are characterized by stagnant laminar films for which the Whitman's model has been developed, we can assume equilibrium conditions at the interface,* and Henry's law can be invoked to relate partial pressure and concentration on that surface:

$$p_i = H_e C_i \tag{7.34}$$

where H_e is the Henry's constant. Table 23.3 reports H_e values for several substances. Since the interface is at equilibrium, no accumulation of molecules occurs and it should be $J_l = J_g = J$. Combining this condition with Equations 7.31, 7.32, and 7.34, one obtains

$$J = K_{\text{net}}\left(\frac{p_g}{H_e} - C_l\right) \tag{7.35}$$

where the net mass-transfer velocity K_{net} across the interface lumps the influence on the transport of both laminar layers:

$$K_{\text{net}} = \frac{K_l H_e}{H_e + RT\left(K_l/K_g\right)} \tag{7.36}$$

According to Equation 7.35, the interphase molar flux depends on the gradient of concentration between the bulk of the gas and that of the liquid. Equation 7.36 can be rearranged as follows:

$$\frac{1}{K_{\text{net}}} = \frac{1}{K_l} + \frac{RT}{K_g H_e} = R_L + R_G \tag{7.37}$$

where $R_L = K_l^{-1}$ and $R_G = RT/(K_g H_e)$. This equation says that the reciprocal of the overall mass-transfer velocity K_{net} is the sum of the reciprocals of the transfer velocities in each of the films (after proper rescaling in the case of the transfer velocity for the gas). Since the reciprocal of a transfer velocity in a film can be considered the resistance that the film opposes to mass transfer, Equation 7.37 highlights an analogy between mass transport across two films and current flowing through an electrical circuit with two resistors in series that are characterized by resistance R_L and R_G. Thus, the resistances to mass transport given by the two films work additively. High values of the ratio $R_L/(R_L + R_G)$ indicate that transport at the interface is liquid controlled, that is, the liquid phase is

* See also Section 18.4.

the bottleneck; conversely, low values are found when mass transfer is controlled by the gas phase. Since this ratio is highly correlated to H_e (Figure 7.7) (Chapra, 1997) for many ecologically important gases, Henry's coefficient can be used to predict the nature of interphase mass transfer for a given substance. The transfer of soluble gases in water encounters most resistance in the gas film, insoluble gases encounter most resistance in the liquid film, and intermediately soluble gases encounter resistance in both films.

The surface renewal model (Danckwerts, 1951) can be applied to mass exchanges between air and flowing waters, such as rivers, for which the presence of a stagnant film at the interface, where air and water are continuously in contact, is not a valid approximation. In this model, which is a modification of the so-called *penetration theory* (Chapra, 1997), it is assumed that turbulence brings many small volumes (or parcels) of water into contact with air for a short period, whose duration is randomly distributed. During this short time, mass transport takes place over the interface. Then, the small volume goes back into the liquid's bulk (where exchange with air is no more possible) and gets mixed with it. Based on such assumptions, the molar flux J across the interface can be derived to be

$$J = \sqrt{D_l r_l}\,(C_i - C_l) \tag{7.38}$$

where r_l (s^{-1}) is the liquid surface renewal rate, and D_l is the liquid diffusivity ($m^2\ s^{-1}$). Therefore, the mass-transfer velocity in Equation 7.38 is $K_l = \sqrt{D_l r_l}$. Analogously, a molar flux across the gas phase can be calculated, assuming that parcels of gas are brought into contact with the liquid for a randomly short time, and the resulting mass-transfer velocity for the gas is $K_g = \sqrt{D_g r_g}$. Therefore,

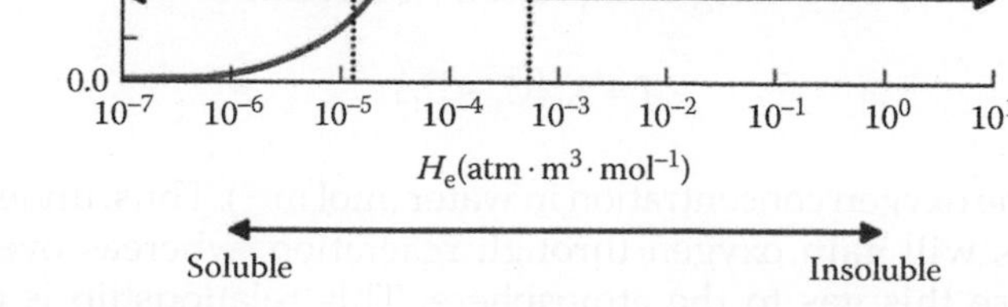

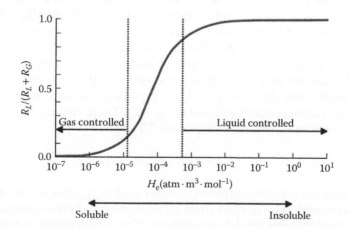

FIGURE 7.7
Fraction of total resistance to mass transfer due to the liquid phase for ecologically important gases as a function of Henry's constant, in lakes. Henry's constants for several gases as a function of temperature can be found in Table 23.3. (Modified from *Fundamentals of ecological modeling*, Jørgensen, S.E. and Bendoricchio, G., Copyright 2001 with permission from Elsevier.)

mass-transfer coefficients in the renewal model depend on the square root of diffusivity; whereas in the two-film model, they depend linearly on it (Equation 7.33), that is, stronger turbulence corresponds to a smaller exponent for diffusivity in the calculation of mass-transfer velocity.

7.8 Reaeration

Reaeration is the transfer of oxygen from air to a water body (natural or artificial) where the dissolved oxygen concentration has been previously reduced, generally by the biological or chemical degradation of organic matter or of other oxygen-consuming substances. Reaeration is an important process in water bodies that are subject to organic pollution, as it contrasts the oxygen deficit created by the bacterial degradation (Section 17.4, Section 18.2) of pollutants. If a large consumption of oxygen is not balanced by reaeration, it can result in hypoxic/anoxic conditions, killing invertebrates and fishes, generating unpleasant odors, and so on. For similar reasons, reaeration is enhanced in wastewater treatment plants through the usage of turbine mixers, aerators, and diffusers. This is done, because key activated sludge processes that treat sewage, such as the oxidation of carbon and nitrogen compounds, need oxygen to take place, but such gas is clearly insoluble, as shown in Figure 7.6 (Henry's constant for O_2 is $H_e = 0.769$ atm m^3 mol^{-1}). The insolubility of oxygen, along with the fact that it is very common in the atmosphere, makes reaeration quite a special case of mass transfer at the interface (Section 7.7). Since oxygen is insoluble, reaeration encounters most resistance from the liquid phase, and thus according to the two-film theory (Section 7.7), its mass-transfer velocity is $K_{net} \sim K_l$; that is, it is determined by the liquid. Finally, due to the stable composition of the atmosphere and the commonness of oxygen, its partial pressure in air is almost constant, and thus based on Henry's law (Equation 7.34), the term (p_g/H_e) in Equation 7.35 can be substituted with the saturation concentration of oxygen C_s. Then, the reaeration flux of oxygen into water according to Equation 7.35 becomes

$$J = K_l(C_s - C_l) \tag{7.39}$$

where C_l is the oxygen concentration in water (mol m^{-3}). Thus, undersaturated water bodies will gain oxygen through reaeration, whereas oversaturated ones will lose this gas to the atmosphere. This relationship is commonly reexpressed as

$$\frac{dO_l}{dt} = K_L(O_s - O_l) \tag{7.40}$$

where O stands for dissolved oxygen concentration (g m^{-3}), the "l" subscript indicates concentration in the liquid, and the "s" subscript indicates saturation concentration.

K_L is known as the *reaeration rate* or *overall mass-transfer coefficient* (s^{-1}):

$$K_L = \frac{K_l}{H} \tag{7.41}$$

where H is mean system depth (m). Temperature T (°C) influences reaeration in two manners. K_L usually increases exponentially with increasing temperature according to the Arrhenius formulation:

$$K_L(T) = K_L(20°C) \cdot \theta^{T-20} \tag{7.42}$$

where θ ranges from 1.015 to 1.040, with a typical value being 1.024. Moreover, temperature also enters C_s (= O_s/M, where M is the molecular weight of oxygen, 32 g mol^{-1}), which clearly decreases in warmer water. Saturation concentration of oxygen is also influenced by substances found in water, such as particulates, and particularly by salinity, with lower C_s found in more saline water. Several empirical relationships exist in the literature that express such dependence of C_s on temperature and salinity. Oxygen saturation is also affected by altitude, as atmospheric pressure enters Henry's law, with the highest C_s found at the sea level (highest pressure). The effects of turbulence and system geometry on oxygen transfer are also important. In the aeration systems applied in wastewater treatment, mixing intensity and tank geometry are considered by multiplying K_L by proper correction factors. Alternatively, K_L is directly measured in pilot-scale or full-scale experimental facilities. In aquatic ecosystems, a large number of theoretical and empirical relationships exist (Chapra, 1997), linking K_L to the type of water body, its geometry, and the characteristics of water motion as well as of the surrounding environment. For example, in rivers and streams, K_L can be expressed as a power function of current velocity (with a positive exponent) and system depth (negative exponent). In lakes, the mass-transfer coefficient is chiefly influenced by wind and increases with higher wind speed; whereas in estuaries, the role of both wind and water velocities should be considered. Reaeration coefficients can also be estimated on the field, based on tracer experiments or measurements of dissolved oxygen concentration over time (Chapra, 1997).

7.9 Volatilization

Volatilization takes place when a substance is transferred from a water body to the atmosphere, partitioning between such liquid and gas phases. As such, volatilization is a special case of transport across interfaces (Section 7.7), and the same theory and models apply, namely the two-film theory in the case of standing water bodies, and the surface renewal theory in the case of flowing waters. In environmental applications, usually, one wishes to describe the volatilization of substances, such as pollutants and volatile organic compounds, that are found in water but are present in a negligible or zero concentration in the air. In this case, the partial pressure of that substance in the air is basically zero, and Equation 7.35 can be modified as

$$J = -K_{\text{net}} C_l \tag{7.43}$$

It should be remembered that the molar flux J is defined as *positive* when going from the gas to the fluid phase; therefore, Equation 7.43 indicates that matter is always purged from the water body for pollutants which are absent from the air. If one multiplies J by the surface A through which volatilization takes place, the relationship mentioned earlier can be rearranged as

$$V \frac{\mathrm{d}C_l}{\mathrm{d}t} = -K_{\text{net}} A C_l \tag{7.44}$$

where V (m^3) is the (constant) volume of the water body modeled. As usual, C_l indicates molar concentration, but it can easily be converted into mass concentration. K_{net} can be estimated according to the two-film theory from Equation 7.36, that is, based both on Henry's constant, which is commonly tabulated in literature (e.g., Fogg and Sangster, 2003), and on mass-transfer velocities in the gas and liquid films (see Section 7.7). High Henry's constants characterize insoluble gases (Figure 7.6) and suggest a large potential for volatilization.

7.10 Transport over Membranes

Membranes are thin layers of matter that separate two fluid phases and which selectively allow the through-flow of only some specific substances, while blocking other ones (e.g., because of a particle size larger than the membrane pores); hence, the name *selective membranes*. The fluids and/or substances that are able to flow through the membrane are called *permeate*, whereas the remaining fraction that is blocked by the semipermeable barrier that the membrane represents is called the *retentate*. Mass transport over membranes takes place because of the presence of a "driving force", which can be a gradient in concentration, as discussed for transport at interfaces (Section 7.7), or in pressure, temperature, gas partial pressure, and electric potential, depending on the type of the process. A classical example of transport over membranes is a selective diffusion process called *osmosis*, which is the net flow of a solvent from an area at a lower solute concentration (hypotonic solution), through a partially permeable membrane, to reach another area at a higher solute concentration (hypertonic solution). The semipermeable membrane does not allow the passage of the solute, and the gradient in the solute concentration is actually smoothed by the flow of the solvent, which tends to diffuse across the membrane, diluting the concentration in the second area and increasing it in the first area. However, osmosis will also increase pressure in the second area (due to added solvent), and this higher pressure will eventually stop the net solvent flow when the so-called *osmotic pressure* is reached. Osmosis plays a key role not only in the transport of molecules across biological membranes, but also in maintaining the turgor pressure of cells, a condition that is exploited by plants to achieve a rigid structure.

Membranes are a common feature of living organisms and their anatomy. They are found around the nuclei of cells as well as around cells and organs. The skin itself is a membrane. Indeed, the very existence of biological systems has been linked to the presence of clear spatial boundaries (see Section 2.2), which distinguish an ordered internal state from the outside chaotic environment, and membranes often represent such boundaries, providing cells, organs, and organisms with protection and structural support. Membranes also perform multiple biological functions by properly regulating flows of chemical elements and compounds through tissues. In biological systems, mass transport over membranes is said to be active if energy has to be consumed for it to take place; otherwise, transport is passive, that is, due to classical diffusion processes (Section 7.2). Due to this energy input, active transport can move a substance in the direction opposite to that which would be caused by a driving force; for example, a solute can move from a phase with low to high concentration.

Natural membranes inspired the creation of artificial ones, which are currently used in many engineering and industrial applications because of their selectivity properties, that is, capability of separating particular substances (which may be valuable or, alternatively, toxic/undesirable). Artificial membranes are mounted in a large number of different geometric configurations (plate and frame, spiral wound, tubular, capillary, hollow fiber, etc.) that depend on trade-offs among costs of production, construction, operation and maintenance, functionality, conditions of applicability, robustness, compactness, duration, and so on. Construction materials can be organic, inorganic (which are not only generally more stable from the chemical and thermal points of view, but also more expensive) (Cano-Odena and Vankelecom, 2008), or a mixture of both in composite membranes. If we limit the list of applications to biological and environmental processes, synthetic membranes are used in disparate fields such as water treatment (e.g., for the treatment of wastewater, desalination, and production of potable water), food industry, and biomedicine; for example, in hemodialysis and tissue engineering (Cano-Odena and Vankelecom, 2008). Membrane transport can be successfully coupled with other processes to form hybrid membranes.

For example, membranes can be integrated with oxidation and adsorption processes to treat wastewater, and with reactions involving biocatalysts (e.g., enzymes) to yield membrane bioreactors, which are used to treat wastewater and gases, enhance fermentation processes, produce pharmaceuticals, biochemicals, and so on. Hybrid membrane processes can also be used to produce energy through fuel cells (Cano-Odena and Vankelecom, 2008).

Synthetic membranes can be classified based on the nature of the driving force, or gradient, whose application on the two sides of the membrane causes mass transport (Table 7.1). The table shows that filtration processes driven by pressure can be classified into subcategories depending on the value of the pressure applied, with higher pressures corresponding to smaller membrane pore sizes, although the precise definition of the values of pressure or pore size ranges can vary according to literature sources (e.g., see Lorch [1987]). Pore size is a key parameter, as it determines the compounds that cannot permeate across the membrane, because they are larger than such size (the membrane acts as a sieve). Other factors that can influence the transport of a solute substance over a membrane include its size, shape, hydrophilicity/hydrophobicity, and potential interactions with the membrane (e.g., charge status of membrane material versus that of ionic compounds, adsorption processes), which can depend on environmental conditions, for example, temperature (Cano-Odena and Vankelecom, 2008).

Membrane- and hybrid membrane-based processes show several advantages with regard to more traditional separation and treatment technologies. In general, membranes (1) tend to produce less or no secondary products; (2) clearly isolate primary products and/or catalysts from raw material and

TABLE 7.1

Membrane Processes, Classified According to the Type of Gradient Driving Mass Transport

Process	Driving Force	Notes
Dialysis	Concentration	
Electrodialysis	Electric potential	Membrane should be conductive, as ions go through it as an electric current
Pervaporation	Partial pressure	A phase transition from liquid to gas takes place at the membrane
Microfiltration	Pressure	Pressure difference of ca 0.1–2 bar
Ultrafiltration	Pressure	Pressure difference of ca 1–10 bar
Nanofiltration	Pressure	Pressure difference of ca 10–35 bar. Also called *hyperfiltration*, along with reverse osmosis
Reverse osmosis	Pressure	Pressure difference of ca 15–100 bar. Also called *hyperfiltration*, along with nanofiltration

Source: Based on Cano-Odena, A., Vankelecom, I.F.J., *Encyclopedia of Ecology*, Elsevier, Amsterdam, 2008.

reaction by-products; (3) are flexible to manage (e.g., it is possible to adjust the residence time within the membrane); and (4) consume less energy (however, this last point is not always true). An example of membrane efficiency is the treatment of water and wastewater: Membrane filtration can remove fine particles, turbidity and bacteria in the case of microfiltration; viruses, colloids, and endotoxins in the case of ultrafiltration; and color, dissolved organics, and even ions in the case of hyperfiltration. On the other hand, it can be expensive to set up membrane processes (the cost of membranes or of treating large fluid volumes can be high, especially if the final product has a relatively low value) and these processes are not always robust, due to the loss/denaturation of biocatalysts in membrane bioreactors and, in general, to the process of membrane fouling (e.g., caused by filtered particles or the formation of biofilms), which can strongly degrade the process performance.

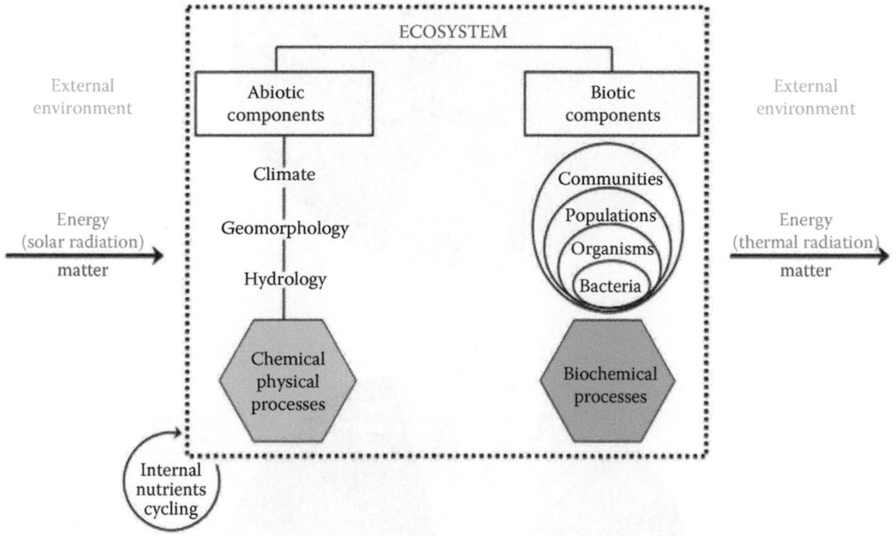

FIGURE 1.3
Ecosystem structure, including biotic and abiotic components as well as the interactions (input and output) with the surrounding environment.

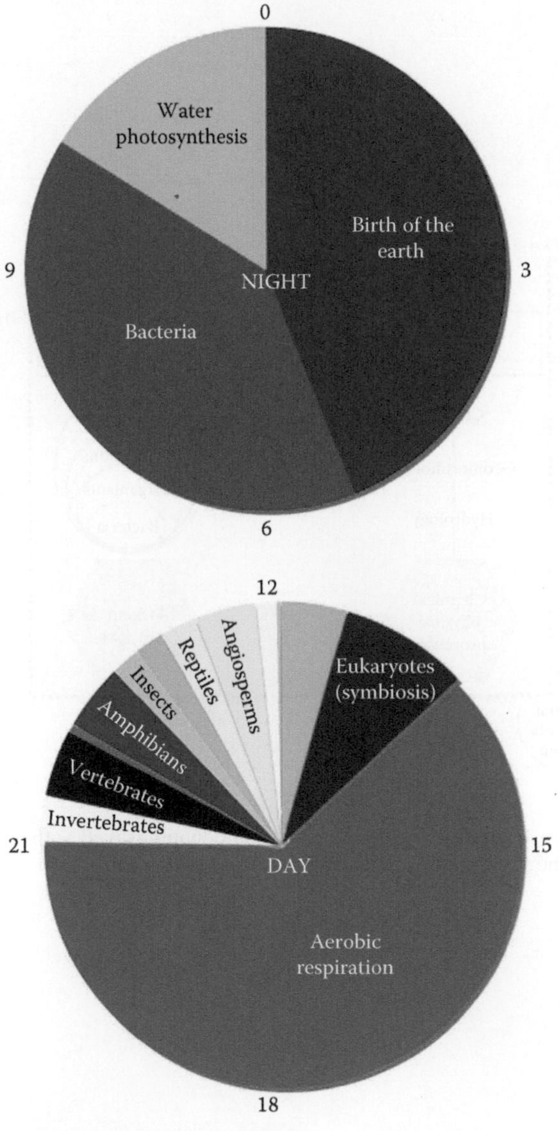

FIGURE 1.7
Evolution on the watch. Astoundingly, a whole day has to pass before the appearance of mammals.

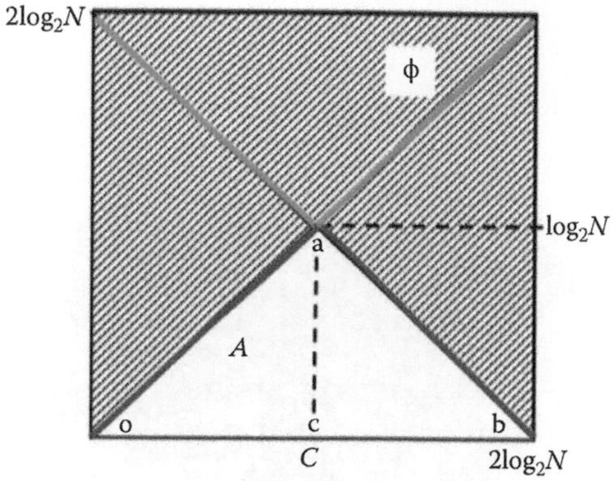

FIGURE 4.5
The domain of AMI. A and ϕ are plotted against C. (o) no connection, (a) linear closed loop, (b) everything connects with everything, and (c) star network where one single compartment feeds all the other compartments.

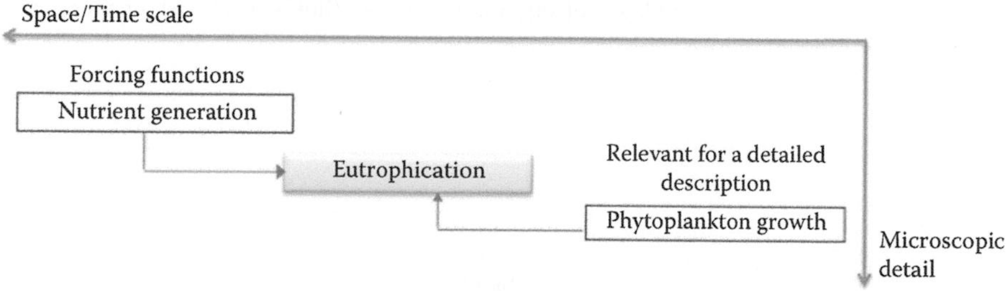

FIGURE 6.2
Hierarchical levels linking scales in time and space and the description's detail. For example, a satisfactory understanding of eutrophication requires the inclusion of forcing loads as well as other relevant microbial processes.

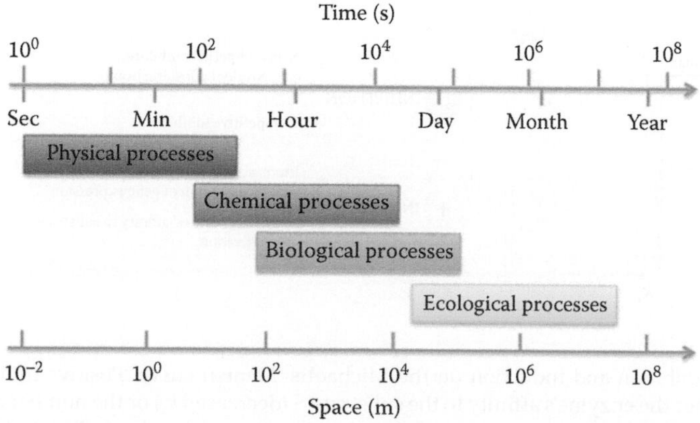

FIGURE 6.3
Typical space and time scale ranges for physical, chemical, biological, and ecological processes. (From Jørgensen, S. E. and Bendoricchio, G., *Fundamentals of ecological modeling* (3rd ed.), Elsevier, Amsterdam, 628 pp, 2001.)

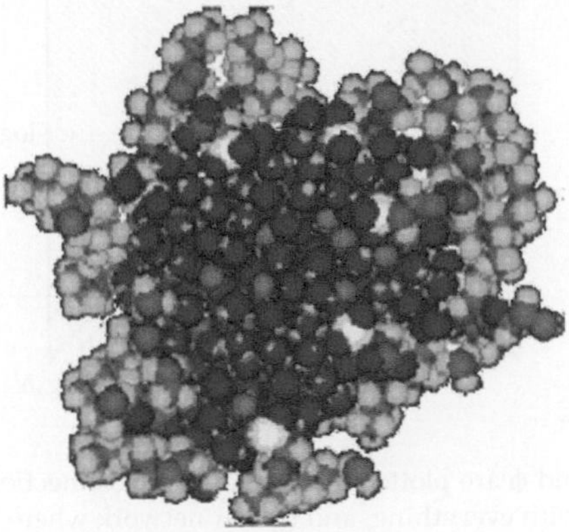

FIGURE 6.4
Membranes appeared as assemblages of organic molecules (lipids, proteins), enabling the onset of stable internal conditions.

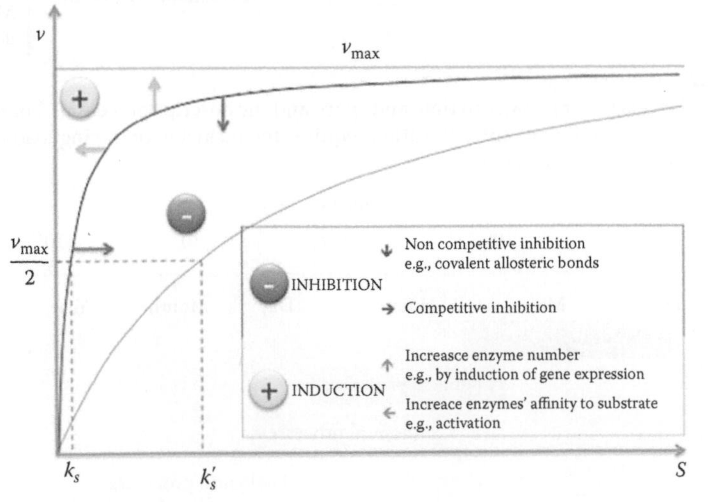

FIGURE 9.3
Effects of inhibition and induction on the Michaelis–Menten curve. Positive induction may increase either the enzyme's affinity to the substrate *S* (decreased k_S) or the number of enzymes itself and, consequently, increase v_{max}. Inhibition lowers enzymes' affinity so that a larger quantity of substrate is needed to reach the half-saturation point.

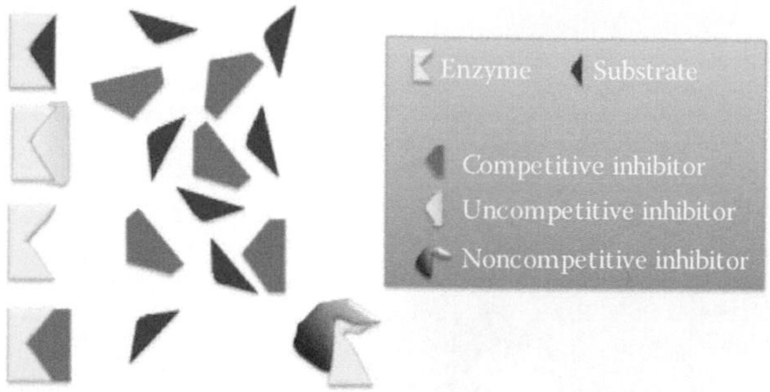

FIGURE 9.4
The lock and key model. Enzymes are plastic structures in which the active site adapts dynamically to the substrate. Inhibitors, which may subtract enzymes from the active pool by behaving similar to a substrate or by definitely reshaping the active site, hinder this adaptation.

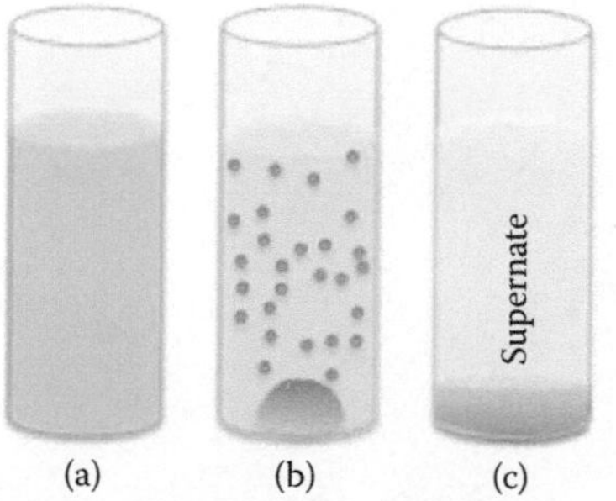

FIGURE 9.5
The mechanism of precipitation. In the solution (a), the formation of crystals by nucleation and growth (b), turns the solution into a suspension that finally undergoes sedimentation (c), *Supernate* refers to the liquid that remains above the solid produced by precipitation.

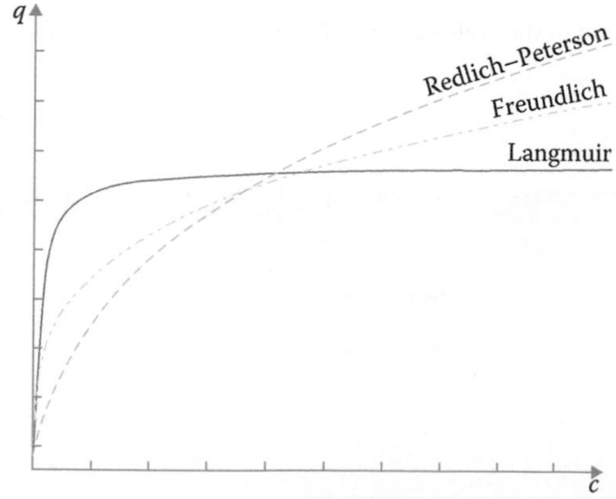

FIGURE 11.1
Plot of adsorption isotherms. If the temperature increases, the curves are displaced toward the bottom of the picture. This corresponds to a shift of the dynamic equilibrium in the direction of desorption.

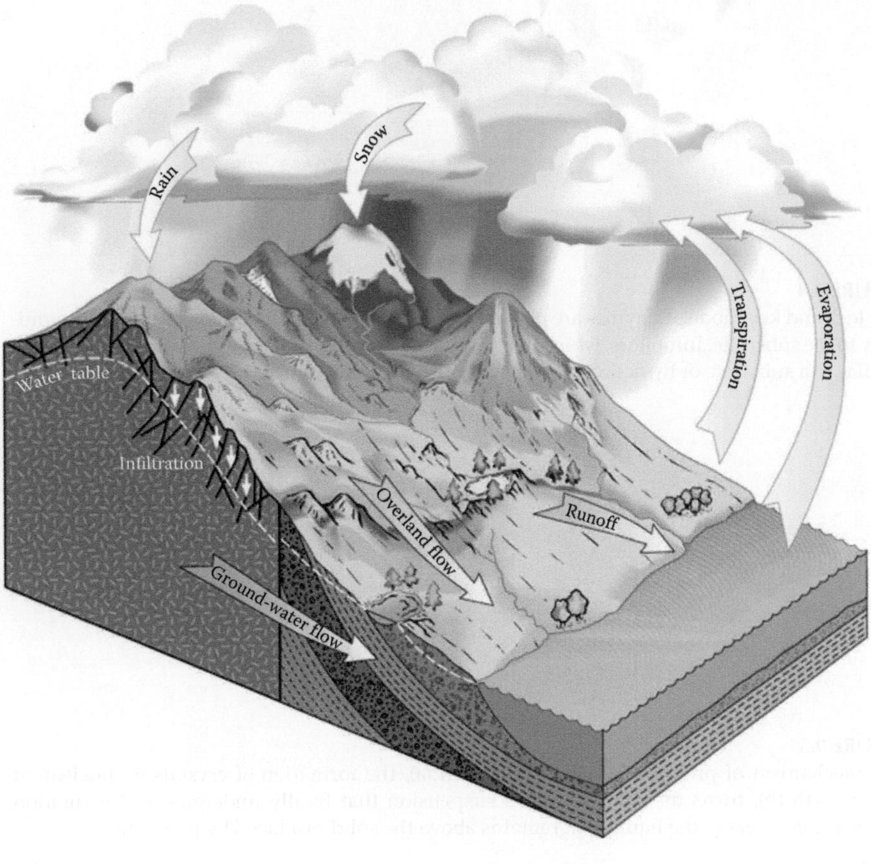

FIGURE 12.1
Schematic representation of the water cycle with indication of the principal pathways.

Eons	10^9 Years ago	Biotechnology		Atmosphere
Archean	3,5	Photosynthesis H_2S	←	CO_2
		CO_2 crisis		
	3	Fermentation	→	CO_2
		H_2 crisis		
Proterozoic	2,5	Photosynthesis H_2O	→	O_2
		O_2 crisis		
	2	Respiration	←	O_2

FIGURE 12.3
The process of biosphere self-regulation.

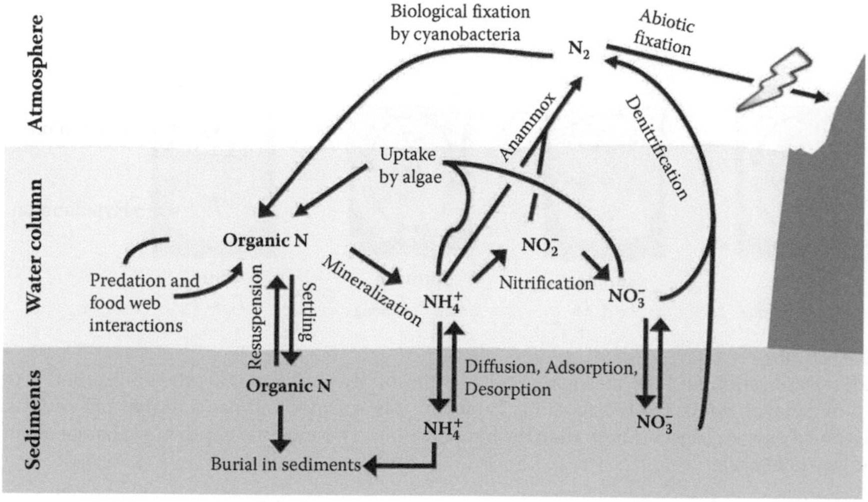

FIGURE 12.5
A schematization of the global nitrogen biogeochemical cycle. Processes and symbols are described in this and other chapters of the book. This figure is mainly focused on aquatic ecosystems, but most processes (fixation, mineralization, nitrification, denitrification, predation) take place also in terrestrial systems.

FIGURE 15.2
Grazing cows on an alpine pasture in the Asiago plateau, Italy.

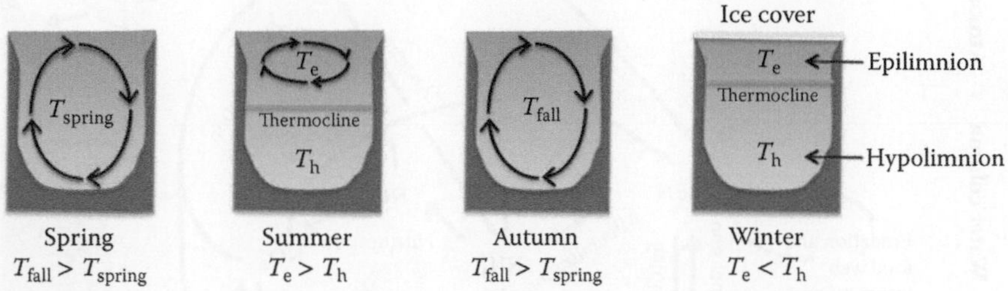

FIGURE 20.1
Lake stratification and the seasonal cycle. In a thermally stratified lake, the epilimnion is the top-most layer, above the thermocline. Typically, it is warmer and has a higher pH and high dissolved oxygen concentration than the hypolimnion. Thermocline depth may change during the seasonal cycle.

FIGURE 20.2
The Brenta river in Nothern Italy, an example of a water body that has been strongly influenced by man since centuries ago. The Brenta river used to flow into the lagoon of Venice, which it had contributed to create, but at the time of the *Serenissima* Republic of Venice its course was diverted into the Adriatic Sea because the sediment load brought by the river was filling up the lagoon.

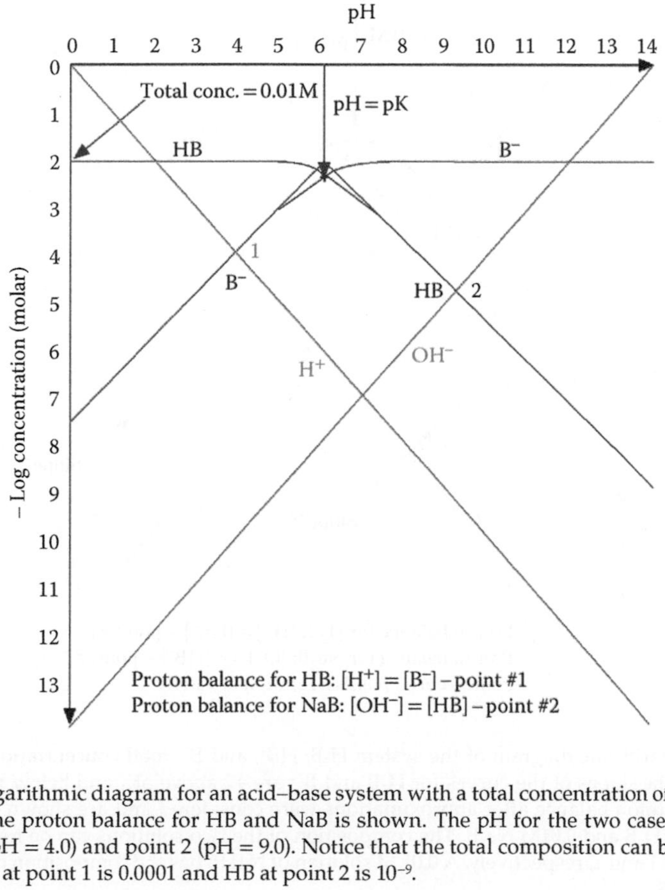

FIGURE A.1
A double logarithmic diagram for an acid–base system with a total concentration of 0.01 m and pK = 6.0. The proton balance for HB and NaB is shown. The pH for the two cases are found as point 1 (pH = 4.0) and point 2 (pH = 9.0). Notice that the total composition can be red at the diagram. B^- at point 1 is 0.0001 and HB at point 2 is 10^{-9}.

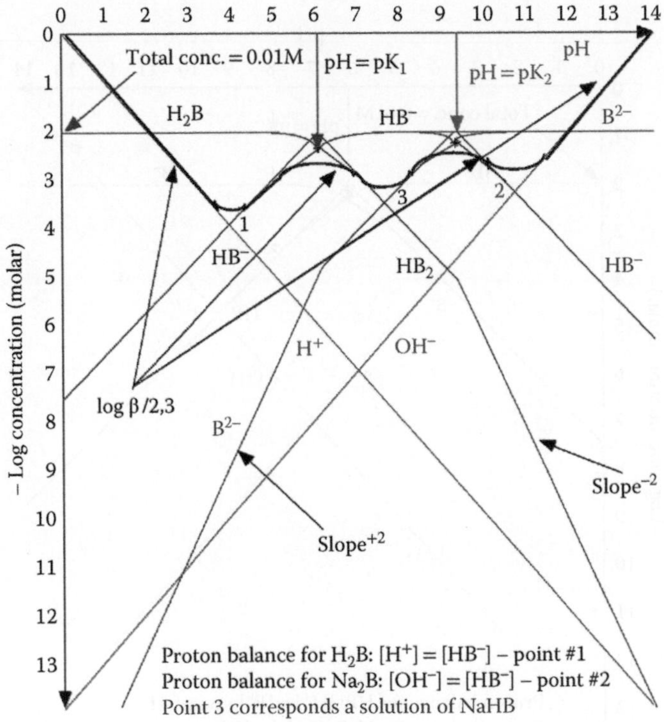

FIGURE A.2
A double logarithmic diagram of the system H_2B, HB^-, and B^{2-} total concentration = 0.01 M. Notice that the slopes of the curves for H_2B and B^{2-} are +2 above pK_2 and below pK_1, respectively. The proton balance after approximations were considered and are shown for the two cases 0.01 M H_2B and 0.01M Na_2B. The composition of the two solutions can be read from the figure points 1 and 2, respectively. A 0.01 M solution of NaHB has the composition corresponding to point 3.

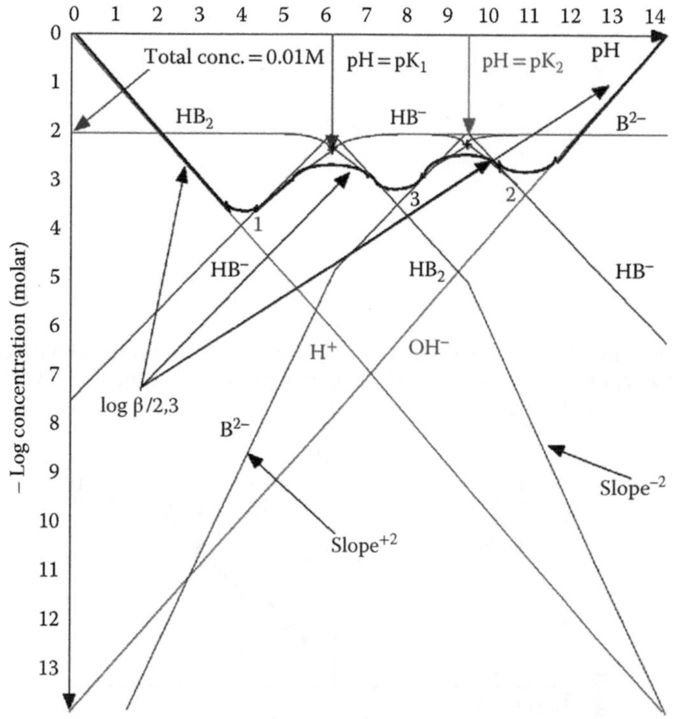

FIGURE A.4
The log ($\beta/2.3$) line for Figure A.2 is shown. At points 1, 2, and 3, the line is 0.3 units above the intersections, and at pK_1 and pK_2, the log ($\beta/2.3$) line is 0.3 below the intersection.

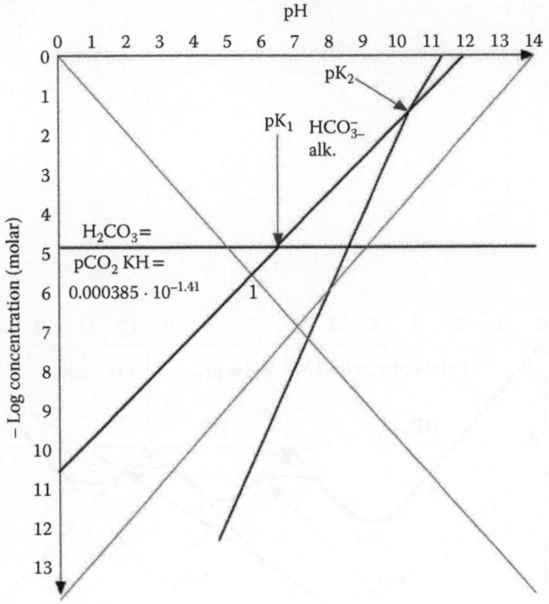

FIGURE A.5
A double logarithmic diagram for an open aquatic system in equilibrium with the carbon dioxide in the atmosphere. A concentration of 385 ppm is presumed. It corresponds to the carbon dioxide concentration during the year 1999–2000.

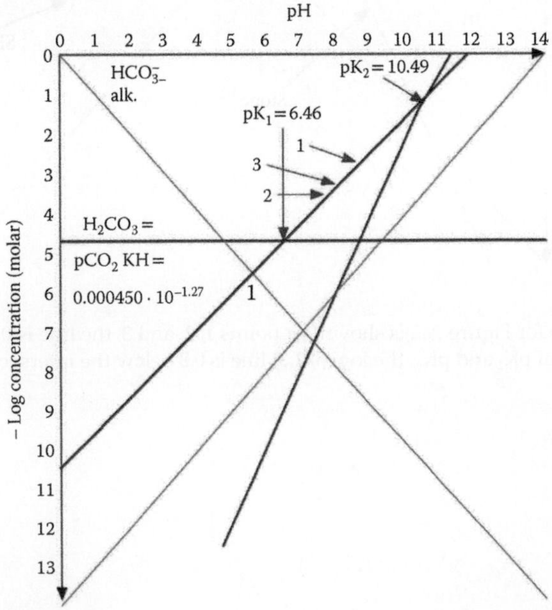

FIGURE A.6
Example 12.1. See the solution.

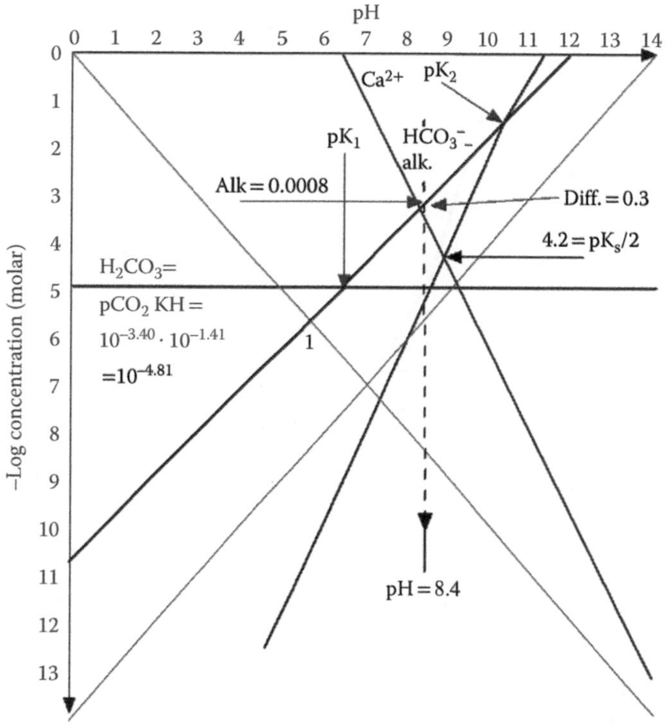

FIGURE A.7
A double diagram for an aquatic system in simultaneously equilibrium with carbon dioxide in the atmosphere and solid calcium carbonate.

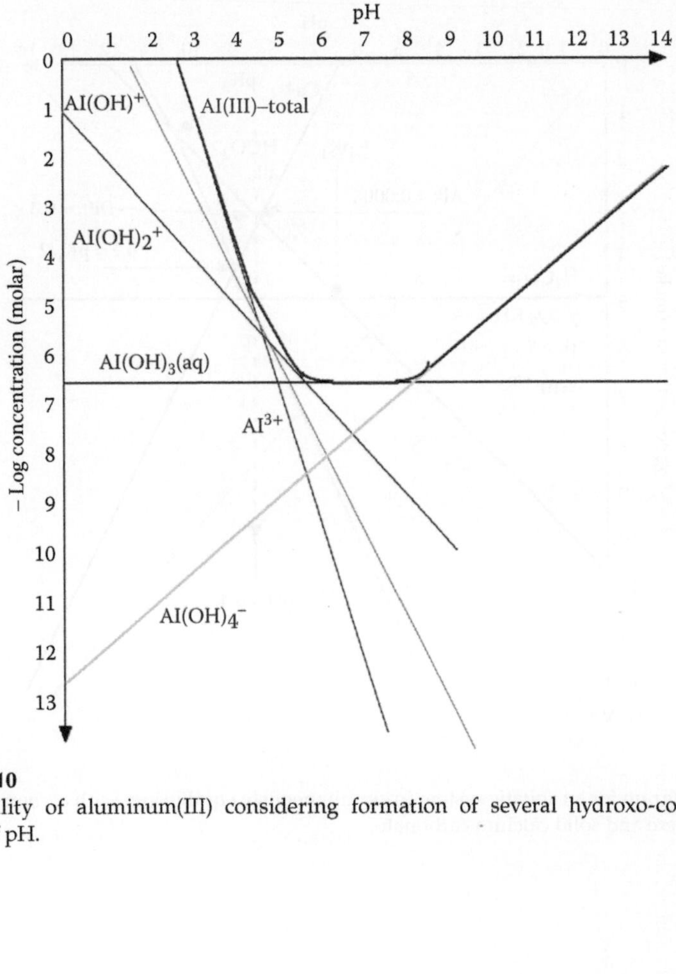

FIGURE A.10
The solubility of aluminum(III) considering formation of several hydroxo-complexes as function of pH.

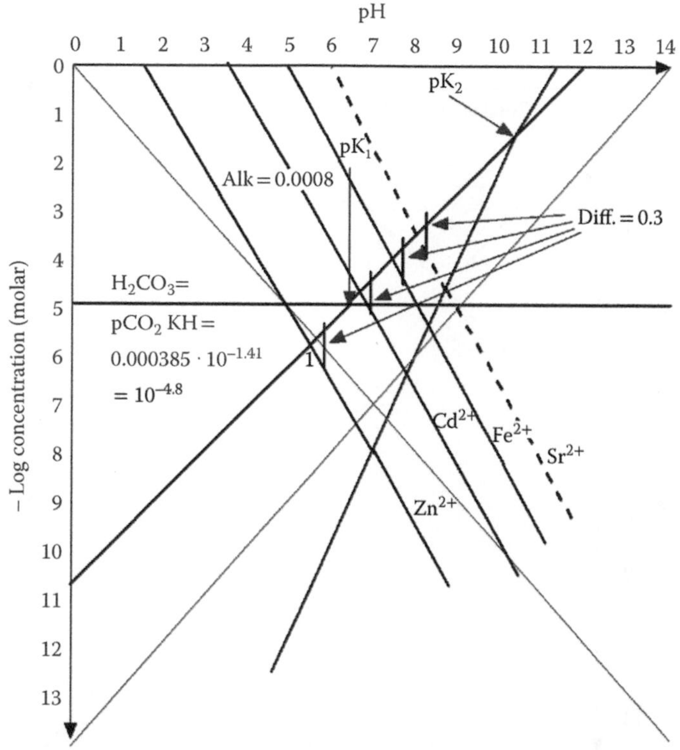

FIGURE A.12
Double logarithmic representation of an equilibrium of solid strontium carbonate, iron(II) carbonate, cadmium carbonate, and zinc carbonate in an open aquatic system. The composition will correspond to the lines where the difference between the concentration of the metal ion and hydrogen carbonate is 0.3, corresponding to the equation $2[Me^{2+}] = [HCO_3^-]$.

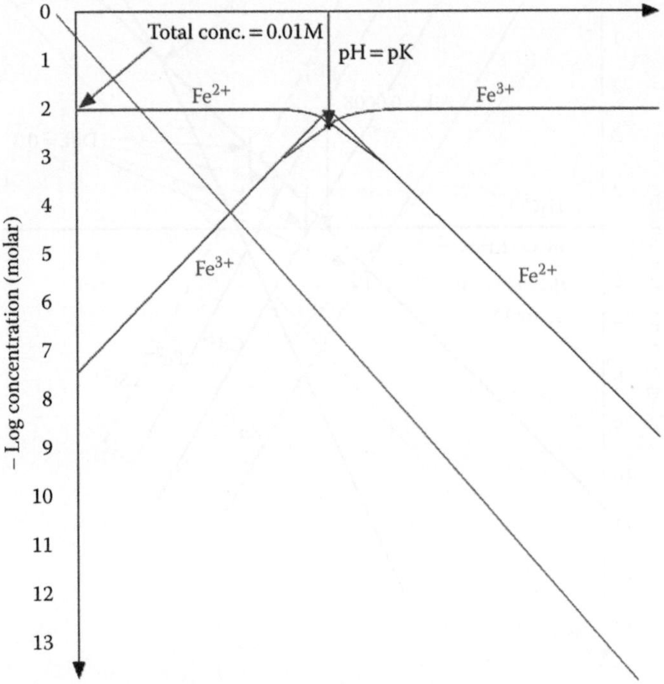

FIGURE A.17
Double logarithmic diagram for the redox process: $Fe^{3+} + e^{-} = Fe^{2+}$.

8

Energetic Factors

8.1 Introduction

Energy has a great influence on most ecological processes. Biotic activity on the earth ultimately originates from the energy provided by solar radiation, which is used by primary producers to produce biomass that fuels all other organisms through predator–prey interactions. Solar radiation is also linked to changes in environmental temperature, a key parameter that is correlated with the rates of most processes in ecosystems. The concept of "dissipative structure" itself (Chapter 3) underlines the close relationship that exists between energy flows and the surprising order that characterizes living systems. In this chapter, we will examine the link between energy and key processes, mostly physical, in ecosystems. The focus will be both on the transport of energy performed by such processes and on its ecological consequences.

8.2 Solar Radiation

Solar radiation is the main energy input to our planet, and it represents a fundamental forcing for many ecological processes. Variations in solar radiation reaching the earth's surface correspond to changes in environmental temperature and weather, with solar radiation being a major component of the energy budget in most terrestrial and aquatic ecosystems, and, consequently, such variations affect the rates of numerous ecological processes. Other important effects of solar radiation on ecological systems include the photosynthesis and photolysis processes, which only take place in the presence of light. Moreover, the amount of light available in ecosystems can shape the behavior and movement patterns of organisms, for example, by making them easier to be spotted by predators.

The revolution of our planet around the sun and its rotation around its tilted axis result in annual (seasons) and daily cycles of incident solar radiation per unit earth's surface. Solar radiation also shows interannual variations, such as 11-year and longer cycles in solar activity. The amount of incident solar radiation per unit earth's surface depends on many factors, such as the time of the day, the season or day of the year, latitude, cloud cover and the presence of light-extinguishing substances in the overlying atmosphere (e.g., due to air pollution or volcanic activity, Section 8.3), local topographic features (shading by mountains or canyons), and elevation above sea level. The incident light irradiance transmitted by the sun and reaching the top of the atmosphere is about 1367 W m^{-2}, known as the *solar constant*, although it is not actually constant and can vary slightly over ecological time scales, as mentioned earlier. As described in Section 1.6, light absorption, diffusion, and reflection processes take place in the atmosphere; they depend on (and can modify) the wavelength of radiation. Eventually, about 46% of total incident radiation directly or indirectly reaches the earth's surface. This fraction is characterized by short wavelengths, that is, ultraviolet and, mainly, visible light.

Direct measurements of solar radiation are often available through field instruments (pyranometers, pyrheliometers, and sunshine recorders) or can be indirectly estimated based on other environmental parameters (cloud cover, etc.) from airports or environmental agencies and weather forecasting systems. Satellite data and other high-resolution datasets available on the Internet can also be used. For the purpose of modeling and extrapolating results to years when measurements are unavailable, empirical relationships can be fitted through measured radiation data, such as sinusoids or other types of relations fitter to model intra-annual changes in solar radiation in locations where cloud cover is very unevenly distributed over the year, for example, in regions characterized by monsoons.

As an alternative to measurements, solar radiation can be modeled directly. Solar irradiance I (W m^{-2}) at time t of the day has a sinusoidal time trajectory from sunrise to sunset, and its value depends on the latitude of the location (Jørgensen and Bendoricchio, 2001):

$$I(t) = \left(1 + \cos\left[\frac{360 \cdot (t - 0.5)}{P(n,\varphi)}\right]\right)\frac{I(n)}{P(n,\varphi)} \tag{8.1}$$

where $I(n)$ is the total solar irradiance on a given day n of the year, and time t can range from $(0.5 - P(n,\varphi)/2)$ to $(0.5 + P(n,\varphi)/2)$ and is expressed in the same units as the photoperiod $P(n,\varphi)$, which is the fraction of light hours in a day. Outside such a range, Equation 8.1 cannot be applied, as it is night and, thus, $I = 0$. The photoperiod is a function of latitude φ and the day n of the year according to (Jørgensen and Bendoricchio, 2001):

$$P(n,\varphi) = \frac{\arccos(-\tan\varphi \tan\delta(n))}{180} \tag{8.2}$$

where $\delta(n)$ is solar declination, that is, the angle between the plane of the equator and the line linking the sun and the center of the earth, which itself is a function of the day n of the year. According to Equation 8.2, since the maximum value of δ is $\approx 23.5°$, which is the earth's tilt relative to its orbit, the photoperiod can be equal to 0, indicating no sun at all during the day (polar night) and to 1, that is, the sun is staying in the sky during the whole day (polar day), for $\varphi \approx 66.5°$ or higher, that is, close to the poles.

To estimate $I(t)$ in Equation 8.1, the total daily radiation should be quantified, though direct/indirect measurements (see above) or sinusoidal models which take into account the effect of the day of the year, latitude, and cloud cover (which can strongly reduce $I[n]$) as well as substances such as aerosols that can attenuate light (Section 8.3; see also Iqbal [1983], Chapra [1997], and Jørgensen and Bendoricchio [2001]). It is important to remember that a fraction of the incident shortwave radiation is reflected by the surface it hits. Such a fraction is known as *albedo* and can vary greatly depending on the type of the surface, from a few percentage points in the case of water bodies when the sun is at the local zenith, to more than 50% in the case of snow.

In order to have a complete picture of the effect of incoming solar radiation on ecosystems, it would be useful to write an energy balance for a control volume and to examine the different exchanges across its surface. The unreflected shortwave solar radiation is not the only input: Atmospheric gases (water vapor, carbon dioxide, ozone, etc.) emit long-wave radiation, which also reaches the earth's surface. The incident long-wave radiation can be described according to a modified version of Stefan–Boltzmann law (radiation is proportional to the fourth power of dry-bulb air temperature) or similar relationships, explicitly accounting for increased emissions due to air humidity and cloud cover, and for the fraction of reflected radiation (Chapra, 1997; Jørgensen and Bendoricchio, 2001). Finally, the long-wave radiation emitted by the earth's surface should be accounted for, again according to Stefan–Boltzmann law, as well as the fluxes of energy related to conduction, convection, and evaporation/condensation (Chapra, 1997; Jørgensen and Bendoricchio, 2001). In the case of ecological systems that are dominated by mass transport phenomena (e.g., rivers), mass fluxes can also be coupled to energy transport; for example, in the case of water advected into a lake where the temperature is different from that of the inflow.

8.3 Light Extinction

Light availability is vital to the functioning of all kinds of ecosystems, as discussed at the beginning of this chapter, because it influences the photosynthesis or plant growth process (Chapters 13 and 15) and affects, directly or indirectly through temperature, the rates of numerous ecological processes.

The solar radiation reaching an ecosystem is only a fraction of the amount emitted by the sun: During its travel, light interacts with matter and experiences several processes that reduce its intensity, which is collectively known as *light extinction* or *attenuation*. For example, light can be extinguished across the atmosphere, in the water column of an aquatic ecosystem, or when penetrating vegetation in terrestrial ecosystems. In ecology, the Lambert–Beer's law, also known as *Bouguer's law*, indicates a common and robust relationship that is used for describing light extinction across a homogeneous medium:

$$I(z) = I(0)\mathrm{e}^{-kz} \tag{8.3}$$

$I(z)$ is the intensity or irradiance of light (W m^{-2}) that is measured at position or depth z (m) along the medium, and k is the light extinction coefficient (m^{-1}). The product kz is called *optical depth*, and it quantifies the capability of a layer of medium with depth z to extinguish light. According to Lambert–Beer's law, the incident light intensity $I(0)$ on the surface of a medium decreases exponentially with the distance traveled across such a medium, due to interactions with matter. Equation 8.3 can be derived from the first-order decay reaction $\mathrm{d}I/\mathrm{d}z = -kz$ when assuming that the medium is homogeneous with regard to extinction, that is, k does not depend on z. Lambert–Beer's law can be used to model light extinction in markedly different media such as air, water, and even vegetation canopies, although it works best in low-concentration media. In spectrophotometry, Equation 8.3 is re-expressed as $A = k \cdot L \cdot C$ and is used to measure the concentration C of a light-absorbing substance in solution, based on the extinction of light throughout a solution sample with length L, as measured by absorbance $A = \ln\left(I(0)/I(L)\right)$.

According to Lambert–Beer's law, the depth of a medium is a key factor that controls light extinction. The extinction coefficient k lumps together all other extinction properties of the medium. The value of k depends on the type of medium and the concentration of light-attenuating substances. For example, turbid water will extinguish more light than does clear water. This coefficient can also depend on the direction in which light travels through the medium, as not all media are necessarily isotropic with regard to light extinction, and on radiation wavelength. The last point is particularly important in ecology: In a medium, light intensity at different wavelengths displays dissimilar degrees of attenuation and, therefore, when modeling light attenuation, the extinction coefficients should be chosen properly depending on the purpose of the study. For example, if one wishes to study the effect of light extinction on the growth of primary producers, k should be determined for the wavelengths covered by Photosynthetically Active Radiation (PAR), mainly in the visible range, which can be exploited by plants (Chapter 13). It is also important to remember that solar radiation is not uniformly distributed across all wavelengths (Sections 1.6 and 8.2).

The causes of solar light extinction in the atmosphere can be classified into two processes, absorption and scattering (or diffusion), both of which are due to atmospheric gases, such as oxygen, nitrogen, water, carbon and sulfur dioxides, nitrous oxide, and ozone, and to aerosols of natural or anthropogenic (pollution) origin. Unlike light absorption, scattering does not entail a net energy loss; that is, it is an elastic process, as photons are simply deviated from their path toward the soil. Rayleigh scattering is caused by particles whose diameter is much smaller than radiation wavelength, for example, O_2 and N_2 for light in the visible range. For very large particles, the geometrical laws of optics are appropriate; whereas Mie scattering is applied to intermediate-sized particles (diameter $< 10 \cdot$ radiation wavelength). In the atmosphere, Equation 8.3 should be modified to account for the effective distance traveled by light toward the ground, when the sun is not at the local zenith. In this case, the light intensity $I(L)$ reaching the soil through an atmospheric layer with vertical height L that is measured from the ground becomes

$$I(L) = I(0)e^{-kmL} \tag{8.4}$$

where $I(0)$ is the light intensity hitting the atmosphere and m is the optical air mass, accounting for the solar angle θ of the zenith. If θ is small ($< 60°$), $m = 1/\cos\theta$; otherwise, complications due to air density, refraction, and the earth's curvature come into play and semiempirical formulas apply. The atmospheric extinction coefficient for a certain wavelength is usually computed as the sum of different terms that separately quantify the role of the scattering and absorption processes mentioned earlier. For example, aerosols cause considerably stronger attenuation of visible light than do gases in polluted areas.

Similarly, the extinction of the incident light on the surface of water bodies is caused by absorption and scattering processes due to the amount (and type) of dissolved and suspended substances in water, in addition to the effect of the water molecules themselves. Equation 8.3 applies again, but now, $I(0)$ is the incident solar radiation on the water surface and z represents water depth. The coefficient k is inversely related to the famous Secchi depth, an empirical measure of water transparency, or it can be estimated in situ by measuring changes in light intensity with depth through photometers. k can also be computed/modeled as the sum of the coefficients for the different processes that cause light attenuation; for example, it can be expressed as a function of turbidity and the concentration of substances in water:

$$k = B + aP + cA \tag{8.5}$$

Background turbidity and the effect of water are represented by coefficient B, whereas a and c are system-specific parameters. Bracket parentheses represent the concentration of particulate material $[P]$ and chlorophyll-a $[A]$, with the latter being a proxy for algal biomass, which can attenuate light during blooms, especially in productive systems; whereas in oligotrophic water bodies, B is typically much larger than the other terms. In aquatic ecosystems, light extinction for the wavelengths covered by PAR (400–700 nm) is usually studied, because phytoplankton and benthic algae need PAR to grow (Chapters 13 and 15), but PAR is not available throughout the whole water column, which is only partially well lit due to light extinction. If PAR is modeled, the dependence of k on $[A]$ indicates the phenomenon of self-shading: Algae limit the growth of their own population by reducing the availability of light in water. PAR can be expressed as W m^{-2} or, alternatively, as photosynthetic photon flux density (number of photons per unit time per unit surface), in the visible range, keeping in mind that the energy carried by a photon can be calculated based on the associated wavelength and the speed of light using the Planck–Einstein equation. A closely related concept to that of PAR is the "euphotic zone," that is, the well-lit part of the water column starting from the surface and down to the depth where the photosynthetic rate equals the respiration rate of primary producers, that is, where net primary productivity is zero (Chapter 15). The depth of the euphotic zone is strongly dependent on system productivity and turbidity, ranging from less than a meter in estuaries and lakes to 200 m in the clear waters of some open oceans.

In terrestrial ecosystems, light extinction by the canopy of vegetation can strongly shape ecosystem functioning and biodiversity, for example in tropical forests, and this process can also be important in agriculture and silviculture, as plants and crops need light to grow and, moreover, light penetration influences evapotranspiration (Section 8.4) and, thus, water balance and management. Light in a canopy can be reflected or (partially or totally) absorbed by the leaves it hits, which is similar to the scattering and absorption processes in water and air. Light extinction can be simulated using a modified version of Lambert–Beer's model:

$$I(z) = I(0)e^{-k \cdot LAI(z)} \tag{8.6}$$

where z is the canopy depth, as measured starting from the top of the canopy (where $z = 0$), and LAI is the cumulated sum of the Leaf Area Index from depth 0 to z. The LAI at a given depth is the ratio of the total one-sided green leaf surface at that depth to the ground surface beneath the canopy. Hence, trees with denser foliage extinguish more light. Since canopies are not homogenous, isotropic media, Equation 8.6 can sometimes be seen as a gross oversimplification of the real process of light extinction and, consequently, more complex models exist, as well as approaches to measuring light attenuation directly within canopies (Barausse, 2008).

8.4 Evapotranspiration

Evapotranspiration is the flow of water vapor from terrestrial ecosystems and water bodies to the atmosphere. As the name suggests, it is the sum of two distinct processes: evaporation and transpiration. Evaporation is the amount of liquid water lost to air by vaporization and that was earlier found on the ground (i.e., soil, hard surfaces such as rocks, concrete, roofs, etc.), the free surface of water bodies, and the surface of plant canopies (i.e., water intercepted by vegetation through branches and leaves). Transpiration represents the amount of vapor leaving the tissue of plants, mainly through the stomata (small pores) found on the leaves. Transpired water was previously uptaken from the soil, along with nutrients, through plant roots. Evaporation and transpiration are difficult to quantify separately, as both flows take place simultaneously when vegetation is present, and, thus, their combination is usually considered. Evapotranspiration is usually measured as the millimeter of water per unit time (e.g., mm d^{-1}) that is needed to balance the vapor flux from a surface.

Evapotranspiration can represent a large fraction of the water balance of an ecosystem on annual timescales. Globally, on average, 57% of the annual precipitation goes back to the atmosphere as evapotranspiration (Irmak, 2008); therefore, this process represents a key component of the global hydrological cycle. Moreover, as the rate of evapotranspiration strongly depends on environmental conditions (see below), the water volumes lost by a land's surface through evapotranspiration can also be particularly significant on seasonal timescales (e.g., in summer) and even during shorter periods, especially in hot, dry, and windy weather. For this reason, the assessment of evapotranspiration fluxes can be extremely important for planning irrigation in agriculture and for the management of water resources in all those areas of the world that are characterized by water scarcity. This point is witnessed, for example, by the publications on this topic by the Food and Agriculture Organization of the United Nations (Allen et al., 1998) and by the American Society of Civil Engineers (Jensen et al., 1990).

A large number of factors can influence evapotranspiration, and they can be clarified by looking separately at the processes of evaporation and transpiration. Evaporation is favored by solar radiation (Section 8.2) and temperature, as vaporization of water from the liquid to the gas state needs energy to take place. Evaporation from a surface is proportional to a driving force that is expressed by the difference between the vapor pressure at the surface and the one in the air (see transport at interfaces in Section 7.7). Hence, if relative humidity is high, such a driving force will be low and, consequently, evaporation will also be weak. Moreover, as evaporation takes place, the air will become more and more saturated with water and, if such air is not replaced with a drier one, the driving force will again tend to zero, reducing the evaporation flux. This consideration clarifies the role of wind, whose speed positively affects evaporation by removing vapor from the evaporating surface, thus increasing the vapor pressure gradient. Examples of quantitative relationships that describe evaporation from a water body such as a lake, as well as further references, can be found in Chapra (1997). In comparison to a free water surface, evaporation from soil is additionally influenced by the availability of water, as well by shading effects due to vegetation. With regard to the former factor, a reduced water availability or supply to the soil (e.g., through irrigation, rainfall or due to shallow water tables) can limit evaporation, which can get close to zero when no water supply to the soil takes place. Instead, when there is plenty of water, the main factor that determines the amount of evaporation will be the meteorological conditions mentioned earlier (Allen et al., 1998). For this reason, excessive irrigation could lead to a waste of water by excessively enhancing evapotranspiration fluxes (Allen et al., 1998).

Transpiration is influenced by meteorological conditions in a similar manner to evaporation. It needs energy to take place, which is provided by solar radiation and temperature, and it depends on the same driving force, the difference in vapor pressure: Thus, transpiration is negatively related to air humidity and positively impacted by wind speed. However, in addition, transpiration can be limited by scarce water availability to plant roots and a low transport rate of water from roots and across stomata. Water availability can depend on soil and groundwater properties such as soil type (e.g., permeability), water table depth, gradients in hydraulic head (Section 7.6), and salinity (salts can reduce the amount of water available to plants due to their affinity with water); on management practices (e.g., irrigation) and meteorological conditions (e.g., rainfall); and on plant characteristics such as root depth and density. Water transport from roots to leaves and transpiration from stomata are influenced by the species of plants/crops, and markedly different evapotranspiration rates can correspond to surfaces that are covered with different plant types (Allen et al., 1998; Irmak, 2008). For example, the vapor exchange surface between plants and air is related to plant density, the number and surface of leaves (i.e., the Leaf Area Index, see Section 8.3), and the stomata density per unit surface of the leaves. The conditions of stomata are also important, as transpiration takes place when stomata are open, and this is linked to environmental factors. If not enough water is available to the roots with regard to the actual transpiration rate, the stomata will close. On the other hand, open stomata are needed to exchange gases in the photosynthesis and respiration processes; therefore, their opening involves a tradeoff between water loss and growth: For example, stronger light intensity not only favors photosynthesis but also provides energy, supporting the transpiration process. Finally, transpiration rates can strongly vary within the same type of plant, depending on its development stage. For example, a recently sown crop or a senescent one will be shorter and/or be characterized by a smaller LAI with regard to a well-developed crop just before the harvest season. Consequently, transpiration and evaporation will display opposite seasonal trends: The former will be enhanced by plant growth (more leaves), whereas the latter will be reduced by the shading effects of the canopy. Since transpiration is positively related to crop production, the management goal in agriculture should be the minimization of water lost as evaporation, which, by definition, does not contribute to plant growth (Irmak, 2008).

Several methodologies that quantify evapotranspiration exist, both on the field (through lysimeters, evaporation pans, atmometers, etc.) and through models, which can be based on energy balances, water balances for the soil, and meteorological data (e.g., Allen et al., 1998; Irmak, 2008). Field measurements are time consuming, expensive, and difficult to perform and extrapolate to other systems; therefore, models that employ empirical or semiempirical equations based on climatic data inputs are widely used nowadays. A common approach is to calculate evapotranspiration (mm d^{-1}) for a given crop as follows:

$$ET = K_c \cdot ET_{\text{ref}} \tag{8.7}$$

that is, the product of the crop coefficient K_c and the reference evapotranspiration ET_{ref}. The reference evapotranspiration represents the evapotranspiration for a standardized vegetation surface (alfalfa or grass) with well-specified characteristics such as height, surface resistance, albedo, and amount of soil water. ET_{ref} can be computed using Penman–Monteith equation based on local meteorological data (Allen et al., 1998; Irmak, 2008), which is for a grass reference surface and daily input data:

$$ET_{\text{ref}} = \frac{0.408(R_n - G) + \gamma \dfrac{900}{T + 273} u_2 (e_s - e_a)}{\Delta + \gamma(1 + 0.34u_2)} \tag{8.8}$$

ET_{ref} is expressed as mm d^{-1}; R_n is the net radiation at the crop surface (MJ m^{-2} d^{-1}); G is the soil heat flux density (MJ m^{-2} d^{-1}); γ is the psychrometric constant (kPa°C^{-1}); T is the mean daily air temperature at 2 m from the soil (°C); u_2 is the wind speed at 2 m from the soil (m s^{-1}); e_s and e_a represent the saturation and actual vapor pressure, respectively (kPa); and Δ is the slope of saturation vapor pressure - air temperature curve (kPa°C^{-1}). This relationship can also be adapted to deal with hourly inputs (Irmak, 2008).

Meteorological conditions are mainly accounted for by ET_{ref}, whereas the crop coefficient not only lumps the influence on evapotranspiration of all other factors, such as the type of crop/plant, its height, albedo, canopy resistance, growth stage (hence, K_c changes with seasons), leaf area, planting date, and several other crop characteristics, but also accounts for the effects of stressful soil conditions for the plant (e.g., salinity and toxic substances, water availability) and evaporation from the soil, management practices, and irrigation (Allen et al., 1998; Irmak, 2008). For this reason, K_c cannot be calculated and should be estimated through experiments. In practical applications, the value of this coefficient that is fit for local crops and conditions can be found in the literature. The strength of the approach of Equation 8.7 lies in its standardization and in the numerous existing literature sources for K_c, making the values of proper crop coefficients available in local contexts where field measurements of ET would not be possible or affordable. The geographical extrapolation of K_c values is possible, because the influence of climatic variability on evapotranspiration is chiefly synthesized by the term ET_{ref} in Equation 8.7, whereas K_c mainly accounts for other factors, such as crop properties and soil conditions.

8.5 Effects of Waves in Aquatic Ecosystems

Besides the direct use of solar energy by photosynthesis, the energy of the sun is converted into mechanical and thermal energy by the Hadley's convection cells mechanism described in Section 1.6. This energy flow sustains the whole planetary atmospheric circulation system.* As stated in Section 7.4 when discussing resuspension, waves on the surface of water bodies get their energy from blowing wind. Depending on the extent of fetch, wind may eventually generate large waves, which then travel until their energy is dissipated by interference or against some obstacle (e.g., shorelines). Under a traveling wave, water moves in vertical circles (with gradually smaller diameters at increasing depth; see Figure 8.1) so that at the end of the cycle of vertical displacement, the water finds itself just in the initial position. What is actually traveling is the energy that causes water displacement. While approaching coastal areas where waters become shallower (when the depth H is less than half the wavelength L), the vertical circles beneath the traveling wave start interacting with the bottom, becoming more and more elliptical. At this point, the waves become taller and steeper (*shoaling*). In proximity of the shore, waves start interacting with the organisms living at the bottom, eventually exerting very strong swinging forces. Waves may also structure the benthic habitat by shaping soft seabed (e.g., sandy bottoms). In shallow habitats, waves can exert a very relevant influence on the life of organisms (Blanchette et al., 2008). In very shallow water, waves can no longer maintain their shape, because the water in the vertical cycles below the wave top is slowed by drag interaction with the bottom. The top portion of the wave will then roll over and begin to break (Figure 8.1). During wave breaking, all of the energy of the wave is dissipated in the turbulent motion of water on the shore. Breaking waves is a violent energy transfer process that results in very high water velocities and high levels of turbulence by resuspension. Certainly, this process may have dramatic consequences on the communities of organisms living in shoreline areas. The most apparent effects of wave sweeping on organisms are mechanical, such as lift and drag. As a consequence, organisms may be removed, displaced, and occasionally destroyed. Measured velocities of the water beneath shoaling and breaking waves can often be larger than 30 m/s. Given that shear stress at the boundary layer

* And, along with the moon orbital position, tidal displacements as well.

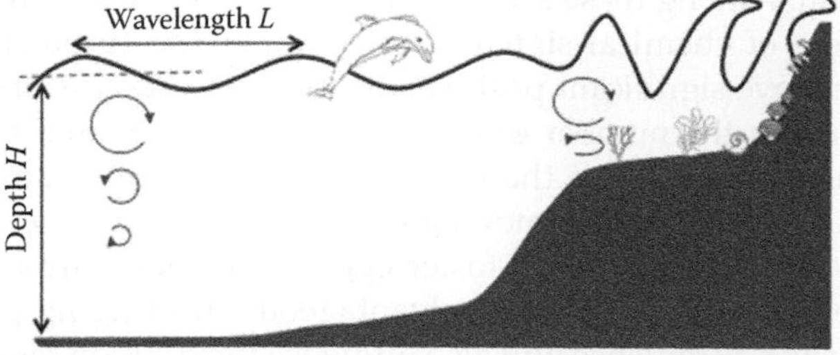

FIGURE 8.1
Traveling waves become shoaling waves approaching the shoreline. (Redrawn after Blanchette, C.A., O'Donnell, M.J., Stewart, H.L., *Encyclopedia of ecology*. Academic Press, Oxford, pp. 3764–3770, 2008.)

depends on the square of the water velocity, forces imposed on biological structures at the sea bottom rise very rapidly with a small increase in water velocity. For a medium-sized organism (let us say some centimeters) with an approximately spherical shape, the forces imposed by sweeping waves can, in normal conditions, be as large as 200 N or 20 kg (Blanchette et al., 2008).

Organisms in such severe environments have developed several survival strategies. Mobile organisms can adjust their behavior in order to minimize perturbation and/or maximize performance. For example, some organism may stay firmly attached to a site under critical conditions and move for feeding or reproduction in periods of relative calm or high tides when waves are less effective. Larger organisms such as seals, dolphins, or killer whales generally do not escape waves, and instead, exploit the energy brought by them for their normal living activities, such as swimming and chasing. Organisms anchored to the benthic environment (i.e., *sessile* organisms), being unable to move, have developed different survival strategies. The strength or the tenacity of attachment to the seabed is the most common one. Flexibility and elasticity, for example in seaweeds, is a cost-effective strategy that is used for energy dissipation. Size is also a relevant parameter. Small organisms (compared with the wavelength) usually experience comparatively less drag force, whereas large organisms (e.g., the long branches of brown algae) can effortlessly absorb water movement while taking advantage of a partial displacement. Finally, fouling (Section 2.2), when mussels, algae, and other organisms are limited in rocky substratum for attachment, is a very widespread survival strategy that is used on wave-swept shores. In fact, wave damping and water slowing can dissipate the energy of water moving inside aggregations. Being the hydrodynamic forces proportional to water speed, this usually leads to lower forces that are exerted on organisms inside aggregations.

Besides being adversely perturbed, aquatic ecosystems may get some benefits as a consequence of wave action. Indeed, not all of the effects of waves are negative; instead, several aspects of biological processes are enhanced by wave action. Among these are matter transport, effects on reproduction, and dispersion of chemical signals. For example, the turbulence induced by waves can have significant positive effects on the success of fertilization and, therefore, on the number of juveniles produced. Sexual reproduction via the release of sperm into the water column is common among aquatic organisms. Given the limited moving capabilities of sperm, energy that is conveyed by water motion may foster egg fertilization. Turbulent mixing due to wave energy can, thus, be advantageous in bringing gametes into contact. However, turbulence will also influence the dilution rate of the gametes. If this inhibits the contact of sperm to eggs or damages the gametes or zygotes, the advantages of mixing are lost. Turbulent water motion can either aid or hinder fertilization depending on both the species and the degree of turbulence. Generally, benthic animals have a planktonic larval stage, and the dispersal and settlement of these larvae are significant factors that affect population dynamics. Larval transport as well is significantly affected by waves. The dispersal of biochemical signals is another process that is strictly linked to water motion. The energy of waves can significantly increment or regulate the dispersal of chemical cues that have important implications on the behavior of organisms. Lastly, feeding and nutrients uptake are fueled by resuspension, mixing, and dispersion consequent of waves. Actually, it may happen that if turbulent mixing is not sufficiently energetic, nutrients may become a limiting factor.

Even if intertidal organisms cannot transform wave energy directly into chemical energy, as photosynthetic plants do with solar energy, several observations indicate that communities that are exposed to high wave action are more productive than similar communities in less wave-exposed areas. In terrestrial ecosystems, photosynthesis is favored by wind, causing pulsing movements of leaves; similarly, aquatic organisms may take advantage of water oscillatory movements.

The enormous quantity of energy carried by waves may give rise to severe ecological perturbations such as coastal erosion or tsunamis. Waves' energy depends on characteristics such as wave amplitude, wavelength, and velocity. In order to understand and quantify the entity of these disturbances, it is, therefore, essential to estimate the energy of waves on the basis of such characteristics. In a given sea state,* the average energy density of waves per unit area of the water surface is proportional to the square of the significant wave height,† h (Phillips, 1980):

$$E = \frac{1}{16} \rho g \, h^2 \tag{8.9}$$

* The sea state in a given time and place is defined by wind direction and strength (Beaufort scale) and by wave height and steepness (Douglas sea scale).

† Defined as the average wave height (trough to crest) of the highest third of the waves.

where E is the mean wave energy density per unit horizontal area (J/m^2), ρ is the water density, and g is the gravity acceleration. As the waves move, their energy is transported with a velocity that is known as the *group velocity*. The wave energy flux per unit width through a vertical plane that is perpendicular to the wave propagation direction is given by

$$P = E \cdot v_g \tag{8.10}$$

where P is the energy flux per unit length of the wave front (J/m/s) and v_g is the group velocity (m/s). The dispersion relation for water waves in a gravitational field links the group velocity to the wave period T (equivalently, to the wavelength L) and generally also to the water depth d. Consequently, the group velocity changes dramatically while passing from deep to shallow water. In deep water, where the water depth is larger than half the wavelength, the dispersion relation is

$$v_g = g\frac{T}{4\pi} \tag{8.11}$$

and, hence, the wave energy flux is given by

$$P = \frac{\rho g^2}{64\pi} h^2 T = \varepsilon h^2 T \tag{8.12}$$

where ε is a constant whose value is approximately 1/2 with units (kW/m^3/s). When the significant wave height h is given in meters, and the wave period T in seconds, P is expressed in kilowatts (kW) per unit length of the wave front. Consider, for example, moderate ocean waves, in deep water, a few kilometers off a coastline, with a wave height of 1 m and a wave period of 4 seconds; from (Equation 8.12), the wave energy flux is approximately 2 (kW/m), meaning that 2 kW per meter of the wave front is the power potential of the wave. During major storms, large waves may be formed offshore being even 15 m high and with a period of about 15 seconds. According to Equation 8.12, the power potential of such waves is around 1.7 MW per meter of the wave front.

When a tsunami reaches the shoreline, it can become a monster wave of 10–20 m height with a period of 10–45 minutes. In shallow water, the dispersion relation changes, becoming relevant to the dependence on water depth; assuming a water depth of 0.1 m, such a wave may reach a destructive power potential of 4 MW per meter of the wave front. Given the ever-growing energy requirements of human societies, the development of wave energy-gathering technologies is of utmost interest nowadays. Wave power plants (or wave farms) of recent design, by the use of floating chambers, can capture as much as possible of the wave energy flux, resulting in waves of lower height in the region behind the wave power device.

Section III

Chemical Processes

9

Chemical Reactions

9.1 Introduction

The term *chemical process* is used to indicate a sequence of one or more operations achieving the transformation of chemicals. Chemical processes fall into three broad categories: processes performed in a chemical laboratory, industrial chemical processes carried out on an industrial scale, and natural chemical processes. Natural chemical processes are carried out in nature ideally without human action. These include, for example, the chemical processes that occur naturally in living organisms (including metabolism and photosynthesis) and the reactions of biodegradation. *Chemical laboratory* and *industrial processes* refer to chemical processes developed and engineered by man.

Generally, a chemical reaction is a transformation of matter that occurs without measurable changes in mass, in which one or more initial chemical elements change their structure and composition to create new products through the formation or breaking of chemical bonds involving an atom's outer electrons. The chemical compounds present at the beginning of the reaction are called *reagents* or *reactants*; whereas those obtained at the end of the reaction are called the *reaction products*. The phenomenon taking place during a chemical reaction can be represented by a chemical equation. Chemical equations are written in a manner that is similar to mathematical equations; they comprise two members: In the first member (i.e., to the left of the arrow or other symbol of the reaction), reagents appear, whereas in the second member (i.e., to the right of the arrow or other symbol of reaction), there are products:

$$aA + bB \rightharpoonup cC + dD \tag{9.1}$$

where lowercase letters indicate the stoichiometric coefficient, that is, the amount of moles (quantity of the element); capital letters A and B, the reagents; C and D, the products (molecules or chemical elements); and arrows indicate the process and direction of transformation.

A reaction cannot occur, is slowed to a halt, or is even reversed if it does not satisfy a number of conditions, such as the presence of reagents in adequate quantity and suitable conditions for the specific reaction; for example, temperature, pressure, and light. Molecules or chemical elements taking part in chemical reactions may belong to different phases: solid, liquid, or gas. In this case, the reaction is said to be *heterogeneous*, and the reaction usually occurs at the phase interface. The phase is indicated in the chemical equation by a g (gas), an l (liquid), or an s (solid) in brackets after the chemical symbol of the element or of the substance. If the reaction occurs in a single phase, it is said to be *homogeneous*. Ecological chemical reactions are often of this latter type. If the inverse of reaction 9.1 can simultaneously take place, the global reaction is said to be *reversible*, and it is written as

$$aA + bB \rightleftharpoons cC + dD \tag{9.2}$$

An *exothermic* reaction generates heat and is, thus, a reaction that involves a transfer of heat from the system to the environment. Similarly, an *endothermic* reaction needs heat to ensue; this is a reaction that involves a transfer of heat from the environment to the system and, therefore, requires external energy to proceed.

Besides the presence of the specific reactants, a minimum quantity of energy is needed for reactions to start and proceed. This quantity of energy is called *activation energy*, E_a. Activation energy is highly specific: For some reactions, this value is low, and, hence, these reactions occur spontaneously; whereas others require high activation energies and cannot, therefore, occur without an external support. For example, supplying more energy to the reagents (e.g., increasing temperature) or adding one or more catalysts to reduce the activation energy may foster and sustain the reaction. Catalysis is a key mechanism in ecological processes, because most of the reactions that involve living organisms need high activation energy. Catalysts for most ecological processes are enzymes, an important class of proteins produced by organisms themselves.

When reactants come into proper contact with enough energy, the reaction can occur and the cross-section of the reaction is said to be *effective*. In this case, the mechanism of the chemical reaction involves the production of activated complexes.* The energetic diagram of a chemical reaction is provided in Figure 9.1.

As anticipated in Chapter 1, all living organisms show approximately a similar chemical composition. This statement becomes more valid when the hierarchical placement of the organism is the highest; indeed: At lower trophic levels, organisms portray a relatively high plasticity in elemental composition (variable stoichiometry); in contrast, higher-level organisms (e.g., multicellular animals) display almost fixed stoichiometry, and they, therefore, can be considered as having a definite chemical composition. Understanding ecological stoichiometry is unavoidable in order to show how the chemical composition

* This term designates a collection of transitional chemical structures in metastable conditions that temporarily persist while old bonds are breaking and new bonds are forming. There is a moment during the reaction when the old bonds are broken and new ones have not yet formed, that is, the transition state in which the energy of the system is maximum (see Figure 9.1).

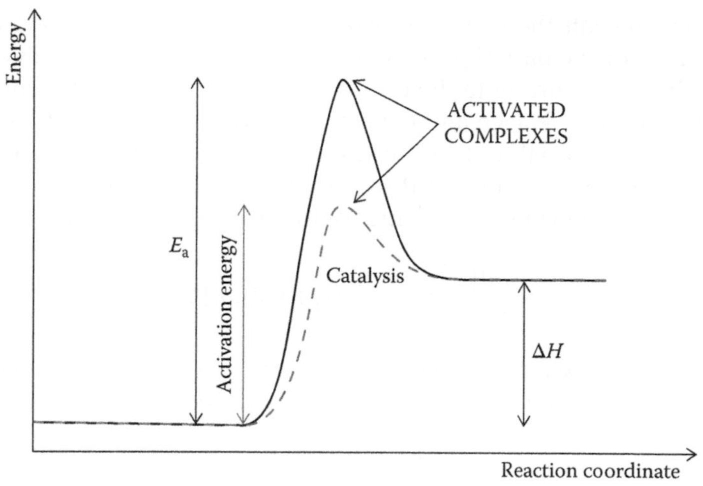

FIGURE 9.1
Enthalpy profile for an endothermic reaction ($\Delta H > 0$). The reaction coordinate represents the degree of progress of the reaction.

of organisms controls biochemical reactions as well as other key factors such as nutrient recycling, resource competition, animal growth, and nutrient limitation patterns in whole ecosystems (Sterner and Elser, 2002). The Redfield ratio (Redfield, 1958) derived for the world's oceans is one very well-known application of stoichiometric principles that has been fundamental for the delineation of the nutrient limitation of growth (see in Chapter 15). Ecological stoichiometry applies both to subcellular-level chemical reactions and to phenomena at the whole biosphere level, such as oxygen and carbon cycles.

In order to unravel reaction kinetics, another important parameter should be considered, which is the reaction rate. Some reactions are very rapid, and they are sometimes even violent, such as explosions. Others are so slow that they can continue for years or centuries. Some are utterly so slow that the reagents involved seem like really stable compounds. To quantify the rate of a reaction, the *degree of advancement* (or progress) of the reaction ξ is customarily used, and it is defined as the proportion of reagents or products that have already been consumed or produced ($\xi = 0$ at the beginning of the reaction, $\xi = 1$ when the reaction is complete). The *reaction rate* is, thus, defined as the derivative of the degree of advancement with regard to time:

$$v = \frac{d\xi}{dt}$$

For example, in Equation 9.1, for all c and d moles of C and D that are formed, a and b moles of A and B disappear. The rate of the reaction v may be written as

$$v = \frac{d\xi}{dt} = -\frac{1}{a}\frac{d[A]}{dt} = -\frac{1}{b}\frac{d[B]}{dt} = \frac{1}{c}\frac{d[C]}{dt} = \frac{1}{d}\frac{d[D]}{dt} \tag{9.3}$$

where [X] represents the activity of element X, that is, the molar concentration of the element or the partial pressure if some of the elements involved are in gaseous phase. According to the law of mass action, which was enunciated for the first time by Waage and Guldberg in 1864, the speed of a chemical reaction is proportional to the concentrations of the participating elements. More precisely, the variation of the quantity of products of an elementary reaction is a function *f* of the activities of the participating elements:

$$v = \frac{1}{c}\frac{\mathrm{d}[\mathrm{C}]}{\mathrm{d}t} = k^{\rightarrow} f([\mathrm{A}],[\mathrm{B}],[\mathrm{C}],[\mathrm{D}]) \tag{9.4}$$

where $k^{\rightarrow}$ is the forward reaction rate constant. $k^{\rightarrow}$ is often assumed depending on T according to the usual Arrhenius equation:

$$k^{\rightarrow}(T) = \mathrm{A}\ \mathrm{e}^{-E_a/RT} \tag{9.5}$$

where A is an experimental factor accounting for effective collision frequency, T is the temperature in °K, E_a is the activation energy, and R is the universal gas constant.

It is important to remark that the term *reversible*, which is associated here with reaction 9.2, is used in this context with a different meaning from that of *thermodynamics*. A reversible reaction is one in which both the reagents and products coexist in a chemical dynamic equilibrium. An irreversible reaction is one in which the equilibrium is completely displaced or to the part of the reagents or to that of the products. To be precise, all chemical reactions are thermodynamically irreversible, therefore exhibiting a non-null entropy production; whereas several of them are chemically reversible. Actually, a truly chemically irreversible reaction is achieved when the settings of products do not allow the reverse reaction to take place; that is, when reaction products do not fulfill one of the following conditions:

1. Products persist, but not one of the elements involved exits the reacting system (e.g., by volatilization).
2. The cross-section remains effective (e.g., proper temperature and products contact are maintained).
3. The energy produced by the forward reaction remains available (in quantity and quality) and provides enough useful energy in order to jump the activation energy barrier for the reverse reaction without external intervention; for example, exogenous energy, reagents, or catalysts supply.

In Table 9.1, a possible classification of chemical reactions in which energy is dissipated is presented. The first two examples, which are of particular interest for ecological processes, encapsulate very complex chains of enzyme-promoted biochemical reactions. In the first case, the chemical energy of glucose is

TABLE 9.1
A Classification of Chemical Reactions According to the Crucial Criteria of Spontaneity and Reversibility

	Symbol	Energy Profile	Sample Prototype Reaction
		Irreversible	
Spontaneous Exoenergetic	$\rightharpoonup$	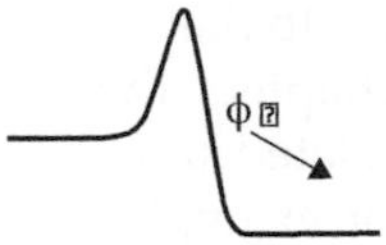	**Consumption** or oxidation of organic matter, e.g., respiration
$C_6H_{12}O_6 + 6O_2 \rightharpoonup 6CO_2 + 6H_2O + \phi$			
Nonspontaneous Endoenergetic	$\rightharpoonup$	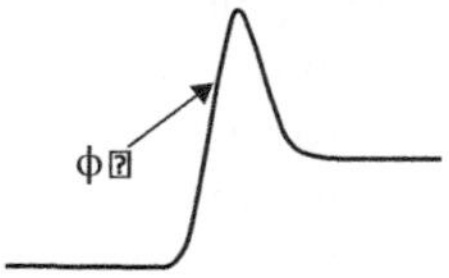	**Production** or reduction of inorganic carbon, e.g., photosynthesis
$6CO_2 + 6H_2O + \phi \rightharpoonup C_6H_{12}O_6 + 6O_2$			
			ϕ = energy (heat, light, and chemical)
Reversible			
Spontaneous	$\rightleftharpoons$		Water dissociation ionic equilibrium
$H_2O \rightleftharpoons H^+ + OH^-$			

Φ Represents energy flows such as inputs (sun or organic matter) or the energy produced by consumption.

converted by respiration into an enormous amount of energy that is available for cellular processes. This is the prototype of the biochemical reaction underlying metabolism. In the second case, the reduction of inorganic carbon is promoted by an input of external energy; for example, in photosynthesis, solar energy is converted into chemical energy and is accumulated by building glucose.

9.2 Chemical Equilibrium

Chemical equilibrium is generally a dynamic steady-state condition (see Section 6.5), which is dependent on the temperature, in which the total concentrations of the chemical species taking part in the reaction do not vary over time. In a chemical reaction, chemical equilibrium is, therefore, the state in which the concentrations of the reactants and products have not yet changed with time. It occurs only in reversible reactions and not in irreversible ones. A different type of equilibrium is attained when a complete conversion of the reactants occurs (referred to as *reaction to completion*). In this case, a progressive, irreversible chemical reaction takes place, which is described by the degree of advancement of the reaction ξ. This reaction does not advance uniformly, meaning that the speed of the reaction v is not constant; instead, it approaches zero when the reactants are almost over and the reaction ends ($\xi = 1$) to an equilibrium. In adiabatic systems, this type of chemical equilibrium corresponds to that of thermodynamic equilibrium.

As discussed in the previous chapters, living systems are necessarily nonadiabatic; instead, they are nonisolated and, to a certain extent, open systems. Living systems are bounded (for example, by a membrane), and it is this boundary that mediates the exchanges of energy/matter with the surrounding environment. A continuous exogenous flow throughout the boundary may turn an irreversible reaction into a reversible one. For example, in an ecosystem, primary production is driven by the solar energy flow and the inputs of inorganic matter; simultaneously, the reverse reaction, consumption, is taking place. By virtue of the second principle of thermodynamics, the amounts of the primary producers and of the consumers are somehow ruled by an evolutional criterion. This condition is similar to a chemical equilibrium in which the forward reaction proceeds at the same rate as the reverse one. The reaction rates of the forward and reverse reactions are generally not zero but, being equal, there are no net changes in the concentrations of the reactants and products.

The more the reaction is shifted to the right (toward products), reagents are more easily transformed into products; the opposite occurs when the reaction is shifted to the left. The process is said to be in *dynamic equilibrium* or *steady state*. This very general definition applies to situations that may arise as a consequence of very different system configurations. Such equilibria happen in phase transitions. For example, if the temperature in a closed system containing a mixture of ice and water is uniformly 273.15°K (0°C), the net amount of ice formed and the melt will be zero. If no vapor escapes from the system (the system is closed!), the amount of liquid water will also remain constant. In this case, three phases, ice (solid), water (liquid), and vapor (gas), are in equilibrium with one another. Similarly, at a particular temperature, equilibrium can also be established between the vapor phase and the liquid phase. Equilibrium conditions also exist between the solid phase and vapor phase.

Four characteristics are common to all the reactions in chemical equilibrium:

1. The spontaneity of chemical equilibrium: The system moves spontaneously toward a state of equilibrium, and if there is a disturbance from the outside, the system will tend to get back into equilibrium (principle of Le Chatelier) (see resilience and resistance in Section 6.5).
2. A process can be said to be in *dynamic chemical equilibrium* if it is reversible, that is, if the nature and properties of the equilibrium are the same regardless of the direction of approach to the equilibrium itself.

3. The steady state or dynamic nature of equilibrium is a situation that is defined by the equal speed of the forward and reverse reactions.
4. The equilibrium state is a thermodynamic condition of optimal energy dissipation, that is, a more favorable compromise between the natural tendency of the system to reach the minimum energy and the minimum entropy production* (see Chapter 3).

Let us consider a reversible chemical reaction similar to that described in Equation 9.2. For such a reaction, the function f appearing in the law of mass action (9.4) is commonly determined experimentally and assumes the general form

$$f = [A]^{\alpha} [B]^{\beta} [C]^{\gamma} [D]^{\delta} \tag{9.6}$$

The exponents α, β, γ, and δ are called *reaction orders* and depend on the reaction mechanism. The index $n = \alpha + \beta + \gamma + \delta$ is the overall order of the reaction and can assume, in general, non-integer values. In elementary reactions, stoichiometric coefficients and reaction orders coincide. These reactions proceed through a single transition state. This is not valid, in general, because rate equations do not always strictly follow the stoichiometry of the reaction. Let us assume that Equation 9.2 represents a simple reversible reaction; then, the speed of forward and reverse reactions may be written as

$$v^{\rightarrow} = k^{\rightarrow} [A]^{a} [B]^{b}$$

$$v^{\leftarrow} = k^{\leftarrow} [C]^{c} [D]^{d}$$

If the forward reaction proceeds with the same speed as the reverse one, the speed of consumption of reagents is equal to that of formation of the same reagents, and this equally applies for products. The equilibrium condition is

$$k_{eq} = \frac{k^{\rightarrow}}{k^{\leftarrow}} = \frac{[C]^{c} [D]^{d}}{[A]^{a} [B]^{b}} \tag{9.7}$$

where k_{eq} is the equilibrium constant of the reaction that depends on pressure and temperature. If k_{eq} is greater than 1, the reaction is shifted to the right toward the products; whereas if k_{eq} is smaller than 1, the reaction is shifted to the left. Given that the chemical reactions involved in ecological processes generally occur at practically constant atmospheric pressure, the only truly relevant dependence of k_{eq} is that on temperature, which is described by the Arrhenius Equation 9.5. The law of mass action provides us with a general method that could be used to write the expression for the equilibrium constant of any reaction. For chemical reactions that are not yet at equilibrium, the ratio calculated from the law of mass action is called the *reaction quotient Q*.

* Which in thermodynamic equilibrium corresponds to maximum entropy.

However, this quantity is not constant. Q values for a reversible chemical reaction have a tendency to reach a limiting value, which is exactly k_{eq}:

$$\lim_{t\to\infty} Q(t) = k_{eq}$$

This is equivalent to saying that a reversible chemical reaction shows the spontaneous tendency to reach a chemical equilibrium state. The law of mass action is universal and applicable under any circumstance. However, for reactions to completion (irreversible), the result may not be very useful. Indeed, in the latter case,

$$\lim_{t\to\infty} Q(t) = 1$$

Chemical reactions may not be as complete as stoichiometry calculations seem to imply. For example, the following reaction is far short of completion:

$$H_2O(g) + CO(g) \rightleftharpoons H_2(g) + CO_2(g)$$

Actually, this reaction is reversible only if the system in which the reaction is taking place is closed. In this case, the ratio

$$\frac{[H_2][CO_2]}{[H_2O][CO]}$$

is a constant. This chemical equilibrium is described by the law of mass action. Of course, when conditions, such as pressure and temperature, change, a period of time is required for the system to establish a new equilibrium. On the contrary, in an open system at room temperature, this reaction is irreversible due to elements in the gaseous phase that are escaping the system, and, in this case, it turns out to be a reaction to completion.

As said earlier, equilibrium conditions may be reached for phase transitions as well as for chemical reactions. Equilibrium is dynamic in the sense that it changes continuously, but the net change is zero. Heat transfer, vaporization, melting, and other phase changes are physical processes that spontaneously tend to equilibrium. If the system is closed and isolated, these changes are reversible. If the system is open, phase transitions are irreversible. For example, a glass containing water is an open system. Evaporation lets water molecules escape into the air by absorbing energy from the environment until the glass is empty. When the glass is covered, it becomes a closed system. The water vapor in the space above water eventually reaches the equilibrium of vapor pressure. Actually, if the glass is nonisolated, when heat transfer between the glass and the environment stops,* equilibrium is attained.

From a thermodynamic point of view, chemical equilibrium is ruled by the second law of thermodynamics. On analyzing the relationship between the free energy G and the degree of advancement of the reaction ξ, an evolutional

* That is, the temperature of the water equals that of the environment.

criterion may be found. From the second principle of thermodynamics, it immediately follows that at constant pressure p and temperature T,

$$\left(\frac{dG}{d\xi}\right)_{T,p} \leq 0 \tag{9.8}$$

where G is Gibbs free energy,

$$\Delta G = \Delta H - T\Delta S \tag{9.9}$$

where H is enthalpy and S is entropy. In exothermic reactions, ΔH is negative and energy is released. Examples of exothermic reactions are precipitation and crystallization (see Section 9.6), in which ordered solids are formed from disordered gaseous or liquid phases. In contrast, in endothermic reactions, heat is provided by the environment. In this case, the entropy of the system increases. Since the entropy increases with temperature, many endothermic reactions preferably take place at high temperatures. On the contrary, many exothermic reactions such as crystallization occur at low temperatures. The evolutional criterion expressed in Equation 9.8 says that at equilibrium, Gibbs energy, as a function of the chemical time* ξ, should be at a minimum and stationary.

Based on Gibbs fundamental Equation 3.4 and by using relations such as Equation 9.9, dG may be written as $dG = V\,dp - S\,dT + \sum_{j=1}^{k} \mu_j\,dn_j$, which at constant T and p becomes

$$dG = \sum_{j=1}^{k} \mu_j\,dn_j$$

μ_j is the chemical potential, and n_j is the molar concentration (the activity) of the jth element. The chemical potential of a reagent A is a function of the activity [A] of that reagent:

$$\mu_A = \mu_A^{o} + \ln[A]$$

$$dn_A = a \cdot d\xi$$

μ_A^{o} is the standard chemical potential, that is, the value of the chemical potential under specified standard conditions (in the standard state as defined in [IUPAC, 1997]), and a is the stoichiometric coefficient. The general concept of chemical equilibrium is expressed by the following thermodynamic evolutional criterion:

$$\left(\frac{dG}{d\xi}\right)_{T,p} = \sum_{j=1}^{k} \mu_j a_j = \Delta_r G \leq 0 \tag{9.10}$$

$\Delta_r G$ is the Gibbs free energy gradient for the reaction.

* The degree of advancement of the reaction may be considered a temporal parameter so that $dG/d\xi$ may be written as $d_\xi G/dt$; see example 3.11.

Several common reactions of interest for the description of ecological processes depend on a single-substance concentration. In this event, the law of mass action (9.4) becomes

$$v = \frac{d[C]}{dt} = -k[C]^n \tag{9.11}$$

n is the order of the reaction. Among these simple reactions, first-order ones are, by large, those that are used in ecological process descriptions. Very seldom, these processes need an order much bigger than three for their description reactions.

9.3 Enzymatic Processes

Enzymatic processes play a crucial role in ecology. Enzymes are special proteins pertaining to the general class of catalysts. These proteins are able to catalyze biochemical reactions by reducing the activation energy E_a of the reaction (see Figure 9.1). The consequence of lowering the activation energy is that of increasing the reaction speed v. Most enzyme reaction rates are millions of times faster than those of noncatalyzed reactions. Indeed, without catalysis by enzymes, most biochemical reactions would not occur in a time that is appropriate for supporting life. The role of enzymes is to facilitate the reaction by forming a complex through binding its active site (the part of the enzyme participating in the reaction) with the substrate (the reagents). The products of the reaction subsequently leave the enzyme, which remains available to start a new reaction. Similar to all catalysts, enzymes are not consumed during the reactions they promote. Actually, these proteins are specific for their substrates and accelerate only those few reactions, from among many possibilities, that are involved in a particular metabolic pathway. This very specificity is, in the general class of catalysts, peculiar to enzymes. Enzymes are able to catalyze around 4,000 biochemical reactions.

The function of a specific enzyme is determined by the spatial arrangement and the three-dimensional conformation of its subunits. Most of the enzymes are considerably larger than the substrates on which they work. Usually, the region involved in the catalytic activity of the enzyme, the active site, is very small (often only 3–4 amino acids). This region is the place of contact with the substrate and is where catalysis occurs. The correspondence between the active site and the substrate is not unique: Different substrates could compete for the same active site in order to achieve diverse reactions. Other substances, inhibitors, may bind to the active site, thus preventing the activation of any reaction. The global consequence is a reduction of the reaction rate. The selective capability of the enzyme is expressed by the *specificity factor*, which is defined as the ratio between the consumption rates of the two competing substances, provided they are in equal quantity. The number of molecules of substrates that can be bound and converted to the product per active site per unit time is called the *turnover number.*

As discussed in Chapter 4, ecosystems may be studied by analyzing fluxes between the different elements or compartments. In this perspective, ecological processes are sustained by flows of matter, energy, and information. Therefore, among the thousands of processes in which biochemical reactions are catalyzed by enzymes, it is worth recalling three of them, as they are really ubiquitous in living systems (Table 9.2). These are primary production (photosynthesis with fixation of inorganic carbon); the synthesis of adenosine triphosphate (ATP), which is the energy fuel for all metabolic processes in the cell; and the transmission of genetic information through DNA replication (Artioli, 2008). These very complex reactions consist of several intermediate stages.

The kinetics of enzymatic reactions is influenced by several factors such as substrate and enzyme concentrations, physical conditions (e.g., temperature), and presence of activators or inhibitors. Typically, enzymatic kinetics is described by the Michaelis–Menten equation (see Chapter 14). This is a simplified but, nevertheless, very practical model that consists of a two-step chemical reaction, which considers the concentrations of both substrate S and enzyme E as reagents. The first step involves the binding reaction that produces the enzyme–substrate complex ES and which is reversible (chemical equilibrium). The second step involves the formation of the final product P. The latter is an irreversible* reaction:

$$\mathrm{S}+\mathrm{E} \rightleftharpoons \mathrm{ES} \rightarrow \mathrm{P}+\mathrm{E} \tag{9.12}$$

The law of mass action for simple reactions is that expressed by Equation 9.11. In the case of first-order reactions such as Equation 9.12, the rates are

$$v^{\leftarrow} = \frac{\mathrm{dS}}{\mathrm{d}t} = k^{\leftarrow}\mathrm{ES} - k^{\rightarrow}\mathrm{S}\times\mathrm{E}$$

TABLE 9.2

Primary Biochemical Reactions That Are Catalyzed by Enzymes

Flow of	Products	Enzymes	
Matter	Glucose	Carboxylase/oxygenase (RubisCO)	Carbon fixation
Energy	ATP	ATP synthase	Storage and transport of chemical energy
Information	DNA	Topoisomerase, helicase, and DNA polymerase	DNA replication, correction, and accommodation

* In other words, the reaction is totally shifted to the right (toward products) so that reagents are straightforwardly transformed into products.

$$v^{\rightarrow} = \frac{\mathrm{d\,ES}}{\mathrm{d}t} = k^{\rightarrow}\mathrm{S}\times\mathrm{E} - k^{\leftarrow}\mathrm{ES} - k_p\mathrm{ES}$$

$$v^{p} = \frac{\mathrm{dP}}{\mathrm{d}t} = k_p\mathrm{ES}$$

where $k^{\rightarrow}$ and $k^{\leftarrow}$ are the reaction rate constants of, respectively, direct and inverse reactions of the enzyme–substrate complex ES, and k_p is the rate constant for P production. Since the first reaction is at chemical equilibrium,

$$\frac{\mathrm{d\,ES}}{\mathrm{d}t} = 0$$

and then,

$$\mathrm{S}\times\mathrm{E} = \frac{k^{\leftarrow} + k_p}{k^{\rightarrow}}\mathrm{ES} = k_s\mathrm{ES} \tag{9.13}$$

Enzymes are not consumed during the reaction; so, the total quantity E_o is constant:

$$E_o = \mathrm{E} + \mathrm{ES}$$

and, along with Equation 9.13,

$$\mathrm{ES} = E_o\frac{\mathrm{S}}{k_s + \mathrm{S}}$$

Finally, the differential equation that describes the Michaelis–Menten model is given by

$$v = -\frac{\mathrm{d}S}{\mathrm{d}t} = \frac{\mathrm{dP}}{\mathrm{d}t} = k_pE_o\frac{\mathrm{S}}{k_s+\mathrm{S}} = v_{\max}\frac{\mathrm{S}}{k_s+\mathrm{S}} \tag{9.14}$$

The maximum rate of the reaction $v_{\max}$ depends on the total amount of available enzyme (E_o) and on the rate at which P is produced from the enzyme–substrate complex. The constant k_s is known as a *half-saturation constant*. This value represents the concentration of the substrate at which the reaction rate is half of the maximum. Equation 9.14, which is plotted in Figure 9.2, describes the relationship between the substrate concentration S and the rate of production v (or consumption of the substrate).

This equation can be generalized by including several factors that influence enzymatic processes. For example, temperature dependence may be easily included by the usual Arrhenius equation 9.5. The ecological process of growth may be described starting by the Michaelis–Menten model. Indeed, by including in the model the fact that products (e.g., microbes) have the ability to increment the quantity of enzymes, a logistic-like dynamic is obtained (see Chapter 14). In this last case, the rate of production will depend on the number of products as well.

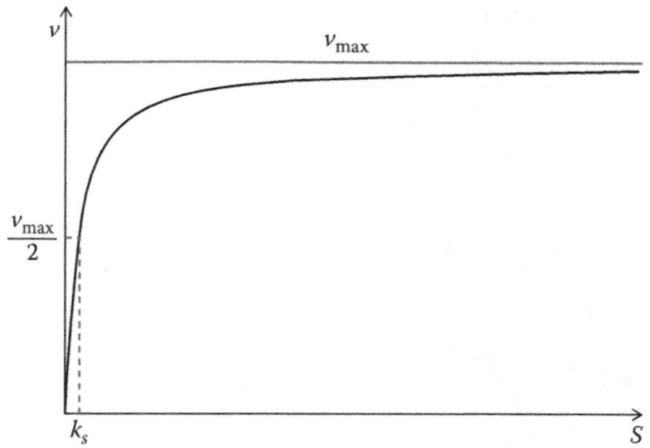

FIGURE 9.2
The Michaelis–Menten model. The maximum reaction rate v_{max} depends on the quantity of available enzymes. k_s is the half-saturation constant.

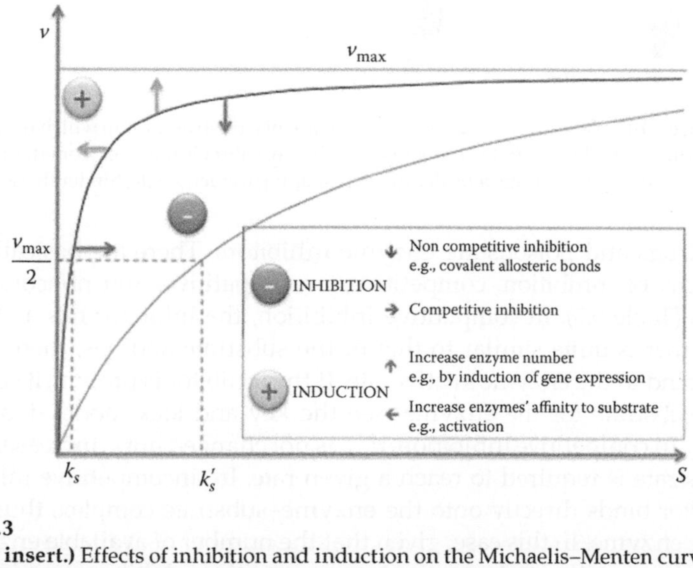

FIGURE 9.3
(See color insert.) Effects of inhibition and induction on the Michaelis–Menten curve. Positive induction may increase either the enzyme's affinity to the substrate S (decreased k_S) or the number of enzymes itself and, consequently, increase v_{max}. Inhibition lowers enzymes' affinity so that a larger quantity of substrate is needed to reach the half-saturation point.

Other molecules such as inhibitors and inducers (or activators) can control enzyme activity (Figure 9.3). Inhibition decreases enzyme activity and can be either competitive or noncompetitive. Induction increases activity either by increasing the number of enzymes (e.g., via fostered gene expression) or by increasing the particular enzyme's affinity to the substrate (e.g., by enzyme activation). Enzyme activators are inorganic compounds that are often involved in the allosteric regulation of enzymes for controlling biologic metabolism.

TABLE 9.3
Inhibition Mechanisms

Class of Inhibitor	k_s	v_{max}
Competitive	Increase	No effect
Uncompetitive	Decrease	Decrease
Noncompetitive	No effect	Decrease

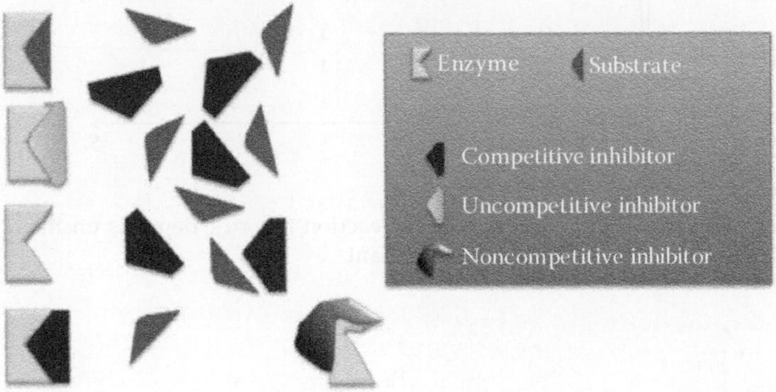

FIGURE 9.4
(See color insert.) The lock and key model. Enzymes are plastic structures in which the active site adapts dynamically to the substrate. Inhibitors, which may subtract enzymes from the active pool by behaving similar to a substrate or by definitely reshaping the active site, hinder this adaptation.

Many drugs and poisons are enzyme inhibitors. There are basically three mechanisms of inhibition: competitive, uncompetitive, and noncompetitive inhibition (Table 9.3). In competitive inhibition, the inhibitor has a chemical structure that is quite similar to that of the substrate and has, therefore, the ability to bind to the enzyme's active site. If the inhibitor is present, it competes with the substrate for the enzyme (see the key and lock model depicted in Figure 9.4). In competitive inhibition, v_{max} is not changed but k_s increases so that more substrate is required to reach a given rate. In uncompetitive inhibition, the inhibitor binds directly onto the enzyme–substrate complex, thus inactivating the enzyme. In this case, given that the number of available enzymes is reduced, v_{max} as well as k_s decreases. Noncompetitive inhibitors can bind to the enzyme at the same time as the substrate, but not to the active site. The enzyme is again inactivated and v_{max} decreases, whereas k_s is unchanged.

The inhibitors described here are said to be *reversible*, because they bind to enzymes with noncovalent interactions such as hydrogen bonding, hydrophobic interactions, and ionic bonding. The multiple weak bonds between the inhibitor and the active site combine to produce very strong and specific interactions. In contrast to irreversible inhibitors, reversible inhibitors can be easily removed by dilution or dialysis. Irreversible inhibitors bind covalently, substantially altering the enzyme's active site and occasionally destroying a functional group that is essential for the activity of the enzyme.

9.4 Redox

In chemistry, the term *redox* (a combination of reduction and oxidation) refers to all those chemical reactions in which the oxidation number* of the atoms changes; that is, all the reactions in which there is an exchange of electrons between chemical species. *Oxidation* refers to an increase in the number of oxidation states of a chemical species (e.g., molecule, atom, or ion), and it is usually accompanied by a supply of electrons by the oxidized species. *Reduction*, instead, is a decrease in the number of oxidation states of a chemical species, and it is usually accompanied by an acquisition of electrons by the reduced species. The two half-reactions of reduction and oxidation always involve a change in the number of oxidation states. This type of reaction spans from simple redox processes, such as the oxidation of carbon that generates carbon dioxide (CO_2) or the reduction of carbon by hydrogen, which produces methane (CH_4), to the most complex oxidation of glucose in living systems through a series of complicated processes of electrons transfer. Redox reactions are the basis of many biochemical processes that are essential to life, as, for example, respiration or photosynthesis, and they are relevant for the production of novel chemical compounds. A useful application of redox reactions is obtained if the reactants are kept separate, but electrical contact is ensured by means of conductive materials. Here, it is possible to intercept the flow of electrons and to use it to produce an electric current, whose potential depends on the chemical nature of the involved species. This effect is exploited in the electrochemical cells of batteries, which are accumulators that act as electrical energy sources in the form of chemical energy. A redox reaction can either occur spontaneously or be forced in the opposite direction through the application of a suitable electrical potential. This phenomenon is widely used in rechargeable batteries as well as in the process of electrolysis.

A redox reaction can be considered a reversible chemical reaction similar to Equation 9.2 and can be represented by the following two half-reactions:

Reduction: Oxidizing species + $n\,e^- \rightharpoonup$ Species reduced

Oxidation: Reducing species $\rightharpoonup$ Oxidized species + $n\,e^-$

where n is the number (or moles) of electrons that are exchanged during the reaction, that is, equal in both half reactions. The overall redox reaction can then be represented as follows:

$$\text{Oxidizing species} + \text{Reducing species} \rightleftharpoons \text{Oxidized species} + \text{Reduced species}$$

* In chemistry, the *oxidation state* (or number of oxidation) of a chemical element in a compound is defined as the number of electrons that are virtually transferred or acquired during the formation of the compound. When a bond joins two atoms, the electrons are considered virtually acquired from that of higher electronegativity.

Chemical species that are directly involved in the process of transfer of electrons are said to be *electroactive,* to distinguish them from any other chemical species present in the system that do not undergo oxidation or reduction. The equilibrium constant is given by Equation 9.7:

$$k_{ro} = \frac{[ox]}{[red]} \tag{9.15}$$

where [ox] and [red] are, respectively, the concentration of the oxidized and reduced forms in the reaction.

Several enzymatic reactions are redox reactions in which one compound is oxidized and the other is reduced. The ability of an organism to carry out redox reactions depends on the redox state of its environment. An indicator for this is the reduction potential, *Eh*. Aerobic bacteria commonly operate at positive *Eh* values, whereas strict anaerobes generally work at negative values. The redox potential can be calculated by Nernst equation. This equation provides the equilibrium reduction potential of a half cell in an electrochemical cell:

$$Eh = E_o + \frac{RT}{nF} \ln k_{ro}$$

where E_o is the standard* reduction potential of the electro-active species involved in the reaction, $R = 8.31\,\text{J}/°\text{K} \times \text{mole}$ is the universal gas constant, T is the absolute temperature in k, n is the number of electrons transferred in the half reaction, and F is the Faraday's constant. The standard reduction potential and the Gibbs free energy (Equation 9.9) are related by the equation

$$\Delta G = -nF \cdot E_o \tag{9.16}$$

If the reduction takes place in a spontaneous way, $\Delta G < 0$ and the standard potential is positive; whereas if the reduction is nonspontaneous, the standard potential is negative.

If the redox potential is positive, environmental conditions are oxidant; whereas if the potential is negative, they are reductant. The main oxidant element is oxygen, the presence of which increases the potential.

The redox potential plays an important ecological role, as aerobic organisms require a high potential, whereas anaerobes require a low potential. For example, with regard to sediments, the values of the redox potential provide important information on possible anoxia conditions. Negative values of the redox potential indicate the presence of poor oxygenation of the sediments with the consequent onset of hypoxia/anoxia conditions, which may, if persistent, lead to the development of methanogenesis with the production of H_2S. Anoxia caused, for example, by eutrophication may provoke the release of phosphorus that is trapped in the sediments. Phosphorus in sediments is

* Measured under standard conditions, that is, at a temperature of 25°C and a pressure of 1 atm.

usually bound as iron^{III}-phosphates, but if the conditions change from aerobic to anaerobic, the following redox reaction will take place:

$$FePO_4(s) + HS^- + e^- \rightharpoonup FeS + HPO_4^{2-}$$

with a further increase of phosphate that is available in the water column for algal growth.

Redox affects the solubility of nutrients, especially metal ions. Many inorganic ions are dominant participants in environmental redox processes. Table 9.4 shows the standard reduction potential of relevant redox processes in aquatic environments with regard to some of these ions.

In an aerobic environment, all organic matter will undergo oxidation if it is present for a sufficiently long time. Certain organisms can adjust their metabolism to their environment, such as facultative anaerobes. Facultative anaerobes can be active at positive *Eh* values as well as at negative *Eh* values in the presence of oxygen carried by inorganic compounds, such as nitrates and sulfates. For example, in an aquatic environment, if reduced material is sufficiently abundant, the oxygen dissolved in interstitial water or at the sediment–water interface will be pumped out and exhausted. The oxidation of

TABLE 9.4

Standard Reduction Potential of Relevant Redox Processes in Aquatic Conditions (Values of Standard Electrode Potentials Are Given in Volts Relative to the Standard Hydrogen Electrode)

Reaction	E_o (25°C)
$Na^+(aq) + e^- \rightarrow Na(s)$	−2.71
$Zn^{2+}(aq) + 2e^- \rightarrow Zn(s)$	−0.76
$Fe^{2+}(aq) + 2e^- \rightarrow Fe(s)$	−0.41
$Co^{2+}(aq) + 2e^- \rightarrow Co(s)$	−0.28
$Pb^{2+}(aq) + 2e^- \rightarrow Pb(s)$	−0.13
$Fe^{3+}(aq) + 3e^- \rightarrow Fe(s)$	−0.04
$2H^+(aq) + 2e^- \rightarrow H_2(g)$	0.00
$Sn^{4+}(aq) + 2e^- \rightarrow Sn^{2+}(aq)$	0.15
$Cu^{2+}(aq) + e^- \rightarrow Cu^+(aq)$	0.16
$AgCl(s) + e^- \rightarrow Ag(s) + Cl^-(aq)$	0.22
$Cu^{2+}(aq) + 2e^- \rightarrow Cu(s)$	0.34
$Cu^+(aq) + e^- \rightarrow Cu(s)$	0.52
$Fe^{3+}(aq) + e^- \rightarrow Fe^{2+}(aq)$	0.77
$Ag^+(aq) + e^- \rightarrow Ag(s)$	0.80
$NO_3^-(aq) + 4H^+(aq) + 3e^- \rightarrow NO(g) + 2H_2O(l)$	0.96
$O_2(g) + 4H^+(aq) + 4e^- \rightarrow 2H_2O(l)$	1.23
$Cr_2O_7^{2-}(aq) + 14H^+(aq) + 6e^- \rightarrow 2Cr^{3+}(aq) + 7H_2O(l)$	1.33
$Cl_2(g) + 2e^- \rightarrow 2Cl^-(aq)$	1.36
$Co^{3+}(aq) + e^- \rightarrow Co^{2+}(aq)$	1.82

organic matter, however, will continue by denitrification and sulfate reduction. All these processes can, in principle, occur by pure chemical oxidation, but, in general, the microbiological oxidation is far more effective.

Redox reactions involving carbon are common to many important biological processes. The carbon atom is found in nature in different oxidation states (Table 9.5).

In the oxidative catabolism,* carbon atoms with different degrees of oxidation are brought to the maximum degree of oxidation (CO_2) with energy recovery; this process is called *respiration*. The reverse process is photosynthesis, which involves the reductive carbon anabolism of carbon dioxide to organic carbon. In many bacteria, the oxidative catabolic pathway is not completed: They do not arrive at the production of CO_2 but instead reach only alcohols. This process is called *fermentation* (see Chapters 12 and 15). Cellular respiration is the oxidation of glucose ($C_6H_{12}O_6$) to CO_2 and the reduction of oxygen to water. The process produces a significant amount of energy as well. The summary equation for cell respiration is

$$C_6H_{12}O_6 + 6O_2 \rightharpoonup 6CO_2 + 6H_2O + \phi$$

For the advancement of the intermediate steps, the process of cell respiration also depends heavily on the reduction of NAD+ to NADH and the reverse reaction (the oxidation of NADH to NAD+). Photosynthesis and respiration are complementary, but photosynthesis is not the reverse of the redox reaction in cell respiration. The summary equation for photosynthesis is

$$6CO_2 + 6H_2O + \text{Light} \rightharpoonup C_6H_{12}O_6 + 6O_2$$

Biological energy is commonly stored and released by means of redox reactions. Photosynthesis involves the reduction of carbon dioxide to sugars and the oxidation of water into molecular oxygen. The reverse reaction, respiration, oxidizes sugars to produce carbon dioxide and water. As intermediate steps, the reduced carbon compounds are used to reduce nicotinamide

TABLE 9.5

Carbon in Ascending Order of Oxidation State

Alkenes	C_2H_4 ethylene
Alcohols	C_2H_5OH ethanol
Alkynes	HCCH acetylene
Aldehydes	CH_2O formaldehyde
Ketones	CH_3-CO-CH_3 acetone
Carboxylic acids	CH_3COOH acetic acid
Carbon dioxide	CO_2

* For the definitions of *catabolism* and *anabolism*, refer to Chapter 14.

adenine dinucleotide (NAD+), which contributes to the creation of a proton gradient, which drives the synthesis of ATP and is maintained by the reduction of oxygen. Mitochondria perform similar functions in animal cells.

The electrons are transferred through these pathways in four modes: as individual electrons, as H atoms, as a hydride ion, or via a direct combination with oxygen atoms. The enzymes that catalyze redox reactions are called *oxidoreductases*. Some important examples of this class of enzymes are dehydrogenases and flavoproteins.

In a living system, the *redox state* refers to the balance of NAD+/NADH and NADP+/NADPH. The redox state is mirrored in the balance of several metabolites (e.g., lactate and pyruvate, beta-hydroxybutyrate and acetoacetate) whose transformation is dependent on these ratios. An anomalous redox state can result in a variety of harmful situations, such as hypoxia, shock, and sepsis. Sometimes, redox reactions assume the role of signal carriers, hence participating in the control of cellular processes.

9.5 Acid–Base

Acid–base are chemical reactions in which there is no variation in the oxidation states of the reagents. The name is derived from the reagents participating in the reaction that are said to be *acid* and *base*. Acid–base reactions differ from redox ones, in which there is variation in the oxidation state of at least one element involved in the reaction. Since the assignment of the oxidation state is conventional,* so also is the change in the oxidation state, and, therefore, the distinction between acid–base and redox reactions are also conventional. In practice, however, oxidation states are assigned with a single conventional method, and then, these two fundamental types of chemical reactions are a well-known and important method of unique classification for chemical reactions.

In an aqueous solution, self ionization of water occurs:

$$H_2O(aq) \rightleftharpoons H^+(aq) + OH^-(aq) \tag{9.17}$$

In conditions of chemical equilibrium, water may dissolve in one hydrogen ion and one hydroxide ion, a water molecule that has lost an atom of hydrogen (a proton). pH, a measure of the acidity and the basicity/alkalinity of a solution in terms of hydrogen ions activity $[H^+]$, is defined as

$$pH = -\log_{10}[H^+] \tag{9.18}$$

* This indicator is defined as the charge that an atom is considered as possessing when electrons are counted according to an agreed-on set of rules.

For pure water at 25°C (standard conditions), the concentration of hydrogen ions is $[H^+] = 10^{-7}$ *moles*, and, thus, pH = 7. According to the acid–base theory for aqueous solutions of Svante Arrhenius (1884), an acid is a substance that somehow increases the formation of hydrogen ions $[H^+]$ in aqueous solution. A base is a substance that somehow promotes the formation of hydroxide ions $[OH^-]$ in aqueous solution. More generally (Brønsted–Lowry theory, 1923), an acid is a substance that is capable of donating one or more hydrogen ions $[H^+]$ that can be accepted by a base. A base is a substance that is capable of accepting one or more hydrogen ions $[H^+]$ that are released by an acid. An acid is a proton donor, whereas a base is a proton acceptor. An acid–base reaction is, thus, a reaction of a chemical species that transfers protons (hence, by charge, balancing electrons too) to another species which is able to accept them. In this type of reaction, the acid is transformed in its conjugate base. The concept of complementarity between acid and base is introduced, as the acid is not such if not in the presence of a counterparty that donates their $[H^+]$ ion, and the base is not such if not in the presence of a counterparty from which to accept an $[H^+]$ ion. A certain chemical substance is not defined as an acid or a base in an absolute sense, but in relation to the substance with which it transfers an $[H^+]$ ion (or an electronic doublet) and, therefore, relatively to a specific reaction. So, even if a substance shows acidic behavior in some reactions (and maybe in its name, the word *acid* appears), it may behave similar to a base in others. Conversely, a substance that shows basic behavior in some reactions may behave an acid elsewhere.

We may indicate a generic acid by the symbol "HA." "A" represents the atoms or group of atoms to which the acidic proton H^+ is bound. When the acid HA reacts in an aqueous solution, the proton transferred to H^+ cannot stand alone but actually binds to a water molecule and produces a second acid, the *hydronium* (H_3O^+). The reaction of HA in aqueous solution may be written as

$$HA(aq) + H_2O \rightleftharpoons H_3O^+(aq) + A^-(aq) \tag{9.19}$$

Concurrent to hydronium, a second base A^- is produced. HA and A^- are called a *conjugate acid–base pair*. Hydronium ion and water also form a conjugate acid–base pair. The hydronium ion, carrying the acidic proton H^+, acts as the acid member of the pair, whereas water is the base. Let us now consider the following reaction of ammonia NH_3 dissolved in water:

$$NH_3(aq) + H_2O \rightleftharpoons NH_4^+(aq) + OH^-(aq)$$

In this reaction, water acts as the acid proton donor, the conjugate base of water is the hydroxide OH^-, and the conjugate acid of NH_3 is $NH_4{}^+$. Water is, therefore, a member of two conjugate pairs. In one case, it is the base member, whereas in the other, it is the acid. Table 9.6 lists a number of common acids and their conjugate bases.

TABLE 9.6

Common Conjugate Acids and Bases

Conjugate Acid (HA)	Conjugate Base (A^-)
HCl	Cl^-
NH_4^+	NH_3
HCO_3^-	CO_3^{2-}
HPO_4^{2-}	PO_4^{3-}

Every proton transfer reaction involves two conjugate acid–base pairs. Such processes may be viewed as competitions for the proton H^+ between two bases. In general, we would expect the stronger base to win this competition. The strength of an acid or a base depends on the number of hydrogen ions or hydroxide ions produced in water solution. The greater the number of ions, the stronger is the acid or base. Examples of strong acids are hydrochloric acid (HCl), sulfuric acid (H_2SO_4), and nitric acid (HNO_3). Strong bases are sodium hydroxide (NaOH), potassium hydroxide (KOH), and calcium hydroxide ($CaOH_2$). Acetic acid ($HC_2H_3O_2$) is a weak acid, whereas ammonia (NH_3) is a weak base. The reaction of acid dissociation (9.19) occurs to varying extents, depending on the strength of the acid, HA. Equation 9.19 is written with a double arrow to acknowledge the fact that often the reaction does not proceed completely to the right.

Instead, a chemical equilibrium may be established between reactants and products. The equilibrium constant can be used as a measure of acid strength. The reaction equilibrium constant is given by

$$k_a = \frac{[H_3O^+][A^-]}{[HA]} \tag{9.20}$$

The concentration of water is not included in the expression for the equilibrium constant, because its value is large and essentially constant during the reaction.

Acids and bases have some distinct and readily observable physical and chemical properties. For example, acids taste sour and are capable of dissolving reactive metals with the production of hydrogen gas (H_2), whereas bases taste bitter and feel slimy. Both acids and bases may lose their characteristic properties during interactions. When coming into contact, acids and bases react with each other. If equivalent quantities of acid and base are combined, the reaction ensuing is generally a reaction to completion of the production of salt* and water.

$$\text{Acid} + \text{Base} \rightharpoonup \text{Salt} + H_2O$$

* A salt is an ionic compound that consists of a cation, which is different from H^+ combined with an anion. Salts usually result from the neutralization reaction of an acid and a base.

This process is known as *neutralization*. For example, when hydrochloric acid and sodium hydroxide react together, sodium chloride and water are obtained:

$$HCl + NaOH \rightarrow NaCl + H_2O$$

Table 9.7 reports examples of acids, bases, and salts that may participate in neutralization.

In Table 9.8, several well-known acids are enumerated in decreasing order of strength, from strong to very weak acids. The equilibrium constants of the forward reaction of dissociation k_a as well as of the reverse one of k_b are indicated.

- $k_a > 1$ for strong acids, including H_3O^+, and for all the acids above it. All the strong acids have large k_a values, meaning that they react essentially to completion with water. These acids are, therefore, strong electrolytes. The large k_a ensues from large concentrations of the products and a very small concentration of the reactant HA. The value of k_a for HCl is practically infinite. In an aqueous solution of HCl, the concentration of nondissociated HCl is null.
- $k_a < 1$ for weak acids. The weak acids dissociate to only a small extent in water. A common weak acid is acetic acid.
- $k_a < 10^{-14}$; these are very weak acids, for example, NH_3 and H_2. As can be seen from the table, these acids lie below and are, thus, less acidic than water. In water solution, therefore, the acidity of these species is negligible (simply cannot be measured).

Strong acids react completely with water. In such cases, conjugate bases are just spectator ions. Strong bases also react completely with water. An acid reacts completely with any base below it in the table and, conversely, a base reacts completely with any acid above it in the table. Very weak acids and bases do not react at all with H_2O.

TABLE 9.7

Acids, Bases, and Salts That May Be Involved in Neutralization

Acids		Bases		Salts
Strong	Weak	Strong	Weak	
HCl	$HC_2H_3O_2$	NaOH	NH_3	NaCl
HNO_3	HCOOH	KOH	CH_3NH_2	K_2SO_4
$HClO_4$	HF	LiOH		NaF
H_2SO_4	H_3PO_4	$Ca(OH)_2$		LiCl
HBr	H_2CO_3	$Ba(OH)_2$		$Mg(NO_3)_2$
HI		$Sr(OH)_2$		

TABLE 9.8
Acids in Decreasing Order of Strength

Categories	Acid, HA	Base, A⁻	k_a(HA)	k_b(A⁻)
	$HClO_4$	ClO_4^-	∞	
Strong acids	HI	I^-		
$K_a > 1$	HBr	Br^-		
	HCl	Cl^-		
	H_2SO_4	HSO_4^-		
	HNO_3	NO_3^-	22	
	H_3O^+	H_2O	1	1.0×10^{-14}
	HSO_4^-	SO_4^{2-}	1.2×10^{-2}	8.3×10^{-13}
	$HClO_2$	ClO_2^-	1.2×10^{-2}	
	H_3PO_4	H_2PO_4–	5.9×10^{-3}	1.7×10^{-12}
	HF	F^-	6.5×10^{-4}	
	HNO_2	NO_2^-	5.1×10^{-4}	
	HCOOH	$HCOO^-$	1.8×10^{-4}	5.56×10^{-11}
Weak acids	CH_3COOH	CH_3COO^-	1.8×10^{-5}	5.6×10^{-10}
$K_a < 1$	$Al(H_2O)_6^{3+}$	$Al(H_2O)_5(OH)^{2+}$	1.4×10^{-5}	
	NH_3OH^+	NH_2OH	9×10^{-7}	
	H_2CO_3	HCO_3^-	4.2×10^{-7}	
	H_2S	HS^-	1.1×10^{-7}	
	$H_2PO_4^{2-}$	HPO_4^{2-}	6.3×10^{-8}	
	HOCl	OCl^-	3.5×10^{-8}	
	NH_4+	NH_3	5.6×10^{-10}	1.8×10^{-5}
	HCN	CN^-	4.9×10^{-10}	
	HCO_3^-	CO_3^{2-}	4.7×10^{-11}	
	$CH_3NH_3^+$	CH_3NH_3	2.3×10^{-11}	4.3×10^{-4}
	HPO_4^{2-}	PO_4^{3-}	4.8×10^{-13}	
	H_2O	OH^-	1.0×10^{-14}	1
	NH_3	NH_2^-	10^{-35}	
Very weak acids	H_2	H^-	2×10^{-38}	
$K_a < 10^{-14}$	CH_4	CH_3^-		
	OH^-	O^{2-}		

Organic acids are a very important class of acids. In general, organic acids are weak acids that do not dissociate completely in water, whereas the strong mineral acids do. Carboxylic organic acids are organic compounds that contain the carboxyl group (COOH). The lighter between the carboxylic acids, up to three carbon atoms, are water-soluble compounds; the heavier ones are oily liquids with a high boiling point, resulting from the hydrogen bonds that the COOH groups exchange with the different molecules. With increasing molecular weight, carboxylic acids become solid low melting waxes. Table 9.9 reports some saturated carboxylic acids (with a linear chain) that are of biological relevance.

Among the many organic acids, several play a relevant role in various biochemical processes. Lactic acid ($C_3H_6O_3$) is constantly produced in fermentation during metabolism. Citric acid ($C_6H_8O_7$) is a natural preservative, and it is also employed in foods and soft drinks for taste processing. Citrate, the conjugate base of citric acid, is an important intermediate in the citric acid cycle, and is, therefore, found in all living things. The oxalic acid's ($H_2C_2O_4$) conjugate base, the oxalate ($C_2O_4^{2-}$), is a chelating agent for metal cations. Uric acid ($C_5H_4N_4O_3$) is a strong organic acid of natural origin. It is produced in living organisms as a byproduct of amino acids and purines metabolism. In its neutralization, ions and salts are produced, and they are known as *urates* and *acid urates*, such as ammonium acid urate. Lower-molecular-mass organic acids, such as formic and lactic acids, are miscible in water; whereas higher-molecular-mass organic acids, such as benzoic acid, are insoluble in molecular (neutral) form. On the other hand, most organic acids are very soluble in organic solvents. Most ecological processes are dependent on pH conditions; therefore, acid–base reactions are of great environmental interest. All biological processes have a pH optimum, which is usually in the range 6–8. This implies that processes, such as primary production, microbiological decomposition, nitrification, and denitrification, are strictly linked to pH conditions. In particular, in aquatic ecosystems, pH plays an important role at all hierarchical levels. For example, the ratio of hydrogen carbonate (HCO_3^-) to carbon dioxide (CO_2) is dependent on pH, and high concentrations of carbon dioxide are toxic for fish. The ratio of ammonium (NH_4^+) to ammonia (NH_3) is conditioned by pH, whereas ammonia is toxic for fish. The fertility of fish and zooplankton is also highly determined by pH. At the water–sediment interface, heavy metal ions are released from the soil and sediment increases very rapidly with decreasing pH. On the contrary, heavy metal ions are precipitated at pH 7.5 or above.

TABLE 9.9

Saturated Carboxylic Acids Listed in Order of Growing Number of C Atoms

n of C Atoms	Name	IUPAC	Formula	Where
1	Formic	methanoic	$HCOOH$	insect stings
2	Acetic	ethanoic	CH_3COOH	vinegar
3	Propionic	propionic	CH_3CH_2COOH	milk
4	Butyric	butanoic	$CH_3(CH_2)_2COOH$	butter
5	Valeric	pentanoic	$CH_3(CH_2)_3COOH$	*Valeriana officinalis*
6	Caproic	hexanoic	$CH_3(CH_2)_4COOH$	goat fat

It is clearly understandable why pH control is a fundamental asset in order to maintain stable suitable conditions for life. pH control is achieved by buffer action. A buffer solution is an aqueous solution that consists of a mixture of a weak acid and its conjugate base or a weak base and its conjugate acid. Buffer action is a property of a weak acid or a base and its salt. Unlike the weak acid that is poorly dissociated, the salt is fully dissociated. This means that the acid's contribution to the total acidity is insignificant compared with that of the salt. As a consequence, the pH of the solution changes very little when a small amount of strong acid or base is added to it. Buffer solutions are used as a means of keeping pH at a nearly constant value in a wide variety of situations. Blood is an example of a buffer solution: A buffer of carbonic acid (H_2CO_3) and hydrogen carbonate (HCO_3^-) is present in blood plasma, to maintain a pH between 7.35 and 7.45. The conjugate bases of organic acids such as citrate and lactate are often used in biologically compatible buffer solutions. The majority of biological samples are prepared and maintained in buffers at pH 7.4. Buffer solutions are necessary to keep the correct pH for enzymes to work. Indeed, several enzymes work only under very specific conditions; if the pH moves outside a narrow range, the enzymes slow or stop working and can denature, that is, they can lose enough three-dimensional structure, which renders them inactive. In many cases, denaturation can permanently disable enzymes' catalytic activity. Table 9.10 reports the components and pH range stabilization for some common buffer solutions.

In order to characterize the effectiveness of a buffer solution, starting from Equations 9.18 and 9.20, it is useful to introduce the alkalinity pK:

$$[H_3O^+] = [H^+] = k_a \frac{[HA]}{[A^-]}$$

$$-\log[H^+] = -\log k_a + \log \frac{[A^-]}{[HA]}$$

TABLE 9.10

Buffer Solutions, Components, and pH Stabilization Range

Components	**pH Range**
HCl, Sodium citrate	1–5
Citric acid, Sodium citrate	2.5–5.6
Acetic acid, Sodium acetate	3.7–5.6
K_2HPO_4, KH_2PO_4	5.8–8
Na_2HPO_4, NaH_2PO_4	6–7.5
Borax, Sodium hydroxide	9.2–11

Alkalinity is, thus, defined as

$$pK = -\log k_a$$

and, hence,

$$pH = pK + \log\frac{[A^-]}{[HA]} \tag{9.21}$$

The alkalinity is the concentration of hydrogen ions that can be taken up by proteolytic species present in the solution. Obviously, the higher the alkalinity, the better the solution is able to maintain a given pH value if acid or base is added. If acid or base is added up to the pH midpoint, in which 50% of the acid or base has been used up, then pH = pK, as log $[A^-]/[HA] = 0$. In this pH condition, the solution is at its maximum buffering efficiency. Further acid or base additions will waste the solution's buffering ability.

The buffering capacity is a measure of the readiness of a buffer solution to counter a change in pH that is caused by the addition of a certain amount of strong acid or base. The change in pH will depend on both the entity of the addition and the characteristics of the solution. Buffering capacity is defined as

$$\beta = \frac{d[\text{base}]}{d\,pH} = -\frac{d[\text{acid}]}{d\,pH} \tag{9.22}$$

where d[base] and d[acid] are the amount of, respectively, strong base and acid that should be added to a liter of solution to produce a variation of d pH.

At very low pH, β increases in proportion to the hydrogen ion concentration. When pH = pK, β is at a local maximum. Buffering capacity is proportional to the acid dissociation constant, k_a. Weaker acids have greater buffering capacities.

Soils and sediments also possess some buffering ability. Ion exchange capacity depends on the content of colloids as well as on their chemical nature; so, the buffering occurs more intensely in soils that are rich in organic matter and, to a lesser extent, in clay soils. Coarse texture soils, being poor in mineral colloids, and often also organic colloids, generally have a milder buffering capacity and are more susceptible to changes in pH.

9.6 Solutions, Precipitation, and Flocculation

Solutions play a crucial role in ecological processes. Indeed, besides enabling chemical species to come into close contact so that chemical reactions can occur, on an ecological scale, solutions transfer molecules and ions between different ecosystem compartments and carry these elements to and from cells (Manahan, 2001). In general, all biological fluids consist of solutions. For example, digestion is a process in which large complex insoluble molecules of food are broken into simple soluble molecules that are then carried by the blood to all districts of the body. On the way back, blood is transported to disposal systems by step products and wastes that are formed through cellular metabolism.

In general, *solubility* refers to the characteristic of a solid, liquid, or gas (the *solute*) to dissolve in a solid, liquid, or gaseous solvent to form a homogeneous solution of the solute in the solvent. The solubility of a substance depends on the solvent properties as well as on environmental conditions such as temperature and pressure. The degree of the solubility of a substance in a particular solvent is expressed by the saturation concentration in standard conditions. This value is the equilibrium concentration at which by adding further solute, the concentration in the solution does not increase anymore. This point of maximum concentration, the saturation point, depends on the temperature and pressure of the solution as well as on the chemical nature of the substances involved.

In ecological processes, solvents are, by large, liquids such as pure substances (e.g., water) or mixtures. Solubility ranges from infinitely soluble solutes (e.g., ethanol in water) to weakly soluble ones (e.g., zinc carbonate in water). The term *insoluble* is often applied to very blandly soluble compounds (e.g., iron [II] sulfide). The solubility of a substance depends on the ecological processes of dissolution, dissociation, and chemical rearrangement in which the substance is involved. The time required for reaching the equilibrium concentration depends on particle size and other kinetic factors: After a sufficient time, even big molecules may eventually dissolve.

When by adding more solute the equilibrium solubility concentration is surpassed, under particular conditions, a supersaturated, metastable solution can be obtained. In this situation, a tiny perturbation of the solution (e.g., gently shaking the solution or adding to it a small amount of some molecules) may result in an abrupt phase shift of the solute.

In chemistry, the term *precipitation* indicates the phenomenon of separation of a solid substance from a solution. An ionic substance leaves the solution and forms an insoluble (or slightly soluble) solid. The final product obtained is called the *precipitate*. Such separation can occur as a result of a chemical reaction or of a variation in the physical conditions of the solution, for example, temperature or pH. For example, precipitation may occur as an exothermic reaction by progressively cooling the solution. Analogously, as said earlier, heavy metal ions start precipitating when pH reaches a value of 7.5 or above. The reverse process of precipitation, the transfer of matter from the solid to the aqueous phase, is the dissolution.

Precipitation results from the natural tendency to equilibrium. For example, settling and sedimentation are processes of physical precipitation. In these processes, a solid is built in a certain amount of time due to physical forces such as gravity or centrifugation. During chemical reactions, precipitation may also occur, in particular, if the density of the solution increases.

Dissolution and precipitation processes are generally slower than chemical reactions among dissolved species. As an example, let us consider a dissolving electrolyte, A_mB_n. According to the law of mass action, Equation 9.6 is written as

$$A_mB_n(s)\Delta m A^{n+} + nB^{m-} \tag{9.23}$$

The opposite process is the corresponding precipitation process. If the equilibrium condition (Equation 9.7) is applied in Reaction 9.23, the solubility product is obtained

$$K_s = \{A^{n+}\}^{m.} \cdot \{B^{m-}\}^{n} \tag{9.24}$$

Solubility equilibriums fall into three general classes: simple dissolution, as in the case of sucrose in water; dissolution by ionic dissociation, such as that of salts; and dissolution by weak acid or base reactions. In every case, the equilibrium constant is indicated by the law of mass action. Let us consider, for example, the following salt dissociation:

$$NaCl(s) \rightleftharpoons Na^{+}(aq) + Cl^{-}(aq) \tag{9.25}$$

If the solution has not yet reached equilibrium, the law of mass action (9.7) gives the quotient or ionic product for this reaction:

$$Q(t) = [Na^{+}][Cl^{-}] \tag{9.26}$$

Once the equilibrium is reached, if k_{sp} is the solubility of NaCl, the solubility of both Na+ and Cl^- will also be k_{sp}. The solubility product is

$$k_{sp} = \lim_{t\to\infty} Q(t) = [Na^{+}][Cl^{-}] \tag{9.27}$$

Increasing salt concentration by further addition of salt, the ionic product Q increases and a new equilibrium is attained. This process is, however, limited by the solubility of NaCl, $k_{NaCl} = 358$ g/L (at 20°C in standard conditions). Once this limit is reached, a precipitate will form. Supplementary salt addition would only increase precipitation. If the limit is approached very slowly, precipitation may not happen immediately. The solution is then supersaturated for some time until precipitation occurs.

Precipitation takes place in two steps: nucleation and growth. The nucleation is the formation of microcrystals of solute. These are the cores of crystallization, which tend to grow by aggregation of the other solute that surrounds the crystal. The crystals, hence, form a suspension. Suspended particles are removed from the suspension by settling or sedimentation (Figure 9.5). In the course of nucleation, the formation of the solid–liquid

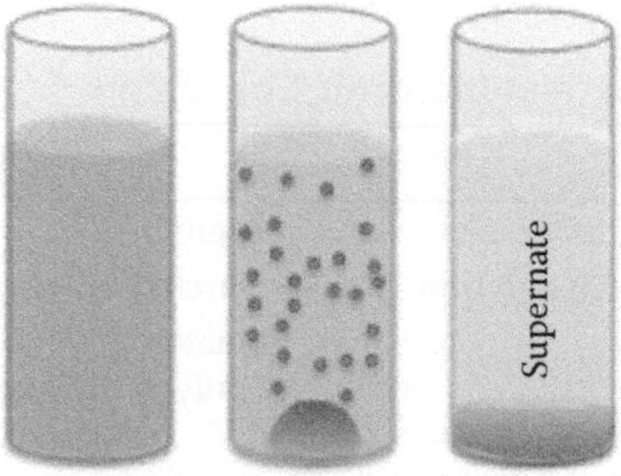

(a) (b) (c)

FIGURE 9.5
(See color insert.) The mechanism of precipitation. In the solution (a), the formation of crystals by nucleation and growth (b), turns the solution into a suspension that finally undergoes sedimentation (c), *Supernate* refers to the liquid that remains above the solid produced by precipitation.

interface requires some energy, in accordance to the relative surface energy of the solid and the solution. If this energy is not available, and no more nucleation surface is available, supersaturation may occur. Concurrent to the phenomenon of precipitation is that of coprecipitation, which may occur by adsorption of foreign ions on the surface of the main precipitate.

In many ecological processes, living organisms use and control the precipitation process. An example is the calcium carbonate in the skeleton of marine coral or in the shell of some mollusks. In this case, the ions of calcium carbonate precipitate from water in the body and form the calcareous part of the exoskeleton. For some solutions (often those involved in biological processes), it is possible that due to supersaturation, the nucleation prevails on growth, forming in this case a colloid, which is hardly separable from the rest of the solution by natural physical means alone. A colloidal solution is quite stable. In a colloid, the microcrystals of the solute formed do not form an agglomerate of a defined geometric shape.

Flocculation or coagulation is the process that occurs when the solid phase tends to separate, forming flakes in suspension in a colloidal solution. The mechanism of adsorption that will be introduced in Chapter 11 is at the base of the process. Temperature, pH, and ionic strength are environmental factors that strongly influence flocculation. Coagulants are chemical or electrochemical agents that, by destabilizing the charges of the colloid, enable flocculation. Coagulants are generally inorganic electrolytes and organic polyelectrolytes. These are chemical compounds of various origins that exert their action, each one in a specific range of pH values. Some common coagulants are listed in Table 9.11:

A sufficiently high temperature, an increase in the ionic strength, and a strong shaking of the colloid are also factors that may foster coagulation.

Flocculation and sedimentation processes play a key role in water purification, as, for example, in sewage treatment systems.

TABLE 9.11

Common Organic and Inorganic Coagulants

Inorganic	Organic
Alum and aluminate	Alginates
Sulfates of aluminum and Iron	Starch derivatives and cellulose
Ferric chloride	Quaternary ammonium polyelectrolytes
Sodium silicate	

10

Chemical Lysis

10.1 Introduction

In biochemistry, the term *lysis* refers to the dissolution or destruction of cells such as blood cells or bacteria. For instance, the disintegration of a cell resulting from the destruction of its membrane by a chemical substance (e.g., an antibody) is often fostered by enzymes (Section 9.3). The action of enzymes results in the chemical subprocess of catalysis (see Figure 9.1), which literally refers to the effect of lowering the process activation energy. In the context of ecological processes, we shall refer to this process as the *biological degradation* of organic matter (see e.g., Chapter 17).

Pure chemical lysis processes are generally based on hydrolysis, which is the breaking of chemical bonds by the addition of water. Generally, hydrolysis is the first step that is involved in the degradation of a substance (e.g., proteolysis).

Photolysis is an ecologically important variant that refers to any chemical process in which a compound is directly or indirectly broken by employing the energy supplied by light.

10.2 Hydrolysis

The term *hydrolysis* refers to the processes involving water, hydrogen ions, and hydroxide ions, and it results in the introduction of a hydroxyl group (OH^-) in the structure of the compound. Hence, this chemical process causes a molecule of water to be added to a substance. Sometimes, this addition causes both the substance and the water molecule to split into two parts. In such reactions, one fragment of the molecule of products (or reagents) gains a hydrogen ion.

A common kind of hydrolysis occurs when a salt of a weak acid or weak base (or both) is dissolved in water. Water spontaneously ionizes into hydroxyl anions and hydrogen cations (see Equation 9.17). The salt also dissociates into its constituent positive and negative ions. Strong acids also undergo hydrolysis; for example, dissolving sulfuric acid in water entrains hydrolysis and gives hydronium and bisulfate.

Metal ions in aqueous solution form hydrated ions whose general formula is

$$M(H_2O)_n^{m+} \tag{10.1}$$

Hydrated ions undergo hydrolysis to a variable extent. The first hydrolysis step may be generically represented as

$$M(H_2O)_n^{m+} + H_2O \rightleftharpoons M(H_2O)_{n-1}(OH)^{(m-1)+} + H_3O^+ \tag{10.2}$$

Usually, the rate of hydrolysis increases with pH, leading, in many cases, to the precipitation of a hydroxide such as Al(OH). These substances are of noticeable ecological relevance, as they are produced by leeching from rocks and subsequent hydrolysis of hydrated metal ions. Consider, for example, aluminum hydroxide ions. Hydrolysis may give rise to the solubility of these ions to form hydrates:

$$Al(OH)_3 + 3H^+ + nH_2O \rightarrow Al(H_2O)_{n+3}^{3+} \tag{10.3}$$

Hydrolysis may also result in the formation of insoluble compounds, often hydroxides, as is the case for Fe in the reaction forming $Fe(OH)_3$, which is highly insoluble and readily precipitates to the sediments:

$$Fe^{3+} + 3H_2 0 \rightarrow Fe(OH)_{3(s)} + 3H^+ \tag{10.4}$$

The increased solubility of heavy metals with decreasing pH, due to the formation of metal hydrated ions (e.g., Equation 10.3), is of great environmental concern as well. For example, as pH decreases as a result of acidic rain in areas with low pH buffer capacities, the toxic effect of metal ions is significantly increased.

Via hydrolysis, metal ions are able to form several species such as hydro-, hydroxo-, hydroxo-oxo-, and oxo complexes. This implies that multivalent metal ions are able to participate in a series of consecutive proton transfers (Jørgensen and Bendoricchio, 2001):

$$\begin{aligned} &Fe(H_2O)_6{}^{3+} \rightarrow Fe(H_2O)_5OH^{2+} + H^+ \rightarrow Fe(H_2O)_4(OH)_2{}^+ + 2H^+ \\ &\rightarrow Fe(OH)_3(H_2O)_{3(s)} + 3H^+ \rightarrow Fe(OH)_4(H_2O)_{2(s)} + 4H^+ \end{aligned} \tag{10.5}$$

The catalytic action of enzymes enables the hydrolysis of proteins, fats, oils, and carbohydrates. The hydrolysis of polysaccharides to soluble sugars is called *saccharification*. Hydrolysis is central for energy metabolism and storage. The energy derived from the oxidation of nutrients is not used directly, but by means of a complex and long sequence of reactions; it is channeled into adenosine triphosphate (ATP), which is a special energy-storage molecule. The ATP molecule contains pyrophosphate linkages that release energy when broken; ATP undergoes hydrolysis in two ways: the removal of a terminal phosphate to form adenosine diphosphate (ADP) and inorganic phosphate or the removal of a terminal diphosphate to yield adenosine monophosphate and pyrophosphate. Adenosine pyrophosphate usually undergoes a final hydrolysis step, resulting in its two constituent phosphates.

10.3 Proteolysis

Proteolysis refers to the breakdown of proteins into smaller polypeptides or amino acids. The process is generally initiated by water supplement and the consequent hydrolysis of the peptide bond (C(O)NH), which is a covalent chemical bond between two molecules in which the carboxyl group of one molecule (COOH) links with the amino group ($-NH_2$) of the other molecule, resulting in the release of a molecule of water (dehydration). The process is not only commonly fostered by protease enzymes, but may also occur by intramolecular digestion, as well as by the intervention of other agents such as heat and acids.

Proteolysis is a fundamental process for all living organisms. For example, during digestion, cellular enzymes break down proteins into food to provide amino acids for the metabolism. After its synthesis, the proteolysis of a polypeptide chain may be necessary for the activation of the protein. Proteolysis is also a key process in the regulation of several physiological cellular functions, as well as in preventing the accumulation of inappropriate proteins in cells.

After protein translation and synthesis, proteolysis of the produced polypeptide occurs in order to obtain a working biochemical agent. Polypeptide proteolytic processing results in several strategic functions such as the adjustment of the protein structure (reshaping of preproteins and cleaning by removal of N-terminal methionine and of the signal sequence), protein activation/deactivation, and protein degradation. Protein degradation is catalyzed by acid hydrolases and protease enzymes that regulate the rate of intracellular protein degradation. Various proteins are degraded at different rates. Abnormal proteins are quickly degraded, whereas normal proteins may vary widely in terms of degradation depending on their functions. The rate of proteolysis may also depend on the physiological state of the cell, such as its hormonal state as well as nutritional status. In time of starvation, the rate of protein degradation increases. Ultimately, proteolysis is central in the regulation of most cellular processes and cycles (e.g., apoptosis, cellular division, and metabolic byproducts elimination).

At higher hierarchical levels, for example in the process of digestion, proteins in food are broken down into smaller peptide chains by digestive enzymes such as pepsin, trypsin, chymotrypsin, and elastase, and further into amino acids by various enzymes such as carboxypeptidase, aminopeptidase, and dipeptidase. It is necessary to break down proteins into small peptides and amino acids so that the intestines can absorb them. The absorbed small peptides are also further broken into amino acids inside cells before they enter the bloodstream.

10.4 Photolysis

Photolysis (or photodissociation) is a chemical process in which a compound is directly or indirectly broken down by light. It may take place wherever light is absorbed by chromophores.* Photolysis may be carried out directly by the light that absorbs chromophores or indirectly as a consequence of the release of reactive species, which are formed via photosensitization.†

The general photolysis of a compound AB may be represented as a chemical reaction

$$AB + h\nu \rightarrow A + B$$

where $h\nu$ is the energy transferred by the photon.

Photolysis is of primary ecological relevance for aquatic ecosystems. Indeed, in these systems, this process can influence many aspects of carbon cycling (Section 12.4). In particular, photolysis is involved in the following:

- Decreasing the average molecular weight of the chromophoric organic substances.
- Changing the optical properties of the water as a result of the changes in chromophoric organic substances. This has a major impact on the penetration depth of harmful ultraviolet radiation and, on the other hand, on the penetration of photosynthetic active radiation, hence controlling the vertical distribution of planktonic primary producers.
- Releasing a complex mixture of reactive oxygen species that can damage organisms and initiate indirect photolysis reactions.

* Chromophores are those parts of a molecule that may absorb photons, hence being responsible for its color.

† This term refers to the process of initiating a reaction through the use of a substance that is capable of transferring energy to the reactants as a consequence of the absorption of photons.

- Producing oxidized C compounds such as CO or CO_2, or short-chain fatty acids, which provide energy and carbon to heterotrophic processes.
- Releasing P and N compounds such as phosphate ions, amino acids, dissolved primary amines, and ammonia, NH_4^+.

In terrestrial environments (soil surface and leaf canopies), photolysis is an important factor of nutrient cycling and trace gas production. In addition to this, the photolysis of water is a central process in photosynthesis (see Chapter 13).

Photolysis may have dramatic effects on organisms. Photodissociation of various chromophoric-dissolved compounds results in the production of reactive oxygen species such as hydrogen peroxide (H_2O_2), which is persistent and may be harmful to several organisms. This molecule easily penetrates biomembranes and, hence, contributes to cells' oxidative stress, inducing and modulating stress-response proteins and enzymes, reducing photosynthetic activity, and increasing membrane (lipid) peroxidation.

Due to photosensitization (indirect photolysis), compounds that do not absorb solar energy directly are subject to photolytic degradation. These processes are important for the degradation of xenobiotic compounds of both anthropogenic and natural origin such as oil spills and some algal toxins. Although photolysis does not lead to complete mineralization of most organic compounds, in several cases, the products may become substrates for heterotrophic bacteria. Photolysis, therefore, should be considered the initial step of biological degradation. In particular, photolysis of dissolved organic matter plays an important role in the global cycle of carbon, and it is involved in the binding and release of nutrients such as P and N (Chapter 12). Photolysis of humic matter, by altering complexion structures (see Section 11.4), may affect the toxicity of heavy metals.

As will be discussed in Section 12.2, photolysis in the atmosphere significantly impacts the oxygen cycle. Indeed, photolysis occurs in the atmosphere as a part of a series of reactions by which primary pollutants such as hydrocarbons and nitrogen oxides react to form secondary pollutants such as peroxyacyl nitrates. The two most important photodissociation reactions in the troposphere are

$$O_3 + h\nu \rightarrow O_2 + {}^*O \qquad \lambda < 320 \text{ nm}$$

which generates an excited oxygen atom (*O) that can react with water to give the hydroxyl radical

$${}^*O + H_2O \rightarrow 2 \cdot OH$$

The hydroxyl radical, thus, initiates the oxidation of hydrocarbons. The second important reaction that is central in the formation of the tropospheric ozone is

$$NO_2 + h\nu \rightarrow NO + O$$

The formation of the ozone layer is also caused by photolysis. Ozone in the earth's stratosphere is produced by ultraviolet light incident on oxygen molecules that contain two oxygen atoms (O_2), splitting them into individual oxygen atoms (atomic oxygen). The atomic oxygen then combines with unbroken O_2 to create ozone, O_3. In addition, photolysis is the process by which chlorofluorocarbons are broken down in the upper atmosphere to form ozone-destroying chlorine-free radicals.

11

Phase Partition

11.1 Introduction

In the context of ecological processes, the term *phase partition* refers to those processes that are involved in the transfer of elements (generally chemical compounds) from one type of environmental matrix to another. The concept of environmental matrix originates in landscape ecology: The environmental phase or the combination of several phases is more representative of the spatial ambit examined. Soil, water, and biota are examples of typical environmental matrices. Usually, environmental matrices are combinations of the elemental phases, for example, suspended solids in the water column. Phase partition is different from the well-known physical process, such as crystallization or fusion, of phase transition. In these processes, matter switches its state amid the three fundamental states: solid, liquid, and gaseous.* Indeed, when dealing with a living system, besides liquid, solid, and gaseous states, an extra phase, which is the biotic one, is often described. The *biotic phase* refers to the environmental matrices pertaining to living systems; for example, physiological fluids, organic tissues, or cytoplasm. The biotic phase populates the biosphere introduced in Chapter 1. This phase may be defined as the totality of all ecosystems or the system that integrates all living components and their relationships as well as the interactions with the elements of the other ecospheres: lithosphere, hydrosphere, and atmosphere. Some of the processes involved in matter transfers among spheres are reported in Table 11.1.

During phase partition, the internal structure of the chemical compounds involved does not change significantly, meaning that the identity of the element is not changed. The chemical bonds that are established or destroyed while a compound associates or leaves a particular matrix depend on the nature of the elements involved. Generally, bonding is caused by

* In physics, there is a fourth fundamental phase, which is that of plasma. However, we shall disregard this phase as being of no practical interest for ecosystems. Actually, in biology, the term *plasma* refers to the liquid component of the blood. This is a fluid solution in water, in which a number of substances are dissolved: proteins; inorganic mineral salts such as sodium and calcium; waste products such as urea; and traces of hormones and antibodies.

TABLE 11.1

Some Processes of Matter Partition among the Ecospheres

Elemental Matrix	Ecosphere	Processes
Gas, air	ATMOSPHERE	volatilization
Liquid, water	HYDROSPHERE	precipitation, dissolution
Solid, sediment	LITHOSPHERE	adsorption, absorption, and desorption
Biotic, organic solutions	BIOSPHERE	bioaccumulation

Note: See also Section 18.4.

electromagnetic attraction between opposite charges, either between electrons and nuclei, or as the result of a dipole interaction. The strength of different chemical bonds changes considerably, spanning from strong bonds such as covalent or ionic bonds to weak ones such as dipole–dipole interactions, van der Waals forces, or hydrogen bonding. The example of a hydrogen bond is that between water molecules. Water is the most ubiquitous solvent in the biosphere, and it is commonly the matrix through which substances of biotic relevance are conveyed. For this reason, water is often used as the reference solvent for the definition of *partition properties* of a given element among the different phases.

A general approach that is used for describing phase partition consists of the definition of a *partition* or a *distribution coefficient*:

$$k_p = \frac{C_1}{C_2} \tag{11.1}$$

as the ratio of the concentrations C_1 and C_2 of a compound in the two phases of a mixture of immiscible solutions at equilibrium. For example, the gas–liquid (e.g., air–water) partition coefficients are the dimensionless constants of Henry's law (see Section 7.7). These constants depend on the specific solute, solvent, and the temperature. They represent a measure of the different solubility of a compound between various solvents. Typically, in order to characterize the solubility of a given compound, two special solvents are employed: water and octanol ($CH_3(CH_2)_7OH$), which is a hydrophobic (nonmiscible in water) solvent. Hence, the partition coefficient is a measure of how hydrophilic or hydrophobic the chemical compound is. In biological matrices, hydrophobic elements with elevated octanol–water (k_{ow}) partition coefficients are preferentially allocated to hydrophobic compartments such as the lipid bilayers of cells membranes. On the contrary, hydrophilic elements with low octanol–water partition coefficients are preferentially located in hydrophilic ambits such as blood plasma.

Adsorption and ion exchange are processes of matter transfer from the liquid to the solid matrix. These processes are of particular interest for environmental systems. When water is in contact with solids (suspended matter, particulate, sediment, etc.) and biota, a significant transfer of matter by adsorption and ion exchange may occur (Jørgensen and Bendoricchio, 2001). In aquatic ecosystems, a significant portion of xenobiotic pollutants is stuck to suspended matter.

11.2 Adsorption and Partition of Organic Compounds

Adsorption is the process in which a molecule or ion of an element in a gaseous or liquid solution, the adsorbate, bonds on the surface of a solid, the adsorbent (e.g., suspended solids in the solution). Adsorbate molecules adhere by creating weak- to medium-strength chemical bonds* on the surface of the adsorbent, possibly forming a number of coating layers. In this sense, adsorption is a surface process, and surface properties (e.g., the surface area, pore size distribution, and surface chemistry) become limiting factors. When the adsorbate diffuses into the structure of the adsorbent, absorption occurs. The reverse of these processes, that is, the element leaving the solid, is known as *desorption*.

In stable environmental conditions, compared with other processes, adsorption is a fast process and a dynamic equilibrium is reached quite quickly with desorption. Similar to Equation 11.1, in this case as well, a partition relation may be introduced:

$$q = k_p \cdot c \tag{11.2}$$

where q (mg/g) and c (mg/L) are, respectively, the solid and dissolved phase concentrations; the equilibrium constant k_p (L/g) is the partition coefficient. Actually, the partition coefficient is not constant: Even in stable conditions—for example, constant temperature—k_p varies with adsorbate concentration in the dissolved phase (c), reflecting the fact that the adsorbent cannot infinitely accept molecules of the adsorbate. From experimental observations of adsorption processes in nature, it can be seen that in isothermal conditions, a general relation is (Jørgensen and Bendoricchio, 2001)

$$k_p = \frac{k}{\delta + hc^{\beta}} \tag{11.3}$$

where k (L/g) is the equilibrium constant (that in isothermal conditions accounts for the process dependence on the temperature through the usual Arrhenius relation, see, e.g., Equation 9.5; h (L/g) accounts for the heat of adsorption; β is a dimensionless coefficient called the *heterogeneity factor* and is related to the surface properties of the adsorbent (theoretical considerations on adsorption provide limit values for this parameter, i.e., $0 \leq \beta \leq 1$); and δ

* Generally, these are van der Waals or occasionally even weak covalent or ionic bonds.

is a dimensionless generalization constant. Relation 11.3 may be adapted to several situations, and it reflects the fact that adsorption is limited. By simplification of Equation 11.3, various well-known adsorption partition equilibria may be derived. These are the so-called *isotherms of adsorption*, and they are obtained by combining Equations 11.2 and 11.3 and assuming limit values of the coefficients δ, *h*, and β, reflecting different adsorbent characteristics. Table 11.2 reports the most used adsorption equilibrium isotherms, from the general Redlich–Peterson to the simpler linear one.

TABLE 11.2

Underlying Hypothesis of Common Adsorption Equilibria Isotherms

Isotherm	Coefficients	Equation	Hypotheses
Redlich–Peterson	$\delta = 1$	$q = \frac{kc}{1+h\,c^{\beta}}$	This is the most general relation that describes multilayer adsorption on a heterogeneous adsorbent. This model is easily calibrated with experimental data by linearization.
Langmuir	$\delta = 1$ $\beta = 1$	$q = \frac{kc}{1+hc}$	This relation is applied to describe monolayer homogeneous adsorption. The theoretical assumptions are as follows: 1. The adsorbate can attach only on particular sites, those displaying the right conditions for establishing the bonds. 2. Adsorption occurs only in a monolayer. 3. All sites of adsorption are energetically equivalent, and the molecules that are adsorbed do not interact with each other. Hence, the already adsorbed materials do not affect the process, and adsorption occurs homogeneously on the surface.
Freundlich	$\delta = 0$ $\gamma = 1-\beta$ $k' = \frac{k}{h}$	$q = k'c^{\gamma}$	This relation describes multilayer adsorption on a heterogeneous adsorbent. The theoretical derivation relies on the following assumptions: 1. Adsorption occurs in a multilayer. The adsorbent is never saturated and would continuously bind to the adsorbate. This isotherm does not apply when the concentration of the adsorbate is very high. 2. The energy required for adsorption exponentially increases with the number of layers.
Linear	$\delta = 0$ $\beta = 0$ $k' = \frac{k}{h}$	$q = k'c$	This form is acceptable only when the adsorbate concentration in solution is small, that is, the adsorbate is very diluted. In this hypothesis, the adsorbtion equilibrium isotherm formally coincides with Henry's law for liquid–gas partition.

In the Langmuir isotherm, the equilibrium constant k is related to the maximum amount of adsorbate that is allowed to stick on to the adsorbent and the enthalpy of the process. The shape of the curve clearly depends on the parameters, but usually, saturation is reached quite fast. In the Freundlich model, the slope of the curve decreases when the concentration of the adsorbate increases, reflecting the fact that the equilibrium of the process changes only slightly with growing concentrations. The adsorption isotherms are plotted in Figure 11.1. The coefficients of the adsorption isotherms vary in response to variations of the temperature. For example, when the temperature rises, augmented vibrational motion causes a shift in the equilibrium toward the desorbed or dissolved phase. Indeed, an increase in thermal vibrational motion makes easier breaking weak van der Waals (sometimes even covalent) bonds. The values of the coefficients of the adsorption isotherms are usually estimated by an interpolation of empirical observations. Experimentally, the Freundlich model seems to be more appropriate to describe adsorption from liquid solutions, whereas the Langmuir model tends to fit better the data on adsorption in gaseous matrices.

As said earlier, adsorption usually reaches a dynamic equilibrium quickly, allowing a description in terms of equilibrium isotherms. However, when the time scale of the processes under study is shorter than the time needed to reach equilibrium, it is possible to use a dynamic description. This is a

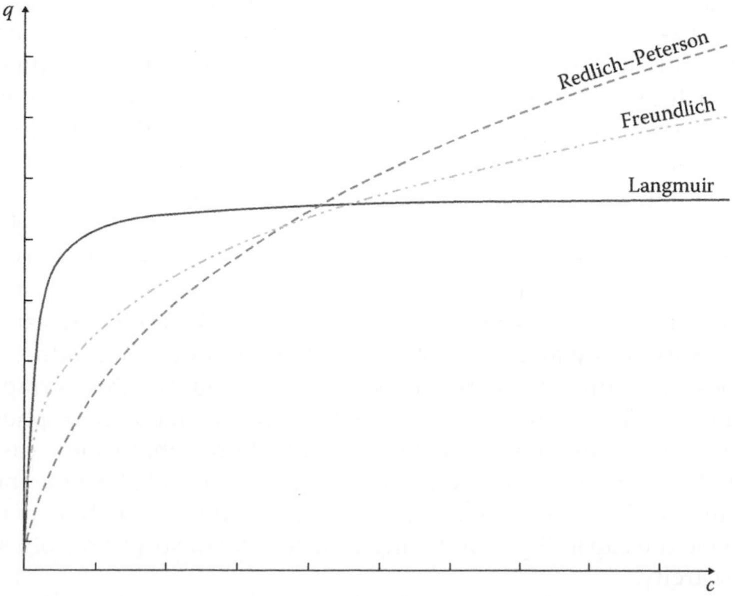

FIGURE 11.1
(**See color insert.**) Plot of adsorption isotherms. If the temperature increases, the curves are displaced toward the bottom of the picture. This corresponds to a shift of the dynamic equilibrium in the direction of desorption.

convenient choice when the contact between adsorbate and adsorbent is small; for example, when rains percolate through an arid soil. In this case, the process may be described with a similar approach to that introduced in Chapter 7 for the description of transport at interfaces. The relation employed for the dynamic description of adsorption is

$$\frac{\partial c}{\partial t} = k(c - q) \tag{11.4}$$

Here, the equilibrium is reached when $c = q$.

Adsorption is a crucial ecological process that drives the exchanges between lithosphere and the other spheres. Besides enabling matter transport among ecosystems' compartments, it is relevant as a start-up for other important processes such as ionic exchange and enzymatic processes (see in Section 9.3). Thanks to adsorption, dissolved compounds can be stuck temporarily or permanently on solids where they can either remain inert or react with solid matrices. In this way, the process participates in the buildup of elements in the lithosphere.

Stunning is the case of phosphorus cycling in ecosystems: Here, inorganic phosphorus is usually found as orthophosphate (PO_4^{3-}) or, more frequently, as its reduced forms HPO_4^{2-} and $H_2PO_4^-$. Phosphorus is a limiting nutrient for several ecological processes, and its availability is fundamental for the maintenance of ecosystem functions such as primary production. Dissolved phosphates are typically adsorbed either on suspended solids or in the soil. This result is achieved by complexation with metal cations, in particular with iron and aluminum hydrous oxides ($Fe(OH)_3$, $Al(OH)_3$), or other metal cations such as Ca^{2+}. Clay also is a good adsorber. As a consequence of these processes, phosphorus dynamic follows mainly that of particulate. Phosphorous solid transport (e.g., erosion) is more relevant than the liquid one (e.g., leaching). Particulate P is not only less mobile than the dissolved form, but also not easily bioavailable. This is one of the reasons that P is more limiting than nitrogen even if it is needed in a lesser quantity by living organisms. As seen in Section 9.4, in aquatic ecosystems, a change in redox potential, for instance as a consequence of the aerobic digestion of a large quantity of organic matter due to eutrophication, may induce anoxic conditions in sediments. In these conditions, phosphorus that is trapped in sediments is released by desorption, with a further increase of phosphate which is available in the water column for algal growth. In some cases, the amount of phosphorus released can be so significant as to become the principal source of P in the system. By changing conditions such as pH, some plants have the capability of inducing P release by desorption from soils in case of scarcity.

Organic surfaces such as those of biological tissues or cell membranes can adsorb some elements. Actually, viruses exploit this process to adhere on the cell membrane and successively inject their genetic nibbles that are interwoven into the cell's DNA. Viruses can also be adsorbed on inorganic particles such as sediments, particularly on sand or clay. In this way, the solid transport becomes a vehicle for viral infections.

Adsorption and ionic exchange are relevant routes for the process of acidification. This process is due to the natural passage of acid substances among ecospheres, for example, atmospheric deposition (see Chapter 23). Sulfates (SO_4^{2-}) produced by fossil fuel combustions are considerably involved in the process. After deposition, sulfates leaching into pore waters in the soil are adsorbed, for example, on oxides such as Fe^{3+} or Al^{3+}. A part of the sulfates is neutralized via ionic exchange by metallic cations in the soil such as Mg^{2+}, Ca^{2+}, or K^+. As a consequence, neutralization subtracts several oligoelements that are valuable for primary production. Moreover, pH in the soil decreases and heavy metal ions release increases very rapidly, which is accompanied by potentially toxic effects. All these processes worsen the ecological soil's conditions. Due to their high productivity, forests are usually the most affected. Adsorption is a reversible process: If the concentration of sulfates in pore water decreases, they can be desorbed and leach back to surface waters.

In aquatic ecosystems, the concentration of organic compounds in the water matrix solution is generally quite low. In the hypothesis of highly diluted adsorbate, a simple linear approach may be acceptable:

$$q = kc \tag{11.5}$$

where k is the partition coefficient. Of course, as said earlier, this simple adsorption equilibrium isotherm has significant limitations. However, for diluted adsorbate, this model can still be used with an acceptable error, being the linear approximation that is consistent with the other isotherm models at very low adsorbate concentrations. The value of the partition coefficient can be estimated via empirical observation, or—for nonionic compounds such as many organic chemicals—as Karickoff et al. (1979) and Karickhoff (1981) demonstrate, k may be expressed as a function of the organic carbon content of the solid phase:

$$k = f_{oc}\ k_{oc} \tag{11.6}$$

where f_{oc} is the weight fraction of total carbon in the solid matrix and has the units gC/g, and k_{oc} is the partition coefficient of organic carbon with units $(mg/gC)/(mg/m^3)$. Given the strong chemical affinity of organic chemicals and organic carbon, it is assumed that adsorption occurs only on the carbonic fraction of the adsorbent. k_{oc} can be estimated by use of the contaminant octanol–water partition coefficient k_{ow} whose units are $(mg/m^3_{\text{octanol}})/(mg/m^3_{\text{water}})$. Di Toro (1985), for example, determines the following relation that is valid for most semivolatile nonionizing organic compounds:

$$\log k_{oc} = 0.00028 + 0.983 \cdot \log k_{ow} \tag{11.7}$$

This relation considers particle interactions and is obtained from analyses on a large set of adsorption–desorption data. k_{ow} values can be found in several publications, as, for example, in EPA (1996), where tables are reported that describe the characteristics of several substances. k_{ow} may eventually be calculated in terms of solubility and molar weight (see, e.g., in Tetko et al., 2005). The octanol–water partition coefficient (k_{ow}) is a measure of the equilibrium concentration of a compound between octanol and water. Elevated octanol–water partition coefficients indicate hydrophobic elements that are preferentially allocated to organic matter rather than to water.

In very productive aquatic ecosystems, the fraction of organic carbon in the solid phase, f_{oc}, should be carefully evaluated. Indeed, the solid concentration in the water column may originate by resuspension of organic elements from sediment, resulting in particulate organic carbon (POC). A significant contribution to this concentration may also come from the decay of organic matter, giving rise to important quantities of dissolved organic carbon (DOC). In this case, DOC should be considered an additional adsorbing phase that may decrease the substance concentration in the water column. Usually, this phase adsorbs toxic substances more easily than does POC. Figure 11.2 illustrates how the adsorption partition coefficient k dependence on $\log k_{ow}$ changes at different suspended solids concentration. Table 11.3 reports the ranges and average values of $\log k_{ow}$ for some organic pollutants.

Adsorption has many industrial applications, as, for example, in water treatment. Besides activating other important processes as enzymatic reactions or ion exchange, adsorption on the solid matrix plays an important

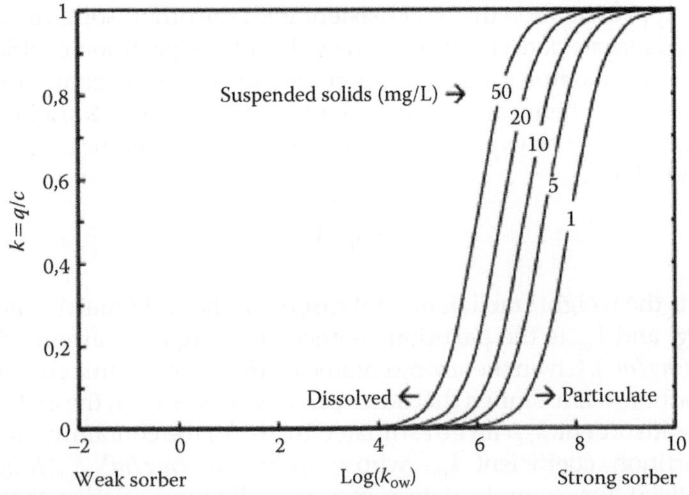

FIGURE 11.2
Adsorption partition coefficient k as function of $\log(k_{ow})$ for different suspended solids concentration. (From Jørgensen, S.E., Bendoricchio, G., *Fundamentals of Ecological Modeling*, Elsevier, Amsterdam, 2001.)

TABLE 11.3

log k_{ow} Values for Several Categories of Organic Pollutants

Organic Pollutant	Range	log k_{ow}	Average
PCB	3.30–6.53		5.60
Phthalate ethers	1.60–9.33		5.50
PAHs	2.67–7.73		5.50
Pesticides	0.53–6.93		3.60
α endosulfan		1.73	
β endosulfan		1.47	
Endosulfan sulfate		1.07	
MAHs	1.60–4.14		3.07
Hexachlorobenzene		6.27	
Phenols	0.93–3.60		2.14
Pentachlorophenol		5.6	
Nitrosamines	0.14–3.33		2.14
Alogenated aliphatics	0.93–2.80		2.00
Hexachloroethane		4.14	
Hexachlorobutadiene		4.80	
Hexachlorocyclopentadiene		5.07	
Ethers	0.14–5.07		1.60

Source: After Jørgensen, S.E., Bendoricchio, G., *Fundamentals of Ecological Modeling*, Elsevier, Amsterdam, 2001.

role in water disinfection. It is indeed an efficient way for virus transport. It should, however, be remarked that viruses in the solid matrix are more resistant and their survival time may increase. If the solid phase is not confined, adsorbed viruses can be transferred along with the solid transport (e.g., via runoff), increasing the risk of infections.

11.3 Ion Exchange

Ion exchange is a reversible chemical process in which ions from a solution are exchanged with similarly charged ions in the solid matrix. In this process, ions in the solution are transferred to the solid matrix, which, in turn, releases ions of the same polarity but of a different element. The ions in the solution are replaced by different ions that were originally present in the solid, and vice versa.

There are many similarities between adsorption and ion exchange. Indeed, the two processes are often analyzed using similar models. Since ion exchange occurs between a solution and the surface of a solid, it can be viewed as a special case of adsorption. Actually, unlike adsorption, ion exchange requires a bidirectional interchange of materials (the ions), and the solution can, hence, maintain its electroneutrality. Ion exchange is typically used for separation and adsorption of ionic compounds (e.g., hydrophilic polypeptides, DNA, or other hydrophilic ionic compounds). In nature, ion exchange is known to occur in several solid matrices, such as clay, humus, metallic minerals, and hydrophilic ionic organic compounds. Clay and other natural materials may be used for water purification by ion exchange.

Ion exchange is a selective process: Binding sites of ions in the solid matrix (the exchanger) are differently preferred by diverse ions. The ions with greater affinity to the exchanger are those that are primarily removed from the water. Ion exchange may be represented by the following reaction (e.g., for cationic exchange):

$$\alpha A^{a+} + \beta B_R \rightleftharpoons \alpha A_R + \beta B^{b+} \tag{11.8}$$

where α and β are stoichiometric and ionization coefficients, respectively; whereas the subscript R refers to the solid matrix or the exchanger. The equilibrium constant of this reversible reaction is given as usual by the law of mass action

$$k_{B,R} = \frac{[A_R]^\alpha}{[A^{a+}]^\alpha} \frac{[B^{b+}]^\beta}{[B_R]^\beta} \tag{11.9}$$

$k_{B,R}$ is called the *selectivity coefficient* of the exchanger R for ions of B type.

Let us consider water softening as an example. This process consists of removing Ca++ or Mg++ ions from water. The presence of these divalent ions results in the so-called *water hardness*. In this case, an ion-exchanger resin can swap Na+ for Ca++ or Mg++ in a reaction such as the following one:

$$Ca^{2+} + 2Na_R \rightleftharpoons Ca_R + 2Na^+$$

The selectivity coefficient $k_{B,R}$ is directly proportional to the preference of a given ion exchanger toward different types of ions. For the same concentration of different ions, the relative preference of an ion-exchanger resin depends primarily on two factors, that is, ionic charge and size. Generally, ions with higher valence are preferred by ion exchangers. Table 11.4 reports the common ionic preferences for different type of exchangers.

Ion exchangers are materials that are built on organic or inorganic matrices which contain ionic functional groups. There are both natural ion-exchange materials (e.g., zeolites that are amorphous hydrated aluminum silicate minerals containing various other elements, which are typically used for water purification) and synthetic ion exchangers. The vast majority of ion exchangers used in industrial wastewater treatment is of synthetic origin. Organic resins constitute the most common type of synthetic ion-exchange materials.

TABLE 11.4

Ionic Preferences for Various Exchangers

Ionic Exchanger	Preference
Cationic resin	$Th^{4+} > Al^{3+} > Ca^{2+} > Na^{+}$
Anionic resin	$PO_4^{3-} > SO_4^{2-} > Cl^{-}$
	$SO_4^{2-} > I^{-} > NO_3^{-} > CrO_4^{2-} > Br^{-}$
General Preference for Cations	
Monovalent cations	$Ag^{+} > Cu^{+} > K^{+} > NH_4^{+} > Na^{+} > H^{+} > Li^{+}$
Divalent cations	$Pb^{2+} > Hg^{2+} > Ca^{2+} > Ni^{2+} > Cd^{2+} >$
	$Cu^{2+} > Zn^{2+} > Fe^{2+} > Mg^{2+} > Mn^{2+}$
Trivalent cations	$Fe^{3+} > Al^{3+}$

Ion-exchange resins are organic compounds that are polymerized and form a porous tridimensional matrix. Synthetic organic resins are predominantly used these days in industrial applications, because their selectivity can be designed for specific purposes. These resins can be prepared by using techniques such as molecular imprinting, which enable an accurate trimming of the selectivity of the exchanger. However, this technique is not applicable for highly hydrophilic molecules such as amino acids, hydrophilic polypeptides, and organic ionic matter. Several natural toxins that frequently occur in the environment—such as cyanobacteria, shellfish, and fish toxins—are highly hydrophilic compounds. For these substances, due to their ionic properties, conventional adsorption based on hydrophobic interaction is not effective. Nevertheless, for such substances, other techniques that are used for constructing suitable exchange resins are available (Kubo et al., 2004).

In typical environmental conditions, many pharmaceuticals, biocides, pesticides, surfactants, and drugs are positively charged. The environmental fate of such cationic species depends partly on sorption to natural sorbents, which are often negatively charged. Ionic exchange with inorganic cations at acidic sites in organic matter, and not mass partitioning, controls the sorption of cationic compounds. Ionic interactions are affected by dissolved concentrations of Na^{+} or Ca^{2+} ions, whereas additional nonionic interactions strengthen the affinity for organic matter. Any ionic compound is soluble in only a polar (e.g., water) or relatively polar solvent, but it cannot be dissolved in an organic solvent, because the forces holding the ionic compound molecules to each other (typically ion-dipole bonds) cannot be replaced by the Van der Walls interactions, which hold up the organic solvent molecules. However, sometimes, an ionic compound with a high molecular weight could have some solubility in an organic solvent, because the large part of the molecule (the part that has high molecular weight) can occasionally dissolve in the organic solvent.

In most cases, ion exchangers are used to treat waters that contain inorganic ions. The sorption of organic species from nonpolar organic solvents by ion exchangers is usually disadvantaged. Moreover, ion exchangers are generally not very effective for large organic molecules, mainly because the size of the molecules dramatically reduces the exchange rate. However, sometimes, ion exchangers may be effectively used in the treatment of specific organic compounds (such as phenols or phthalates). In this case, the ion exchanger does not act as such but more as a conventional adsorbent.

The capacity of an ion exchanger is the amount of ionic species that can be exchanged per unit mass of dry exchanger. The exchange capacity is a function of the number of available exchange sites in the resin and is commonly expressed in milliequivalents of ionic species per gram of dry weight of ion-exchange particles (*meq/g*).

The capacity of most ion-exchange resin is in the range 2–10 *meq/g*. Operating factors such as pH or temperature may affect the capacity of the exchanger. The capacity of the exchanger resin is consumed with time and should be periodically regenerated. This result is accomplished by employing other chemical processes (involving acid, bases, or washing salts solutions) in order to extract and concentrate the exchanged ions (e.g., by precipitation), hence restoring the capacity of the exchanger.

Customarily, in dynamic equilibrium conditions and at fixed temperature, experimental observations on ion-exchange processes are confronted with adsorption models such as the Langmuir or Freundlich equations.

11.4 Complex Formation

Complex formation is a reaction by which two or more components form a (more complex) compound. This reaction type is of particular ecological relevance when metal ions (named *central atom, M*) are chemically bonded to various organic compounds (named *ligands, L*). Frequently, more than one ligand can be complex bound to the central atoms; for instance:

$$M + L \leftrightarrow ML \qquad K_1 = \frac{\{ML\}}{\{M\}\{L\}} \tag{11.10}$$

$$ML + L \leftrightarrow ML_2 \qquad K_2 = \frac{\{ML_2\}}{\{ML\}\{L\}} \tag{11.11}$$

$$ML_2 + L \leftrightarrow ML_3 \qquad K_3 = \frac{\{ML_3\}}{\{ML_2\}\{L\}} \tag{11.12}$$

$$ML_i + L \leftrightarrow ML_{i+1} \qquad K_i = \frac{\{ML_{i+1}\}}{\{ML_i\}\{L\}} \tag{11.13}$$

The process through which two or more ligands react simultaneously with the central atom is, of course, possible:

$$M + L_2 = ML_2 \qquad \beta_2 = K_2K_1 = \frac{\{ML_2\}}{\{M\}\{L\}^2} \tag{11.14}$$

As seen, the equilibrium constant where *i* ligands simultaneously react with the central atom is

$$\beta_I = K_1K_2K_3 \ldots K_i.$$

Reactions can also take place by the addition of protonated ligands:

$$M + HL \leftrightarrow ML + H^+ \qquad {}^*K_1 = \frac{\{ML\}\{H^+\}}{\{M\}\{HL\}} \tag{11.15}$$

A parallel expression is used for a reaction with the second HL, the third HL, and so on. By multiplication by {L} in the nominator and denominator of the expression in Equation 11.15, it is seen that $^*K_1 = K_1K_{aL}$.

Here, we have used the more generally expressed rule that the equilibrium constant for a process, K*, which consists of *i* steps, is equal to the product of the *i* equilibrium constants of the steps, K_1, K_2, K_3 ...K_i.:

$$^*K = K_1 \ldots K_2, \ldots K_3 \ldots K_i . : \tag{11.16a}$$

or

$$\log {}^*K = \log K_1 + \log K_2 + \log K_3 + \ldots\ldots \log K_i \tag{11.16b}$$

11.5 Activities and Activity Coefficients

As seen in Chapter 9, an activity can be written as the product of the concentration [A] and the activity coefficient *q*:

$$\{A\} = q[A]$$

The activity is basically defined in such a way that the activity coefficient q = {A}/[A] approaches unity as the concentration of all solutes approaches zero. This means that for a water solution, the activity coefficient becomes unity as the solution approaches the pure ionic medium; that is, when all concentrations other than the medium ions approach zero. The activity coefficient *q* can be found for individual ions by empirical expressions as presented in Table 11.5, where I is the ionic strength

$$I = 0.5\ \Sigma C_i Z_i^2$$

TABLE 11.5

Equations for Individual Activity Coefficients

The Name of the Approximation	Equation: log q =	Valid at I <
Debye–Hückel	$-AZ^2\sqrt{I}$	0.005M
Extended Debye–Hückel	$-AZ^2\dfrac{\sqrt{I}}{(1+Ba\sqrt{I})}$	0.1M
Güntelberg	$-AZ^2\dfrac{\sqrt{I}}{(1+\sqrt{I})}$	0.1M
Davies	$-AZ^2\left(\dfrac{\sqrt{I}}{(1+\sqrt{I})}-0.21\right)$	0.5M

TABLE 11.6

Parameter *a* for Individual Ions

Ion Size Parameter *a*	For the Following Ions
9	H^+, Al^{3+}, Fe^{3+}, La^{3+}, Ce^{3+}
8	Mg^{2+}, Be^{2+}
6	Ca^{2+}, Zn^{2+}, Cu^{2+}, Sn^{2+}, Mn^{2+}, Fe^{2+}
5	Ba^{2+}, Sr^{2+}, Pb^{2+}, CO_3^{2-}
4	Na^+, HCO_3^-, $H_2PO_4^-$, acetate, SO_4^{2-}, HPO_4^{2-}, PO_4^{3-}
3	K^+, Ag^+, NH_4^+, OH^-, Cl^-, ClO_4^-, NO_3^-, I^-, HS^-

and Z_i is the charge of the ion. "A" in the table is = $1.82 \times 10^6(\varepsilon T)^{2/3} \approx 0.5$ for water at room temperature (ε is the dielectric constant of the solvent, and T is the temperature). B ≈ 0.33 for water at room temperature, and ***a*** is an adjustable parameter that corresponds to the size of the ion (see Table 11.6). log *q* for ions is negative, which implies that *q* is less than one and that it decreases with increasing ionic strength and charge of the ion. The activity is less than the concentration, because negative ions form a shield around positive ions and positive ions form a shield around negative ions. The more ions the solution contains, the stronger the shield is and the higher the ionic strength is. The electrical force is proportional to the square of the charge, which at least explains that the effect increases more than proportionally to the charge of the ion. The models presented in Table 11.5 can, therefore, be understood as a consequence of the electrical forces in solutions. They can, however, not be proved, but are derived from observed empirical correlations.

It is clear from Section 11.2 that by application of Equations 11.1 through 11.3, we can find the equilibrium constant that presumes activities or fugacities. Equilibrium constants taken from handbook tables are also based on activities and fugacities, whereas we are most often interested in concentrations. Introduction of the activity coefficient, *q*, makes it possible, however, to set up the following relationships:

$$K = \frac{\{C\}^c\{D\}^d}{\{A\}^a\{B\}^b} = \left(\frac{[C]^c[D]^d}{[A]^a[B]^b}\right)\left(\frac{q_C^c q_D^d}{q_A^a q_B^b}\right) \quad (11.17)$$

Concentrations can now be determined, provided the activity coefficients are known.

Usually, many equilibrium calculations are carried out for the same solution with a well-defined ionic strength. It would, therefore, be beneficial to find an equilibrium constant K′ that is valid for concentrations for the considered solution. From Equation 11.17, we obtain

$$K' = \frac{K\, q_A^a q_B^b}{q_C^c q_D^d} \quad (11.18)$$

In accordance with the International Union of Pure and Applied Chemistry (IUPAC)'s convention for determination of pH, we should consider pH = −log {H⁺}. It is, therefore, suggested to use a so-called *mixed acidity constant*, K'_{am}, which can be found from K_a for the process HA ∫ $A^- + H^+$ (Figure 11.3)

$$\begin{aligned} K_a &= \frac{\{A^-\}\{H^+\}}{\{HA\}} \\ &= \frac{\{A^-\}\, q_{A-}\{H^+\}}{[HA]q_{HA}} \\ &= \frac{K'_{am}q_{A-}}{q_{HA}} = \frac{K'_a q_{A-}q_{H+}}{q_{HA}} \end{aligned} \quad (11.19)$$

q_{HA} is, of course, 1 if HA has no charge; see the equations in Table 11.5. Calculations of pH presume the application of K'_{am}. In environmental calculations, where the accuracy is generally lower than for laboratory calculations, activity coefficients is recommended for salinities above 0.1%; whereas it is usually not necessary to apply activity coefficients for aquatic ecosystems with a salinity below 0.1%.

Illustration 11.5.1

Find the mixed acidity constant and the concentration quotient (equilibrium constant) for all dissolved carbon dioxide regardless of whether it is hydrated or not and for the ammonium ion in a marine environment with a salinity of 2.6% (assume that it is sodium chloride), when logarithm to the acidity constant in distilled water for all dissolved carbon dioxide at the actual temperature is known to be 6.2 and for the ammonium ion at the actual temperature is 9.2.

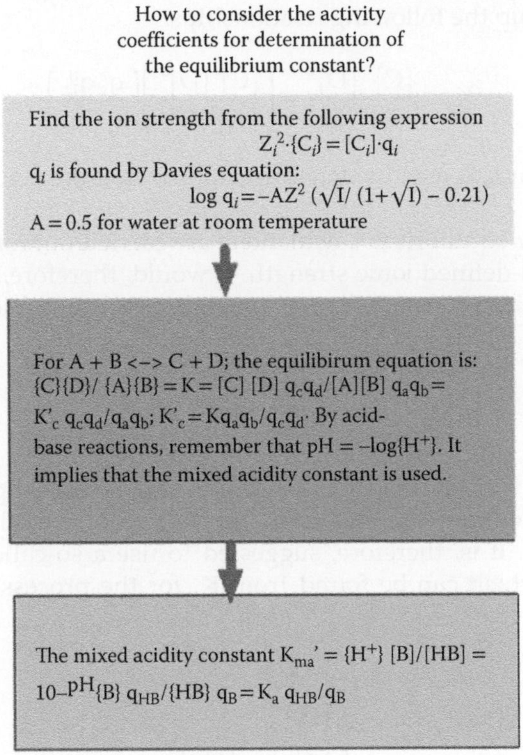
How to consider the activity coefficients for determination of the equilibrium constant?
Find the ion strength from the following expression $Z_i^2 \cdot \{C_i\} = [C_i] \cdot q_i$
q_i is found by Davies equation: $\log q_i = -AZ^2 (\sqrt{I}/(1+\sqrt{I}) - 0.21)$
A = 0.5 for water at room temperature
For A + B <-> C + D; the equilibirum equation is: $\{C\}\{D\}/\{A\}\{B\} = K = [C][D] q_c q_d/[A][B] q_a q_b = K'_c q_c q_d/q_a q_b$; $K'_c = K q_a q_b/q_c q_d$. By acid-base reactions, remember that pH = −log{H^+}. It implies that the mixed acidity constant is used.
The mixed acidity constant $K_{ma}' = \{H^+\}[B]/[HB] = 10^{-pH}\{B\} q_{HB}/\{HB\} q_B = K_a q_{HB}/q_B$

FIGURE 11.3
The stepwise calculation of the mixed equilibrium constant and the concentration quotient is shown. For more details, see Weber and DiGiano (1996).

SOLUTION

$$I = 0.5\left(\frac{26}{(23+35.5)}\right)+\left(\frac{26}{(23+35.5)}\right) = 0.445 \text{ M}$$

Davies equation is applied:

For the hydrogen carbonate ion, the ammonium ion, and the hydrogen ion:

$$\text{Log } q = -0.5\left(\frac{\sqrt{0.445}}{(1+\sqrt{0.445})} - 0.2.\ 0.445\right) = -\ 0.156$$

$$q = 0.698 \approx 0.7$$

The acidity constants for all dissolved carbon dioxide:

$$\log K_a' = \log K_a + 2 \log q = 6.2 - 0.312 \approx 5.9$$
$$\log K_{am}' = 6.2 - \log q = 6.2 - 0.156 \approx 6.0$$

The acidity constant for the ammonium ion:

$$\log K'_a = \log K_a + \log q - \log q = 9.2$$
$$\log K'_{am} = \log K_a - \log q = 9.2 - (-0.156) = 0.9356 \approx 9.4$$

Many processes occur simultaneously in aquatic ecosystems. The appendix at the end of the book will treat the types of processes presented earlier and will show how to provide an overview of many simultaneous processes of the same reaction type. The overview makes it possible to distinguish between processes of importance and processes that are negligible in the context. The double logarithmic representation will provide an excellent tool to overview the processes, and this is very important in the management of aquatic ecosystems, as the chemical processes rules the biological processes in such ecosystems. The appendix presents an interesting paradigm on how to apply in practical cases environmental chemistry methodologies for an ecosystem's state appraisal.

Section IV

Biological Processes

12

Biogeochemical Cycles

12.1 Introduction

The term *biogeochemical cycle* refers to the set of pathways involving biological, geological, and chemical processes, by which a chemical element or substance moves through different environmental matrices (as discussed in Chapter 11). Common synonyms of biogeochemical cycle are turnover or cycling of substances (e.g., P or N cycling). Cycling means that, after a specific amount of time, the substance is found back in the starting phase and the set of pathways is then repeated. For example, circulation of chemical elements, such as carbon, oxygen, nitrogen, phosphorus, calcium, and water, through various ecospheres is referred to as biogeochemical cycle. Depending on the particular environmental conditions, various elements show characteristic *cycling periods.* While cycling throughout the ecospheres, an element may be accumulated in specific places, called *reservoirs.* For example, in aquatic ecosystems, phosphorus is often accumulated in the bottom sediment. Generally, reservoirs hold substances for times significantly longer than the cycling period. The length of time that a chemical is detained in one place is called its *residence time.* When chemicals are seized for a short period, they belong to *exchange pools* (e.g., plants and animals). Cycling time and storage time are critical factors defining the behavior and role of a particular substance or element throughout its biogeochemical cycle.

Ecologically relevant biogeochemical cycles are water, oxygen, carbon, nitrogen, phosphorus, and sulfur. Climate change and human impacts are drastically changing the speed, intensity, and balance of some relatively unknown cycles (e.g., the mercury cycle as well as the cycle of atrazine). Due to the troubling ecological implications, these cycles are increasingly studied. Indeed, while our planet is constantly subject to an energy flow from the sun, its chemical composition is essentially fixed, and additional matter may come only occasionally from meteorites. Thus, since matter is not replenished like energy, material cycling is essential for the persistence of ecological processes. These cycles involve all the ecospheres from the living biosphere to the nonliving lithosphere, atmosphere, and hydrosphere. All chemical elements in living things are part of biogeochemical cycles. For example, all the nutrients such as carbon, nitrogen, and phosphorus, that are essential for all living organisms, are part of a closed cycle. On the global ecosystem scale, these chemicals are recycled instead of being constantly replenished such as in an open system. In general, any type of matter is passed through the biogeochemical cycles between organisms and amid the various environmental matrices of the biosphere. Sulfur is continuously recycled in the sulfur cycle that may be used in ecosystems as a source of energy through the oxidation and reduction of sulfur compounds. For instance, very specialized organisms living near hydrothermal vents in the deep oceans, such as the giant tubeworm can exploit hydrogen sulfide as an energy source. This way ecosystems where no sunlight can penetrate, such as the deep sea, can find sustenance.

As seen in the previous chapter, water is the most ubiquitous solvent in the biosphere and usually carries the substances of biotic relevance. Water is recycled through the water cycle that is exemplified in the Figure 12.1 referred to as the *hydrologic cycle*, which involves interactions between all the ecospheres and pathways above and below the surface of the earth. Although the overall water balance of the ecospheres is constant over time, water moves from one reservoir to the other (e.g., from river to ocean and from the ocean to the atmosphere). The principal pathways of the water cycle are evaporation, transpiration, precipitation, infiltration, runoff, overland, and groundwater flows (Figure 12.1). Water undergoes phase transitions at various places in the water cycle switching its state between solid, liquid, and gaseous. Phase transitions implicate the exchange of heat and consequently temperature variations. For example, water takes up energy from the surroundings during evaporation refrigerating the environment, while in condensation process, water releases energy warming the environment. Water in solid, liquid, or vapor form is moved through the atmosphere by advection. Therefore, climate and meteorology are strongly influenced by the water cycle.

Some relevant processes and pathways governing water movements are described in Table 12.1.

Approximately 5×10^{14} m^3 of water arrive on the surface of the planet each year as precipitation, 80% of which falls over the oceans.* Total annual evapotranspiration is roughly in balance with precipitation. The great part of water on the earth is salt water and only 2.5% of the totality is fresh water, of which around 70% is frozen in the icecaps of Antarctica and Greenland, and only the remaining 30% (corresponding to only 0.7% of total water resources worldwide) is available for consumption. Of the available fresh water, approximately 87% is employed for agriculture (IPCC, 2007).

Water reservoirs play an important ecological role. For example, as a consequence of water cycling, water is purified by transfer from one reservoir to the other. Terrestrial reservoirs (e.g., groundwater, lakes, and rivers) are filled during rainy times and release freshwater to the land. Given that water transports

* This is not surprising, though, as stated in Section 1.6, seas account for 71% of the planet's surface.

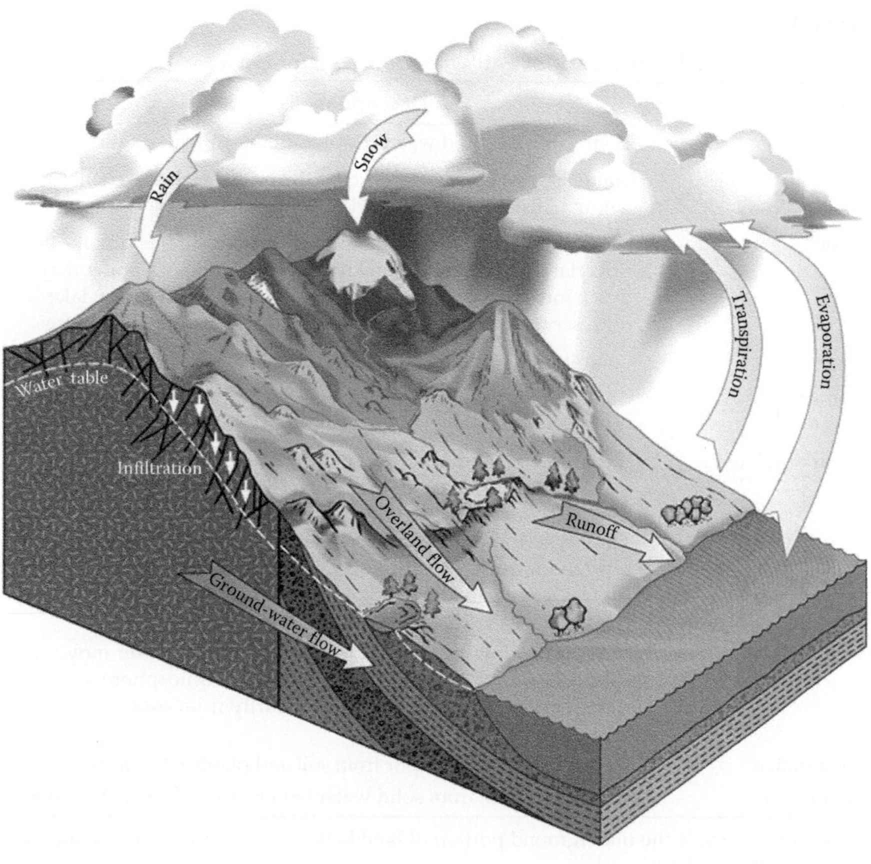

FIGURE 12.1
(**See color insert.**) Schematic representation of the water cycle with indication of the principal pathways.

chemicals to the different matrices in the biosphere, and the quality of water in reservoirs and potential impacts on this are critical ecological factors.

On a global planetary scale, residence times in reservoirs span from days to months, when the seasonal climatic cycle is considered. Yet, reservoirs such as glaciers and groundwater can exhibit residence times as long as thousands of years. This is a relevant issue particularly when confronting with cases of groundwater contamination. In such cases, the impacts can emerge very late in time, being groundwater pathways (especially deep groundwater) characterized by transiting times up to some millennia. In Table 12.2, the residence times for several reservoir of ecological relevance are reported.

Water being a persistent element, cycling times in the case of clean water are in general comparable with residence times. Water quality instead, as a measure of water conditions in relation to the requirements of biotic species or consumption needs, depends critically on residence times in the different reservoirs.

TABLE 12.1

Processes Involved in the Various Pathways of the Water Cycle

Process	Description
Condensation	From vapor to liquid water droplets in the air (clouds and fog).
Precipitation	Condensed water vapor falls onto the planet's surface (rain, snow, hailstorm, fog drip, drizzle, and sleet).
Runoff	Water moving across the land including both surface runoff (overland flow) and channel runoff. While flowing, water may leach into the ground, evaporate into the air, be stored in lakes or reservoirs, or extracted for agricultural or other human uses.
Snowmelt	Contribution to runoff by melting snow.
Infiltration	Water flowing from the surface into the ground (soil moisture, groundwater).
Subsurface flow	Water flowing underground, in the *vadose zone** and the aquifer. Subsurface water may emerge to the surface (springs or wells) or eventually flow into the oceans. Gravity or gravity induced pressures push water to the surface at lower elevation than where it infiltrated. Generally, groundwater moves slowly and its residence time in aquifers may be of thousands of years.
Evaporation	Phase transition from liquid to gas resulting in water moving from the ground or water bodies into the atmosphere. Energy for evaporation comes primarily from solar radiation.
Transpiration	The release of water vapor from soil and plants into the air.
Sublimation	Phase transition, from solid water (snow or ice) directly to vapor.

* The vadose zone is the underground portion of land between the surface and the top of the groundwater table.

TABLE 12.2

Residence Times for Some Reservoirs Involved in the Water Cycle

Reservoir	Residence Time	
Antarctica	20×10^3	Years
Deep groundwater	10×10^3	Years
Oceans	3×10^3	Years
Shallow groundwater	100–200	Years
Lakes	50–100	Years
Glaciers	20–100	Years
Seasonal snow cover	2–6	Months
Rivers	2–6	Months
Soil moisture	1–2	Months
Atmosphere		Days
Biota		Hours

It should be remarked that, through erosion and sedimentation, water is also involved in reshaping the geological appearance of the landscape.

The following sections provide synthetic descriptions of the essential biogeochemical cycles of oxygen, carbon, nitrogen, and phosphorus.

12.2 Oxygen Cycle

During the microcosm (see Section 1.7), bacteria have drastically changed the composition of the atmosphere adjusting the level of oxygen to 21% of the air. Nowadays the atmosphere is the main reservoir of free oxygen on the earth. This element is widely present in both the hydrosphere and the lithosphere, where oxygen is immobile being bonded to rocks and constitutes the predominant reservoir of oxygen. The largest part of oxygen production comes from the biosphere by photosynthesis. Oxygen is chemically consumed by oxidative processes such as those taking place in biological respiration. Figure 12.2 illustrates the oxygen cycle with arrows' size indicating the relative importance of flows and percentages the relative size of reservoirs. Oxygen is transferred across all the ecospheres. Reaeration is the process governing oxygen exchanges between the atmosphere and the hydrosphere (as described in Section 7.8). Wind and water as well as plants and animals, are the major causes of *weathering* that produces chemical or mechanical breaking down and loosens the earth's surface minerals, so that these can be transported away by erosion due to water, wind, and ice. Oxygen is transferred from the lithosphere to the hydrosphere and biosphere through this process. Burial is the process governing the transfer of oxygen from the biosphere to the lithosphere. Here, oxygen is immobilized and stored to form the largest reservoir of this element on the earth. The storage capacities as well as the residence times of the most relevant oxygen reservoirs are reported in Table 12.3.

Photolysis (see Chapter 10) is the principal process responsible for the oxygen and ozone balance in the atmosphere. The atmosphere layer between 10 and 50 km altitude is called the *stratosphere*, while the layer underneath, up to 10 km, is referred to as the *troposphere*. This layer subdivision is due to the stratification in temperature, with warmer layers higher up and cooler layers farther down in the stratosphere, in contrast to the troposphere near the earth's surface, where the higher parts are cooler than the lower ones. The boundary between the troposphere and the stratosphere is called the *tropopause*. The following chain of reactions describes the dynamic equilibrium between oxygen and ozone in the stratosphere,

$$O_2 + h\nu_{<240\,nm} \rightarrow 2O$$
$$O_2 + O \rightarrow O_3$$
$$O_3 + h\nu_{200-310\,nm} \rightarrow O_2 + O$$
$$O_3 + O \rightarrow 2O_2$$

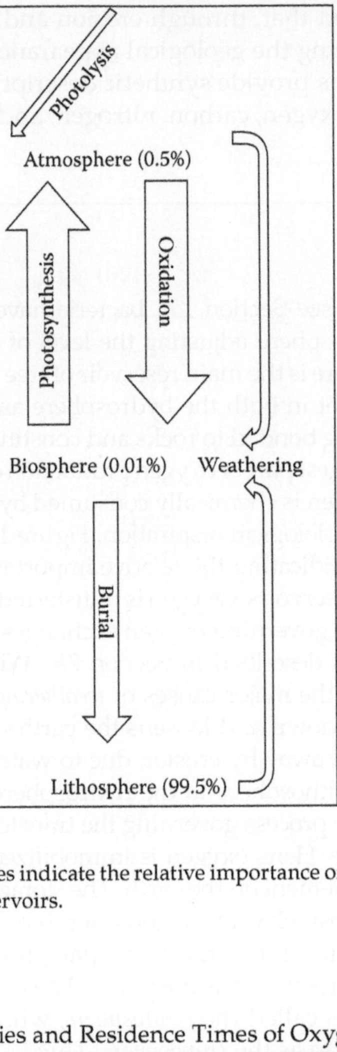

FIGURE 12.2
The oxygen cycle. Arrows' sizes indicate the relative importance of the flows while percentages the extent of the different reservoirs.

TABLE 12.3
Storage Capacities and Residence Times of Oxygen Reservoirs

Reservoir	Capacity (kg O_2)	Residence Time (years)
Atmosphere	1.4×10^{18}	4500
Biosphere	1.6×10^{16}	50
Lithosphere	2.9×10^{20}	5×10^{8}

Ozone is created by UV radiation hitting oxygen molecules and splitting them into individual oxygen atoms. The atomic oxygen then combines with unaffected O_2 producing ozone, O_3. Thus the ozone layer is formed as a result of oxygen photolysis. The reverse reaction is ozone photolysis, fired by UV radiation, but wider spectrum. This way the upper atmosphere is turned into a filter that protects the earth's surface from harmful ultraviolet rays. However, photolysis is also the process by which CFCs are broken down in the upper atmosphere to form ozone-destroying chlorine-free radicals (see Chapter 23).

Large part of tropospheric ozone formation occurs when nitrogen oxides (NO_x), carbon monoxide (CO), and volatile organic compounds (e.g., methane, CH_4) react in the atmosphere in the presence of sunlight. These chemicals are often referred to as ozone precursors. In the troposphere, the key reaction chain in the formation of tropospheric ozone is

$$NO_2 + h\nu \rightarrow NO + O$$
$$O_2 + O + M \rightarrow O_3 + M$$
$$NO + O_3 \rightarrow NO_2 + O_2$$

where, M is a useful molecule capable of absorbing energy (e.g., gaseous nitrogen, N_2) and making stable ozone. This dynamic equilibrium is altered by the presence of pollutants that potentially prevent the dissociation reaction. Air pollution may result as a consequence of chemical reactions in the atmosphere of nitrogen oxides and volatile organic compounds with sunlight, which leaves airborne particles and ground-level ozone. This toxic mixture of air pollutants can include aldehydes, nitrogen oxides, volatile organic compounds as well as tropospheric ozone. All of these chemicals are usually highly reactive and oxidizing. Industrial and urban emissions as well as motor vehicle exhaust and chemical solvents are the major anthropogenic sources of these chemicals.

The consumption of oxygen in the ecospheres is ascribed to different ecological processes as described by the following oxygen demand indicators:

- Biochemical oxygen demand (BOD)
- Chemical oxygen demand (COD)
- Nitrogen oxygen demand (NOD)
- Sediment oxygen demand (SOD)

The BOD concentration (in mg/l) is a measure of the quantity of oxygen required for the removal of dissolved organic pollutants by biological oxidation, that is, carried by aerobic bacteria, which is an indirect measure of the quantity of dissolved organic pollutants in the matrix examined. As stated in chapter 9, organic matter is composed by nearly fixed proportions of the basic elements C, O, P, N, H, and so on. According to the proportions established by the Redfield ratio, the degradation of organic matter carried out by microorganisms can be described by the following conceptual redox reaction (Jørgensen and Bendoricchio, 2001):

$$C_{106}H_{253}O_{116}N_{16}P_1 + R(O) + \text{Decomposers}$$
$$\rightarrow aCO_2 + bNH_4^+ + cHPO_4^{2-} + dH_2O + eH^+ + \text{Energy}$$

where, *a*, *b*, *c*, *d* and *e* suitable stoichiometric coefficients and *R(O)* the oxidizing chemical agent. Of the various oxidizing agents reported in Table 12.4, the one that gives the highest amount of energy (in the form of ATP) will be used first. Therefore, if oxygen is present, oxygen will be used, nitrate will be used after oxygen depletion, and so on.

The COD concentration measures the quantity of dissolved organic pollutants that can be removed by chemical oxidation. The global reaction transfers each carbon atom of organic matter in a molecule of CO_2. BOD refers to that part of COD needed by aerobic biological organisms to break down organic material so that the BOD/COD ratio gives an indication of the fraction of organic pollutants that are biodegradable. COD is determined by the stoichiometry of the chemical oxidation reactions. COD and BOD are typically used as indicators characterizing the organic pollution load in water bodies. The oxygen cycle in aquatic ecosystems, where very simplified water quality models such as the Streeter and Phelps equation are employed, are described in Chapter 20.

An important part of BOD is that employed in the process of *nitrification* (NOD), that is, the oxidation of reduced forms of nitrogen, as described by the following reactions:

$$NH_4^+ + \frac{3}{2}O_2 \rightarrow NO_2^- + H_2O + 2H^+$$

$$NO_2^- + \frac{1}{2}O_2 \rightarrow NO_3^-$$

Organic nitrogen is hydrolyzed to $N\text{-}NH_4$ without oxygen consumption. The ammonium produced is then nitrified using bacteria like *Nitrosomonas* and *Nitrobacteria* in two steps: first to nitrite and subsequently to nitrate. The rate constants are affected by various factors such as temperature, pH, and bacterial strain.

Oxidation of organic matter in sediments and biota respiration are accounted by SOD (g $O_2/m^2/day$), which may represent in some water bodies, a large fraction of the oxygen consumed. The two main sources of

TABLE 12.4

Common Oxidizing Agents (in Order of Growing Energy Production)

Common Oxidizing Agents	
Molecular oxygen	O_2
Nitrate ion	NO_3^-
Manganese dioxide	MnO_2
Iron oxides	For example, FeO(OH)
Sulfate ion	SO_4^{2-}
Carbon dioxide	CO_2

SOD are the degradation of organic matter settled on the bottom and the respiration of benthic biota. SOD is a consequence of surface processes and is strongly affected by oxygen diffusion into the sediment induced by bioturbation. Degradation of organic matter in sediments is influenced by the diffusion of dissolved oxygen from the water column to pore water, and the diffusion of mineralized reduced forms of organic matter from pore water to water column. Bioturbation of sediments by benthic organisms increases the interface exchanges and is considered as an increment of the active exchange surface. For example, the labyrinth of small tubes created by worms in marine sediment may provide an exchange surface up to four times larger than the corresponding horizontal surface of the bottom.

Respiration of primary and secondary producers causes oxygen consumption. However, this oxygen demand is often neglected being considerably less important compared to other sources, which tend to be preponderant. In the absence of significant organic loads and reareation, respiration can become an important factor affecting water quality.

12.3 Aerobic and Anaerobic Conditions

As seen in Section 9.4, biological energy is commonly stored and released by means of redox reactions. For instance, chlorophyll photosynthesis involves the reduction of carbon dioxide into sugars and the oxidation of water into molecular oxygen (autotrophs organisms). Respiration is a reverse reaction, in which sugars oxidize to produce carbon dioxide and water (heterotrophs organisms). The equilibrium of these redox reactions is regulated by the redox potential that in turn is stringently linked to oxygen availability. The ability of an organism to carry out redox reactions depends on the redox state of its environment described by the reduction potential, *Eh*. Aerobic bacteria are commonly operating at positive *Eh* values, whereas strict anaerobes work at negative ones. The different habitats within an ecosystem are characterized by the biogeochemical conditions of the matrices involved. These conditions are determinant in establishing which of the possible final electron acceptors will dominate biochemical functional reactions of organisms.

The possible redox regimes as a function of the redox potential *Eh*, which in the oxidized and reduced forms will prevail, and the related equilibrium conditions depend on the complex system of interacting microbial catalytic functions present in the habitat under focus is depicted in Table 12.5. In standard conditions, the reactions in the upper part of the table proceed rightward, in reverse direction in the lower part. The quantity of ATP produced by the reaction is proportional to the vertical distance of the involved oxidized and reduced forms in the table. For example, when glucose is oxidized via oxygen, respiration will give the largest yield of biochemical energy.

TABLE 12.5

Redox Regimes as a Function of the Redox Potential, *Eh*

Oxidizing Agents				
e- Acceptor	Oxidized		Reduced	Redox Conditions
O_2	O_2	$\rightharpoonup$	H_2O	*oxidizing*
NO_3^-	NO_3^-	$\rightharpoonup$	N_2	
Mn^{4+}	MnO_2	$\rightharpoonup$	$MnCO_3$	
Fe^{3+}	$FeO(OH)$	$\rightleftharpoons$	$FeCO_3$	*Eh*
SO_4^{2-}	SO_4^{2-}	$\leftharpoondown$	HS^-	
CO_2	CO_2	$\leftharpoondown$	CH_4	
CO_2	CO_2	$\leftharpoondown$	$C_6H_{12}O_6$	*reducing*

Organisms that can only survive in aerobic conditions are known as strictly aerobes. These organisms use organic matter, nutrients, and oxygen to produce carbon dioxide and replicate. When molecular oxygen is depleted, microorganisms using chemically combined oxygen such as that of nitrite and nitrate will take over. This condition is referred to as *anoxic*. Anoxic organisms coexist easily with both aerobic and anaerobic organisms. Often anoxic organisms are facultative, that is, they possess the ability of selecting and performing the appropriate redox process depending on oxygen availability. These organisms use organic matter, nutrients, and combined oxygen to produce nitrogen gas, carbon dioxide, and novel organisms.

In anaerobic conditions, living systems are able to obtain energy throughout a complex series of biochemical reactions carried out in multiple steps by specialized microorganisms that require little or no oxygen for their metabolism. This process produces *biogas* that is mainly composed of methane and carbon dioxide. Anaerobic energy production requires four fundamental steps: *hydrolysis*, complex organic matter is decomposed into simple soluble organic molecules; *fermentation* (acidogenesis), the chemical decomposition of carbohydrates in the absence of oxygen that carried on by enzymes, bacteria, or yeasts; *acetogenesis*, the conversion of fermentation products into acetate, hydrogen, and carbon dioxide; and *methanogenesis*, production of methane and hydrogen/carbon dioxide. Fermentation, acetogenesis and methanogenesis are accomplished by highly specialized bacteria. Acetogenic bacteria grow in close association with methanogens. Anaerobic decomposition takes place only under strict anaerobic conditions.

Although energy-yielding metabolism often uses oxygen as the most efficient electron acceptor (in terms of ATP production), as recalled in Table 12.5, several other elements can be used as electron acceptors. To appreciate which electron acceptor is selected, it is critical to define the physiological and biochemical setting of redox processes that are relevant for microbial cycling of carbon, nitrogen, sulfur, and so on. Under aerobic conditions (availability of O_2), a set of physiological reactions regulate the geochemical state of waters, sediments, soils, and the atmosphere. In the absence of oxygen (anaerobic conditions), different group of processes take place, which are usually catalyzed by specialized organisms. Anaerobic respiration produces less energy yields compared to aerobic respiration. The very different aerobic and anaerobic conditions considerably affect ecological matrices. The presence of critical biogeochemical compounds like for instance O_2, CH_4, SO_4^{2-} and NH_4, depends on the changing equilibrium between aerobic and anaerobic conditions. Predicting the dynamic of such equilibrium requires a detailed and deeply integrated biochemical and ecological description.

12.4 Carbon Cycle and Biosphere Stability Conditions

The carbon cycle is the most important of all the biogeochemical cycles. All organic matter contains carbon; this element is essential to all forms of life. The carbon and oxygen cycles are intimately linked. In order to better understand the role these cycles have in maintaining stable conditions in the biosphere, it is interesting to trace the evolutionary history of life on our planet as these cycles are intimately linked. As discussed in Section 1.7, it is during the era of microcosm, extending for nearly two billion years from 3.5 to 1.5 billion years ago, that bacteria "invented" all the biotechnologies employed for biosphere self-regulation. Particularly during this era, bacteria drastically changed the composition of the atmosphere, adjusting the levels of humidity, hydrogen, CO_2, oxygen (21%), and ozone (Margulis and Lovelock, 1974).

A hypothetic but plausible sequence of the relevant steps in the process of biosphere self-regulation that took place during the Archean and Proterozoic eons is briefly depicted in Figure 12.3.

In an environment characterized by an atmospheric prevalence of carbon dioxide, the first biotechnology developed with the appearance of life was the autotrophic production of organic matter using solar energy and atmospheric CO_2 as a source of carbon. Earlier photosynthesizers* used hydrogen sulfide (H_2S), a very toxic and inflammable gas of volcanic origin, as electron donor. This process continued until atmospheric carbon dioxide was severely depleted. Around three billion years ago, a first dramatic condition emerged that might be called CO_2 crisis. This condition resulted in a general temperature decrease of the planet, as CO_2 is a greenhouse gas. At that time, a new family of bacteria developed the ability of anaerobic energy production by fermentation. This process produces biogas such as methane and carbon dioxide that is returned to the environment. These bacteria may have significantly increased hydrogen

* Typically this process is performed by the purple and green sulfur bacteria.

Eons	10^9 Years ago	Biotechnology		Atmosphere
Archean	3.5	Photosynthesis H_2S	←	CO_2
		CO_2 crisis		
	3	Fermentation	→	CO_2
		H_2 crisis		
Proterozoic	2.5	Photosynthesis H_2O	→	O_2
		O_2 crisis		
	2	Respiration	←	O_2

FIGURE 12.3
(See color insert.) The process of biosphere self-regulation.

loss to space, hence initiating a net oxidation process of the atmosphere/ocean/sediment system (the hydrogen crisis), making the atmosphere rich in methane. It is reasonable to assume that organic carbon immobilization (e.g., by precipitation and burial) and oxidation of reduced organic matter and iron were in balance until atmosphere oxidation began. When water photosynthesis by cyanobacteria is more efficient biological oxygen production process, oxygen concentration in the atmosphere become critically high (the oxygen crisis dated about 2.4 billion years ago), and methanogen communities, being strictly anaerobes, would have been replaced by aerobic respiring organisms. Cyanobacteria, by producing oxygen that was toxic to anaerobes, were basically responsible for the largest extinction in the history of the earth.

This self-regulation process was accomplished through unavoidable feedback mechanisms between the biosphere and the atmosphere and resulted in today's atmosphere (see Table 12.6). In particular, oxygen concentration is maintained at the critical value of 21%. Aerobic respiration is not efficient below this concentration, while larger oxygen content would imply a dangerous oxidative condition in which self-combustion events become significantly possible.

The concentration of the greenhouse gas CO_2 is also regulated. This value stayed constant slightly below a value of 280 ppm during the last millennia (this value is referred to as the preindustrial concentration). However, since last two centuries began a sharp increase, that in the recent years reached an estimated rate of 2 ppm/year and brought this concentration to 383 ppm in year 2007.*

On the earth, the largest stock of carbon is found in the rocks in inorganic forms such as carbonates and bicarbonates. Another important sink is materials of biologic origin in which carbon is fixed in organic forms. The atmosphere also stores carbon in the form of gases like CO_2, CH_4, and CO.

* See in sources by Dr. Pieter Tans, NOAA/ESRL (www.esrl.noaa.gov/gmd/ccgg/trends/) and Dr. Ralph Keeling, Scripps Institution of Oceanography (scrippsco2.ucsd.edu/).

TABLE 12.6

Composition of Atmosphere Today

Gas	
Nitrogen, N_2	78.1%
Oxygen, O_2	20.9%
Argon, Ar	0.9%
Minor constituents (ppm)	
Carbon Dioxide, CO_2	383
Neon, Ne	18.18
Helium, He	5.24
Methane, CH_4	1.7
Krypton, Kr	1.14
Hydrogen, H_2	0.55

The carbon cycle may be tracked by subdividing it into a set of interconnected processes (Figure 12.4):

1. Photosynthesis is the process by which atmospheric CO_2 is chemically converted into organic compounds (carbon-based sugar molecules). Each year, photosynthesis by plants moves about 111 billion tons of carbon from the atmosphere to the biosphere. This process is carried out by photosynthetic or chemosynthetic autotrophs (e.g., cyanobacteria or methanogens) and is often referred to as CO_2 fixation.
2. Carbon is returned to the atmosphere as CO_2 by the processes of respiration and fermentation (oxidation of organic matter to CO_2). Plants and animals transform carbon-based organic molecules into CO_2 and other byproducts to obtain energy. Fifty billion tons of carbon is released each year into the atmosphere through respiration by organisms other than detritivores. The detritus cycle accounts for 60 billion tons of yearly carbon emissions coming from organic matter decay in soils. Together with respiration, these flows account for a large part but not all of the carbon fixed by photosynthesis: approximately 1.4 billion tons per year are permanently stored in soil, rocks, and sediments.
3. The global carbon cycle involves oceans, atmosphere, and lithosphere exchanges. Atmospheric CO_2 enters the oceans diffusing into surface waters and reacts with water by a reversible reaction producing carbonic acid, $CO_2(aq) + H_2O \rightleftharpoons H_2CO_3$. This process achieves a dynamic steady chemical equilibrium, maintaining a relatively stable ratio of CO_2 to H_2CO_3. The value of this ratio depends on factors such as water temperature and alkalinity. Until the nineteenth century, the flow of CO_2 diffusing into the ocean and that released from the ocean into the atmosphere was maintained approximately in balance

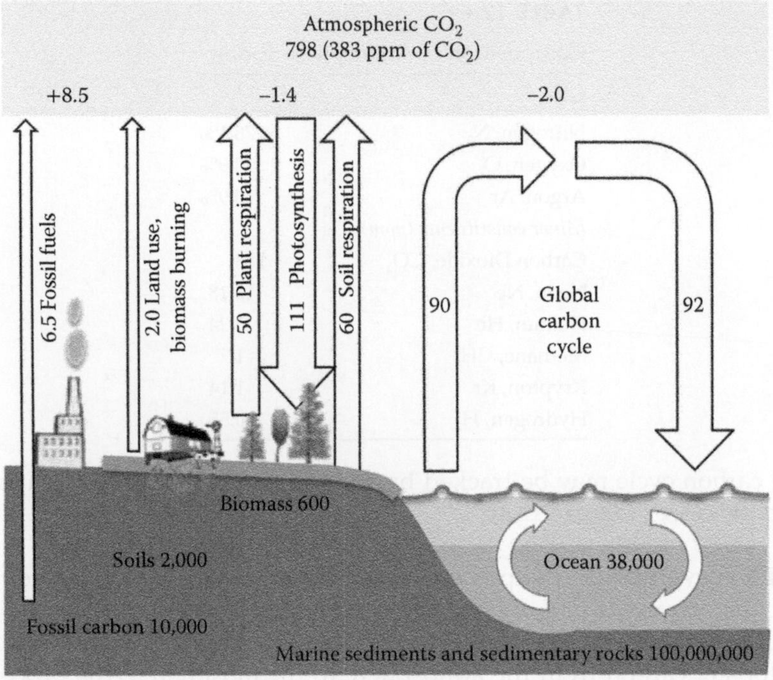

FIGURE 12.4
The global carbon cycle. Flows are expressed in gigatonnes of carbon per year (10^9 tC/year). Figures in white represent stocks (10^9 tC). The flows are estimated by employing the best knowledge available of the processes involved. These calculations provide an estimation of the net carbon flow to the atmosphere of 5.1×10^9 t/year. However this value is not consistent with measured values (approximately 3.2×10^9 t/year); that is, there is less carbon in the atmosphere than expected. The missing carbon is associated with a yet unknown carbon sink or to an underestimation of some flow.

by geochemical and biological processes. However, this equilibrium has been displaced by supplementary unbalanced CO_2 emissions of anthropogenic origin into the atmosphere, leaving a residual net carbon inflow into the oceans of about 2 Gt. This phenomenon leads to ocean *acidification* (see Chapters 20 and 23), that is, the ongoing decrease in pH of the oceans that has resulted in the last two centuries in a 30% increase in acidity of seawater. The important reaction that controls the acidity is $H_2CO_3 \rightleftharpoons H^+ + HCO_3^-$. This reaction produces hydrogen ions and bicarbonate and buffers seawater against large changes in pH (see Section 9.5). Living things fix bicarbonate with calcium (Ca^{2+}) to produce calcium carbonate ($CaCO_3$) used to build shells by organisms such as coral, clams, oysters, some protozoa, and some algae. After death, this material is accumulated in the bottom of the ocean as carbonate-rich deposits and then in sedimentary rocks. Ocean deposits are the largest carbon reservoir on the planet (Figure 12.4).

4. Flows of anthropogenic origin have affected the carbon cycle since the Neolithic revolution that 10,000 years ago saw human tribes changing in agricultural communities. With the industrial revolution, the flow of carbon stored in fossil fuels to the atmosphere becomes substantial: burning coal, oil, and natural gas returns CO_2 to the atmosphere moving it from a reservoir that until recently represented an end point for the carbon cycle. In the year 2000, the amount of carbon emissions to the atmosphere due to fossil carbon combustion was estimated to be around 6.5 Gt. Land use changes induced by humans over the last several hundred years have drastically changed landscapes by reducing the area covered by forests, a process known as *deforestation*, or more in general by replacing natural areas with agricultural or urban lands. By reducing the number of trees through agricultural practices such as slashing and burning, the photosynthetic capability of the biosphere has been impaired, while the increased decay or combustion of organic material has generated a supplementary CO_2 flow to the atmosphere. Overall these factors may be accounted for by a net carbon flow to the atmosphere estimated to be around 2 Gt per year. Altogether, anthropogenic carbon emissions into the atmosphere are approximately 8.5 billion tons per year.

Measured values of CO_2 levels in the atmosphere give currently a result about 3.2 billion tons. The value provided by the budget illustrated in Figure 12.1 is much larger, that is, 5.1 Gt. A probable explication for this difference resides in some unidentified feedback mechanism between atmospheric CO_2 concentration and biota fixation ability.

Besides biological activity and anthropogenic impacts, mechanisms of control over the carbon cycle include volcanic activity as a prominent one. Long-term historical records of CO_2 concentration in atmosphere are strongly correlated with the volcanic rock formation rate. Rocks weathering is a significant process for CO_2 transfer from the lithosphere to the atmosphere and the hydrosphere.

12.5 Nitrogen Cycle

Nitrogen is a key chemical element to life, found in large quantities in fundamental biochemical molecules and compounds, such as amino acids and proteins, nucleic acids (DNA, RNA), and chlorophyll. Although nitrogen is the most common element in the earth's atmosphere, about 78% of the total volume of dry air, its gaseous molecule N_2 is markedly inert and cannot be uptaken by most organisms. For these reasons, nitrogen is often a limiting factor for the growth of primary producers, representing the energetic basis of ecosystems, and is central in key human activities such as farming (fertilizers contain N): consequently, its biogeochemical cycle is very well-studied. But, an excess input of this nutrient to coastal waters, for example, through rivers carrying nitrogen loads generated by wastewater treatment plants, industries, and agricultural/farming activities, can be negative and lead to eutrophication (Chapter 20). Eutrophication of water bodies is often connected to hypoxic and anoxic phenomena, blooms of toxic algae, fish kills, and biodiversity loss, but, sometimes, also to increased fishery landings through enhanced system primary productivity. The study of the nitrogen cycle is important to understand the fate of common N-based molecules, such as ammonia (NH_3), ammonium ion (NH_4^+), nitrate ion (NO_3^-), nitrite ion (NO_2^-), mono-nitrogen oxides (NO_x, including nitric oxide NO and nitrogen dioxide NO_2), nitrous acid (HNO_2), and nitric acid (HNO_3), whose effects can include toxicity for man (e.g., nitrates can lead to methemoglobinemia or blue baby syndrome in infants), animals or plants, or changes in environmental pH (e.g., acidification through acid rains) (van den Berg and Ashmore, 2008), which control other ecological processes. NO_x is commonly produced by combustion, for example, by industries or car engines, while NH_3 is involved in many industrial processes.

The nitrogen cycle (schematically represented in Figure 12.5) is a complex biogeochemical cycle, which involves several transformations and biological, physical, and chemical processes. Man greatly impacts the nitrogen cycle through the industrial production of fertilizers causing eutrophication,

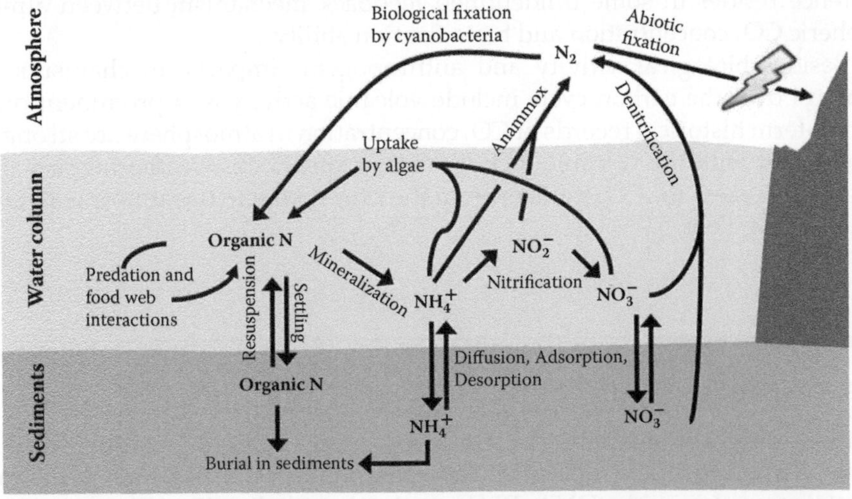

FIGURE 12.5
(See color insert.) A schematization of the global nitrogen biogeochemical cycle. Processes and symbols are described in this and other chapters of the book. This figure is mainly focused on aquatic ecosystems, but most processes (fixation, mineralization, nitrification, denitrification, predation) take place also in terrestrial systems.

and by causing/favoring the emissions of N-containing pollutants and greenhouse gases into the atmosphere. Examples of such a complex cycle can be appreciated in wetlands (Kadlec and Knight, 1996; Jørgensen and Bendoricchio, 2001), where most nitrogen cycle processes take place.

As mentioned, the largest reservoir of nitrogen (about 90% in weight) (Widdison and Burt, 2008) on the earth is the atmosphere, where nitrogen is mainly found as an inert diatomic gas (N_2) and as traces of nitrous oxide (N_2O, a greenhouse gas with great global warming potential), NO_x and volatilized ammonia. The other large reservoir on the earth (10% in weight) is represented by rocks and sediments, while fast exchange pools such as terrestrial and aquatic biota, water and soils contain negligible nitrogen amounts (Widdison and Burt, 2008).

Inert atmospheric nitrogen mainly reaches lands and oceans through the key process of biological *fixation* of N_2, which is uptaken and transformed into ammonia by particular N-fixing bacteria and archea (diazotrophs), including both free living organisms (e.g., cyanobacteria, or blue-green algae) in soils and water bodies, and those bacteria forming a symbiotic relationship with legumes (symbiotic bacteria, such as those belonging to the *Rhizobium* genus, live in nodules of their roots) and few other plants (e.g., water fern *Azolla*). Atmospheric N_2 can also be fixed into nitrogen oxides by high-energy processes such as lightning, ultra-violet radiations, combustion (e.g., in car engines, industries), and electrical sparking: those oxides, together with NO_x emitted into the atmosphere by combustion of fuels containing bound-nitrogen, are then washed by rain and dew, and reach the soil. In general, nitrogen deposition from the atmosphere to earth's ecosystems can be "wet," for example, NO_x or NH_3 can be removed from the air by rain, fog, snow, and so on, or "dry," that is, when nitrogen-containing molecules settle or hit the surface of soil and water bodies without being carried by precipitation. "Acid rain" is usually meant to include both wet and dry deposition of nitric acid. Globally, biological fixation of N_2 is higher than non-microbial atmospheric N-inputs (at least four times as large) (Postgate, 1998). Instead, a quantitatively important human pathway of N_2 fixation is represented by the Haber (or Haber-Bosch) industrial process for the synthesis of ammonia, based on elemental nitrogen and hydrogen gases. The Haber process made intensive agriculture possible at the beginning of the twentieth century, since the ammonia obtained could then be used to produce fertilizers (nitrogen often limits plant growth) and other nitrogen-based industrial products. The process reaction, which requires high pressure (150–200 atm) and temperature (about 500°C), and the presence of a catalyst (usually iron), is

$$N_2 + 3H_2 \rightleftharpoons 2NH_3 \tag{12.1}$$

According to the N-cycle paradigm, the atmosphere is the only nitrogen reservoir taking part in the cycle, from which this nutrient can reach earth's ecosystems and exchange pools. Rocks are basically inert and uninvolved in the global cycle. Recent research, however, shows that trees can uptake nitrogen directly from weathered sedimentary bedrocks rich in nitrogen, and suggests that this pathway could be quantitatively important in the global cycle of nitrogen (Morford et al. 2011).

Once nitrogen has reached terrestrial and aquatic ecosystems from atmospheric and land reservoirs, it can cycle through its different forms in abiotic and biotic exchange pools. Several processes are involved, which are detailed below.

Biologically fixed ammonia is used by microorganisms to build up organic compounds containing nitrogen, and can be uptaken by plant roots or phytoplankton when fixing microbes die and decompose; it can even be uptaken directly in plants hosting symbiotic bacteria. Similarly, nitrogen in fertilizers can be uptaken by plants. *Uptake* of nitrogen by plants and algae mainly exploits two bioavailable inorganic forms of nitrogen, ammonium, and nitrate ions, although other such forms exist (e.g., nitrites or organic compounds such as urea). N-uptake can also be performed by heterotrophic bacteria and fungi. Primary producers, N-uptaking organisms as well as nitrogen-fixing micro-organisms can be predated by other organisms, so that the organic nitrogen found in their tissues is assimilated by the predators, entering the *food web* and indirectly sustaining all non-primary biological production (Chapters 16 and 19). When living beings die or release feces (and exudates in the case of primary producers), they are decomposed by bacteria and fungi, and *mineralization* of organic matter and nitrogen takes place, and the latter is transformed into the bioavailable ammonium (hydrolysis also comes into play, Chapter 10). Nitrogen pools, which are commonly involved in these processes include particulate organic nitrogen (PON), representing both living micro-organisms and dead N-containing particulate organic matter, and dissolved organic nitrogen (DON), representing N found in dissolved organic compounds (e.g., amino acids, amines, but also larger macro-molecules).

In addition to being uptaken, ammonium can undergo *nitrification*. Nitrification is a two-step oxidation of ammonium into nitrites and then nitrates performed by bacteria:

$$NH_4^+ + 1.5O_2 \rightarrow NO_2^- + H_2O + 2H^+ \tag{12.2}$$

$$NO_2^- + 0.5O_2 \rightarrow NO_3^- \tag{12.3}$$

The first step (ammonium to nitrite) is performed by bacteria such as *Nitrosomonas* species and is usually much slower than the second (nitrite to nitrate; performed by bacteria such as *Nitrobacter* species), so that nitrite concentration is generally lower than that of nitrates. As shown in Equations 12.2 and 12.3, nitrification is an aerobic process requiring oxygen and taking place in well-aerated waters or near the soil surface.

The nitrogen cycle is finally closed by bacterial *denitrification*, which is the transformation of nitrates into gaseous N_2, and into N_2O in lower amounts.

This process requires particular conditions to take place: an electron donor should be available to bacteria, such biodegradable organic matter, and anoxic conditions should be achieved, since oxygen inhibits denitrification: denitrifying bacteria are facultative; that is, if oxygen is present, they will choose it instead of nitrate to oxidize organic matter.

Denitrification is a key ecological process removing nitrogen from ecosystems in an irreversible manner. Another important pathway is the recently discovered *anammox* process (Arrigo, 2005), the microbial anaerobic oxidation of ammonium to N_2 using nitrite as an oxidant. Anammox can be used to remove nitrogen from wastewater, and is observed both in the sediments and the water column of oxygen-poor natural water bodies (e.g., productive marine ecosystems such as coastal shelves and upwelling systems). Other flows of nitrogen to the atmosphere exist, such as volatilization of ammonia, which is pH-controlled based on the equilibrium between NH_3 (which can be toxic for living organisms) and NH_4^+, or the above-mentioned NO_x fluxes produced by combustion.

Several other processes are involved in the complex nitrogen cycle, and are described elsewhere, such as settling of PON as particulate matter (including phytoplankton cells) in aquatic ecosystems, vertical transport through diffusion in stratified water bodies, and adsorption-desorption equilibrium in soil and sediments (Chapter 11). Finally, it is important to notice that most of the processes which involve nitrogen cycling in fast exchange pools, such as mineralization, nitrification, denitrification, and so on, are biological processes, and thus are usually strongly influenced by environmental factors such as temperature, pH, moisture, oxygen and resource availability (see specialized texts, for example Kadlec and Knight, 1996, regarding wetlands). Most of these processes can be modeled as first-order reactions describing the decay of the concentration C of a particular nitrogen form over time t:

$$C(t) = C_0 \cdot e^{-kt} \tag{12.4}$$

where $k = k_{20} \cdot \theta^{(T-20)}$, T(°C) is temperature, and θ is a parameter slightly greater than 1, describing the Arrhenius-like dependence of the reaction rate k on temperature. Table 12.7, adapted from Jørgensen and

TABLE 12.7

N Cycle, Relative Speed of the Different Processes

Process		k_{20}—Decay Rate at 20°C (d^{-1})
Mineralization of DON	$DON \rightarrow NH_4^+$	0.002
Mineralization of PON	$PON \rightarrow NH_4^+$	0.01–0.03
Oxidation of ammonium to nitrite	$NH_4^+ \rightarrow NO_2^-$	0.1–0.5
Oxidation of nitrite to nitrate	$NO_2^- \rightarrow NO_3^-$	0.5–2.0
Oxidation of ammonium to nitrate	$NH_4^+ \rightarrow NO_3^-$	0.1–0.2
Denitrification	$NO_3^- \rightarrow N_2$	0.1
Release of NH_3 from sediments		0.001–0.01

Bendoricchio (2001), reports parameter values highlighting the relative speed of different processes in the biogeochemical cycle: mineralization appears to be the limiting step of the nitrogen cycle compared to other "fast" nitrogen transformations.

12.6 Phosphorus Cycle

The biogeochemical cycle of phosphorus is well studied due to the important role of this element in supporting life of all organisms. Phosphorus is found in nucleic acids (DNA, RNA) storing genetic information, in the molecule ATP used as an energy carrier in metabolism, and in several other important biochemical molecules and organic tissues of living beings. For example, cellular membranes contain phosphorus (as phospholipids), similarly to bones and teeth. The cycling of this element is central in ecology because it often limits the growth of primary producers, similar to nitrogen. These two nutrients are often studied together and both are generally measured in algal growth studies (Chapter 15). Typically (but there are many exceptions) phosphorus is the limiting nutrient in freshwater ecosystems, while marine water is N-limited. The P-cycle is strongly impacted by man (similarly to nitrogen), as phosphorus is contained in fertilizers, detergents and several other chemicals, and wastewater of human and livestock origin. The great boost in P-emissions from land to water over the past decades has often led to the process of eutrophication of water bodies (Chapter 20) (Nixon, 1995).

The peculiarity of the phosphorus cycle (Figure 12.6), as compared, for example, to that of nitrogen, is that it basically lacks a gaseous phase, and an atmospheric reservoir. The bulk of the earth's reservoirs of phosphorus are found in its crust, mainly as phosphate rocks such as apatite. Such phosphorus can be involved in the fast exchange pools of the cycle (cycling through waters, sediments, biota, etc.; see Figure 12.6) only after rocks have been weathered. Weathering is typically a very slow process compared to the cycling of this element among fast exchange pools, and this difference in rates can explain why phosphorus is a key limiting nutrient. Human needs for phosphorus are great, as it is used to produce fertilizers sustaining global agriculture and to satisfy the requirements of various industrial processes to some extent (Liu and Chen, 2008), and the slow weathering flux of phosphorus from reservoirs is not sufficient to meet such needs. Thus, phosphorus is currently extracted from rocks at an intense rate through mining of phosphate rocks (Nixon, 1995; Liu and Chen, 2008). Given the intense exploitation of rock reservoirs of phosphorus, which are nonrenewable on short time-scales, current global P-reserves may be depleted in less than a century (Cordell et al., 2009).

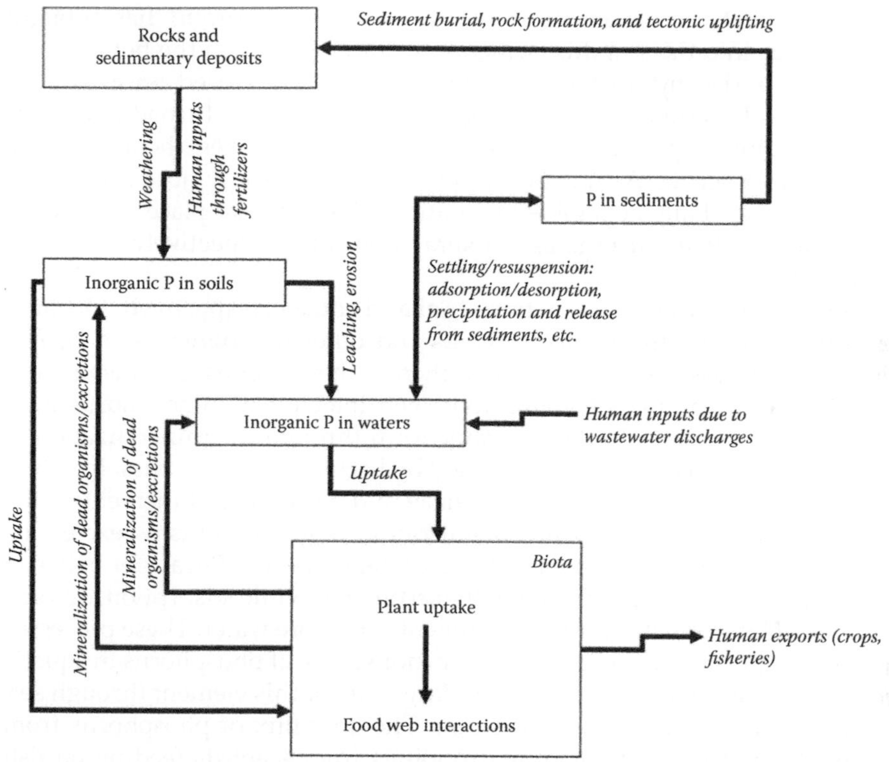

FIGURE 12.6
A schematization of the global phosphorus biogeochemical cycle.

Inorganic phosphorus generally occurs as phosphate ions or salts, and the latter can be soluble or insoluble depending on environmental conditions (e.g., pH, temperature and redox potential, Liu and Chen, 2008; e.g., anoxia can lead to P-release from sediments). The orthophosphate ion PO_4^{3-} is the main bioavailable P-form (but, for example, algae can also uptake lower P-amounts as colloidal organic labile form). The equilibrium of this dissolved inorganic form with other forms, such as HPO_4^{2-}, $H_2PO_4^-$, and phosphoric acid H_3PO_4, is regulated by pH. Phosphorus has also a strong tendency to be adsorbed (Chapter 11) to soil and sediment particles, which is a key fact: adsorbed phosphorus is often abundant, but generally not bioavailable and, moreover, particulate phosphorus can be the most important P-form transported by rivers to water bodies through erosion (see below). Organic P-forms include particulate (POP) and dissolved organic phosphorus (DOP) and colloids.

The P-cycle can be thought of as the superimposition of one inorganic and two organic cycles, and the interactions between them (Liu and Chen, 2008). In the inorganic cycle, weathering moves phosphorus from rocks to soils, and then into freshwater bodies through leaching and soil erosion (e.g., from croplands that receive fertilizer inputs). Once river runoff has brought phosphorus into the sea, this element can settle and reach the bottom, and finally be buried by further sediments. The cycle is closed on geological time-scales (hundreds–million years) once sediments are turned into rocks by the sediment pressure, and tectonic uplifting makes them available again to weathering. As mentioned, phosphorus is nearly absent from the atmosphere: globally-negligible inorganic exchanges take place from water bodies and soils to air (e.g., as sea spray and dust, respectively) and back (as P-deposition).

The two organic P-cycles are terrestrial and aquatic, respectively. On land, phosphorus cycles from soils to plants and other microorganisms (*uptake*, primarily of dissolved phosphorus), then to herbivorous organisms, and finally to other animals through food-web interactions. Phosphorus goes back from the organic to the inorganic form through the mineralization of dead animals and excretions performed by bacteria. In water, dissolved bioavailable phosphorus is uptaken by algae and bacteria, and then enters the food web. As on land, bacterial mineralization of dead organisms, excretions and exudations closes the cycle. Diffusion regulates the P-transfer between the water column and pore water in the sediments, while adsorption/desorption control the transfer between sediments and pore water. These processes are very important, as sediments are a major source of phosphorus in aquatic ecosystems, since sediments receive a large flux of this element through settling organisms, excretions, and so on. The only flux of phosphorus from oceans to land is represented by droppings from seabirds feeding on fish (guano deposits in Pacific coasts can be large) (Liu and Chen, 2008), and by the anthropic flux of phosphorus contained in fishery catches. As discussed, the influence of man on P-fluxes reaching soils and water bodies, and hence on the two organic P-cycles, is great, due to phosphorus contained in fertilizers (based on phosphorus extracted from rocks, and to a much lower extent on recycled organic phosphorus, e.g., manure application) and wastewaters. Crop harvesting is a significant anthropogenic P-output from cultivated soils; man can also indirectly favor P-emissions from soils, by accelerating soil erosion (Liu and Chen, 2008).

13

Photosynthesis

Photosynthesis is the production of organic molecules (*synthesis*) based on carbon dioxide, water, and sunlight's energy (*photo* is the Greek word for light), which is performed by plants, algae, and some bacteria, such as cyanobacteria and green–purple sulfur-oxidizing bacteria. Photosynthetic organisms, termed photoautotrophs because they can capture light and use it to fix the inorganic carbon from CO_2 into high-energy organic compounds stored in their bodies (autotroph means *self-feeding*), are fundamental in ecology because they represent the energetic and carbon basis of all ecosystems. No other organism can synthesize organic carbon compounds based on inorganic carbon, with the exception of chemoautotrophs using chemical reactions as a source of energy instead of light, such as bacteria living nearly thermal vents under the sea (whose global productivity is however negligible). Therefore, predation upon photoautotrophs and the consequent incorporation of organic carbon in the predator tissues are the ways by which newly synthesized organic carbon from CO_2 can enter the food web. Similarly, photosynthetic organisms represent the ultimate source of energy for ecosystems, since they can fix the energy coming from the sun, that is, outside of the earth. Most of biological productions on the earth can be traced back by means of predator–prey interactions to the energy and carbon fixed during photosynthesis and hence to the sun's energy (see Chapter 15; notice that also the so-called photoheterotroph organisms use sunlight as a source of energy, but they use organic and not inorganic carbon as their carbon source).

Besides being fundamental for the growth of plants, algae, and related food webs, photosynthesis is a key ecological process for another reason: it can release oxygen. As detailed below, photosynthesis requires an electron donor to take place; in the most common type of photosynthesis, such electron donor is water, and this leads to the production of oxygen according to the following overall reaction:

$$6\ CO_2 + 6\ H_2O + \text{Photon energy} \rightarrow C_6H_{12}O_6 + 6\ O_2 \qquad (13.1)$$

Equation 13.1 shows how photosynthesis modified the chemical composition of the earth's atmosphere, over billions of years, by producing free oxygen whose concentration allows the survival of aerobic organisms representing the most common life forms on earth. It also highlights that photosynthesis is tightly connected to the biogeochemical cycles of oxygen and carbon (Chapter 12) and thus becomes a key process to understand the ongoing climate change. Besides the oxygenic (aerobic) photosynthesis exemplified by Equation 13.1, anoxygenic (anaerobic) photosynthesis is also possible, although it is less common and can be inhibited by the presence of oxygen. For example, green–purple sulfur-oxidizing bacteria can use H_2S as an electron donor instead of H_2O, and thus they do not release oxygen during photosynthesis.

According to Equation 13.1, water and sun's energy are combined during photosynthesis to transform carbon dioxide into glucose (*carbon fixation;* then, more complex organic carbon compounds can be synthesized by the photosynthetic organism), yielding oxygen as a by-product. Thus, this process requires water and sunlight, however, not all solar radiation is available to perform photosynthesis. Only photons characterized by wavelengths in the (near) visible range of 400–700 nm, which is called photosynthetically active radiation (or PAR), can be taken up by plants through peculiar pigments, the most famous being the green pigments termed chlorophyll. Indeed, chlorophyll-*a* concentration is commonly used as a (rough) proxy for algal biomass in aquatic ecosystems. Chlorophyll is found in chloroplasts, which are cellular organelles specialized in taking up light and then performing photosynthesis. In plants, most photosynthetic activities take place on leaves.

The chemical energy stored by photosynthesis into glucose is continuously used by the metabolic process of *cellular respiration*, which consumes oxygen and takes place both in the dark and in the light, unlike photosynthesis. During respiration, glucose is broken down using oxygen as oxidant, while water and carbon dioxide are released, making its overall effects opposite to photosynthesis.

Equation 13.1 sums up the net effect of photosynthesis, but is actually the overall result of two sets of enzymatic reactions (Lambers et al., 2008) the light-dependent (or "light") and the light-independent reactions, also known as "dark" reactions. In light-dependent reactions, sunlight's energy is captured by chlorophyll and is used to split water molecules into molecular oxygen (which is then released), hydrogen protons, and free electrons. This drives a flow of electrons and protons that ultimately lead to the conversion of sun's energy into chemical energy in the form of the molecules adenosine triphosphate (ATP) and nicotinamide adenine dinucleotide phosphate (NADPH), which is obtained by the reduction of the $NADP^+$ molecule. The light-independent reactions take place both in the light and in the dark (actually, this is not entirely true as some of the enzymes involved in these reactions need light to be activated) (Lambers et al., 2008). The energy contained in ATP and NAPDH molecules is used to fix the inorganic carbon from CO_2 into glucose and other organic compounds, through a set of reactions called carbon reduction or Calvin cycle, which involves enzymes such as RuBisCO. ATP provides energy to the reaction, while NAPDH molecules provide the electrons needed to reduce CO_2, being transformed back into $NADP^+$.

Several environmental factors can influence the rate of photosynthesis, as discussed in Chapter 15. Equation 13.1 shows that a scarce supply of water,

CO_2, and light can slow down the photosynthetic process. For example, CO_2 supply can be limited by the rate of its diffusive dynamics in water or into the leaf, and by the resistance that diffusive motion encounters (Sections 7.2 and 7.7). In plants, CO_2 is up taken through leaf pores termed stomata through which transpiration of water from the plant to the external environment takes place; thus in dry environments, a trade-off arises for the plant, which has to balance the process of inorganic carbon uptake and that of water loss. In some primary producer species, scarcity of CO_2 can cause O_2 to bind with the RuBisCO enzyme instead of CO_2 in a process called *photorespiration*, which reduces CO_2 fixation, lowering the efficiency of photosynthesis, and also decreases O_2 production. This process is enhanced by the high availability of oxygen and light and low availability of carbon dioxide, for example, by dry and hot climate, when stomata close to stop water loss, and all internal carbon dioxide is used by photosynthesis.

Light can only be absorbed in the visible spectrum and changes across the day and the year (Section 8.2), hence its availability can be a limiting factor for photosynthesis. However, the relationship between photosynthetic rate and light is not always positive but rather has the shape of an optimum-curve, because excess solar radiation can damage the photosynthetic organism. A number of strategies to dissipate or avoid excess light exist (e.g., Lambers et al., 2008; Li et al., 2009), for example, chloroplasts moving to avoid incoming photons or algae changing their position in the water column. Photosynthesis is a set of enzymatic reactions so that its rate is positively influenced by temperature, but, similarly to light, the photosynthesis–temperature relationship has an optimum shape, since at too high temperature enzymes deactivate and, in the case of terrestrial plants, water availability decreases and water loss through stomata increases. Response to pH can be complex: low environmental pH values correspond to an increased availability of CO_2 in water (the equilibrium is shifted away from the carbonate ion), meaning more inorganic carbon supply for photosynthesis; on the other hand, strongly acidic pH and, in general, pH outside the typical environmental range could negatively influence photosynthetic rate, for example, by affecting the activity of enzymes, which are characterized by an optimum pH range. Finally, nutrient limitation typically affects the growth of photosynthetic organisms (Sections 12.5 and 12.6; Chapter 15), since for example chlorophyll is high in nitrogen while ATP contains both phosphorus and nitrogen.

It is important to realize that the relationship between photosynthesis and environmental conditions is not constant over time, but organisms can react to changed external conditions and stressful environmental states, such as reduced incident light and CO_2 availability, by undergoing morphological and physiological adjustments within their own lifetime—a phenomenon known as *acclimation*—and by showing evolutionary responses (that is, genetic changes) in their population—a phenomenon known as *adaptation* (Lambers et al., 2008).

14

Growth

14.1 Introduction

In ecology, the term *growth* of an organism or population of organisms refers to the chain of metabolic processes resulting in the increase of organism weight or number of individuals. If x is a variable representing the concentration of organic material (biomass) of a given type, the speed of growth is expressed by $\mathrm{d}x/\mathrm{d}t$ (see, e.g., Section 6.4). Unlimited unconstrained growth, with constant growth rate μ, may be described by a linear growth model as

$$\frac{\mathrm{d}x}{\mathrm{d}t} = \mu$$

Of course, such an oversimplified model has little ecological meaning. Indeed, growth depends on several species-specific characteristics and environmental factors, such as age, temperature of the environment, ration of the food ingested, and body size (see Figure 14.1). In general, μ includes several of these factors, for example,

$$\mu = \mu(T, L, S, \ldots)$$

with T the temperature, L the light, and S the substrate (e.g., nutrients). This function is usually expressed as a composition of various limiting factors (see Chapter 15).

Typically, the limiting effect of the substrate availability is described as an enzymatic interaction (Section 9.3) and is taken into account with a Michaelis–Menten approach:

$$\frac{\mathrm{d}x}{\mathrm{d}t} = \mu(S) = \mu_{max} \frac{S}{k_s + S} \tag{14.1}$$

The limitation's characteristic is provided by the substrate's semisaturation concentration k_s and the maximum growth rate μ_{max} (Figure 9.2). In the case of primary production, a similar approach is employed for including the effect of light limitation on growth. As discussed in Section 9.3, the kinetics

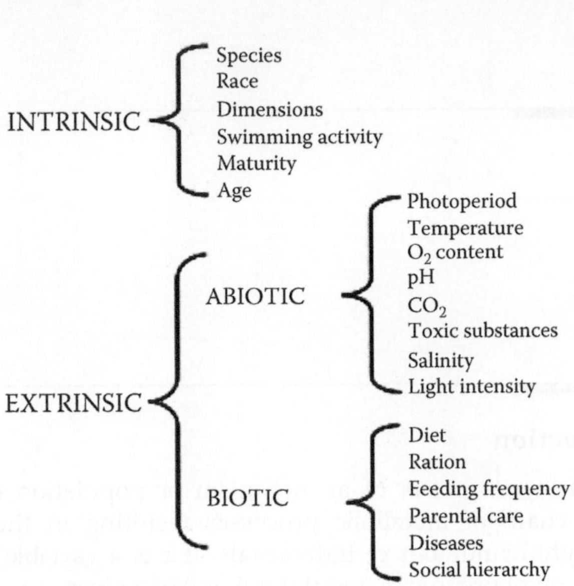

FIGURE 14.1
Intrinsic and extrinsic factors affecting the growth of an organism.

of enzymatic reactions is influenced by several factors, such as substrate and enzyme concentrations, physical conditions (for instance the temperature), and presence of activators or inhibitors. Besides these factors, which may be referred to as extrinsic, biological systems growth rate depends on intrinsic factors as well (Figure 14.1), of which the size of the body or population is a fundamental one (being often the other factors amenable to this one). Indeed, by including in the model the fact that the rate of production will depend on the quantity of products too (e.g., bacteria usually reproduce by simply splitting in two), the exponential or Malthusian growth model is obtained:

$$\frac{dx}{dt} = \mu x \tag{14.2}$$

This model describes an unlimited exponential growth. However, even if this model may be applicable to the early stages of growth, from experimental observations on microbial communities it is evident that unlimited growth is unrealistic. Actually in real populations, there is an upper limit to the number of individuals the environment can support. Ecologists call this limit the *carrying capacity* of the system, x_0. The carrying capacity represents the mathematical formalization of the biological phenomenon of intraspecific competition, which accounts for the fact that the members of the same species usually strive for the same resource in an ecosystem, such as food, light, nutrients, space, and so on. A practical way of describing this phenomenon is by considering the growth rate dependent on the number of

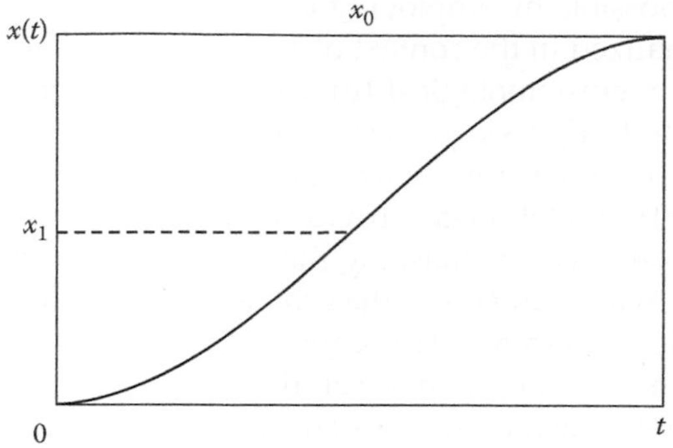

FIGURE 14.2
A sigmoid function describing limited growth. x_0 is the carrying capacity while the point of inflection $x_1 = x_0/2$ corresponds to the maximum growth rate.

individuals (alias the biomass), in such a way that the rate slows to zero when the population approaches the carrying capacity:

$$\mu = r\left(1 - \frac{x}{x_0}\right) \tag{14.3}$$

where the intrinsic growth rate r includes all those other relevant factors that are not accounted for by the carrying capacity, for instance the temperature. From Equations 14.2 and 14.3 the *Logistic growth model* is obtained:

$$\frac{dx}{dt} = rx\left(1 - \frac{x}{x_0}\right) \tag{14.4}$$

This latter model shows a substantial agreement with experimental data obtained while studying microbial population dynamics. The general solution of Equation 14.4 is the classical *sigmoid* curve represented in Figure 14.2.

$$x(t) = \mathrm{k}\frac{1}{1 + e^{\phi_r(t)}} \tag{14.5}$$

where $\phi_r(t)$ is a generic parametric function of the growth rate (e.g., $\phi_r = -rt$) and k a constant encompassing the carrying capacity x_0 and the initial condition $x(0)$. In particular if, $x(0)$ is smaller than x_0 the shape of the curve is that of Figure 14.2.

14.2 Metabolic Growth Model

The logistic growth model (Equation 14.4) is a general growth model that can be constructed on the basis of metabolic energy flow analysis. This approach, as discussed in Chapter 4, allowed von Bertalanffy to propose in the 1930s its revolutionary mathematical model of metabolic growth for living organisms.

This result was possible by employing energetic flow analysis that may be consistently formalized in the context of system theory. The very interesting aspect of this new epistemological thinking is the generality of the metabolic paradigm. Indeed, basic principles of physics, chemistry, and biology are all encompassed in a common framework by the concept of metabolism (Brown et al., 2004). Metabolism provides a conceptual connection for the physiological processes characterizing the biology of all living things, spanning from individual organisms to the ecology of populations, communities, and finally whole ecosystems. This common framework allows investigating analogous processes taking place at different scales, making evident the underlying general principles or laws governing the nature of life. The sake for general relationships of body size with anatomical structures, physiology, and furthermore behavior, in mathematical biology translates to search scale invariance for universal properties (or, as anticipated in Section 6.2, *allometric principles*).

Metabolism refers to a complex network of biochemical enzymatic reactions regulating the concentrations of substrates and products and the rate of energy conversion. An indicator of metabolic activity is the *metabolic rate*, which is defined as the energy consumed by a living organism (or a population) per unit time (watts). This quantity is a fundamental biological one, because it encompasses all the flow rates of energy uptake, transformation, and allocation.

Metabolism is usually subdivided into two separate contributions, *anabolism* and *catabolism*:

$$\text{Metabolism} = \text{Anabolism} + \text{Catabolism}$$

Catabolism accounts for the energy consumed for sustaining the metabolic processes that are essential for life, for example, to harvest energy in cellular respiration. Instead, anabolism refers to the energy converted to construct cell components such as proteins and nucleic acids. This quota of energy is used for growth. If x is a variable representing the concentration of organic material (biomass) of a given type, growth may be written as

$$\frac{\mathrm{d}x}{\mathrm{d}t} = \text{Anabolism} = \text{Metabolism} - \text{Catabolism} \tag{14.6}$$

In the realm of the energy/mass conservation principle (Section 2.3), Equation 14.6 is the continuity equation analogous of Equation 2.1. As stated in Section 2.4, writing such a relation is useful only if the assumption of the mass energy equivalence is accepted. This assumption may be revised in terms of equal chemical composition of the organic matter, as discussed in Chapter 9 (Redfield, 1958). In these hypotheses, the flow paradigm (Section 6.4) is straight forwardly applied to describe metabolic processes. The continuity Equation 14.6 may be used to build a general metabolic description of the energy flows through a living system. Stated that Anabolism = $\mathrm{d}x/\mathrm{d}t = G$ and the energy of consumed food C is a proxy for the metabolic rate, applying Equation 14.6 to fish growth, anabolism may be rewritten as

$$G = C - F - U - R = \text{IN} - \text{OUT} \tag{14.7}$$

$$R = R_s + R_d + R_a$$

where the symbols are exemplified in Figure 14.3, that describes a likely conceptualization of the energy flowing through an organism (e.g., fish).

The term IN in Equation 14.7 represents the system's energy intake (consumed food), and OUT represents the energy output that is subdivided into three terms: undigested food, feeding catabolism, and fasting catabolism. While fasting catabolism accounts for the losses due to the metabolic processes sustaining an organism that is not feeding, the other two terms represent the energy lost for the processes of feeding (e.g., digestion, hunting, or discarded food). From Equation 14.7:

$$\frac{dx}{dt} = \text{IN} - \text{OUT} = C - [\text{FAST} + \text{FEED} + \text{ND}] \tag{14.8}$$

A significant record of experimental observations shows that all those terms are differently proportional to a power of the quantity of biomass x. In particular, those related to nutrition have the same dependence, that is, $C \sim x^n$, $\text{FEED} \sim x^n$ and $\text{ND} \sim x^n$, while fasting, catabolism is generally described by a different power, $\text{FAST} \sim x^m$. These relations may be explained by assumptions concerning the system's structure and topology. For example, the food-adsorbing surface is often assumed proportional to x^n (Jørgensen and Bendoricchio, 2001). Hence, Equation 14.8 becomes a generalization of the logistic model (14.4):

$$\frac{dx}{dt} = hx^n - kx^m \tag{14.9}$$

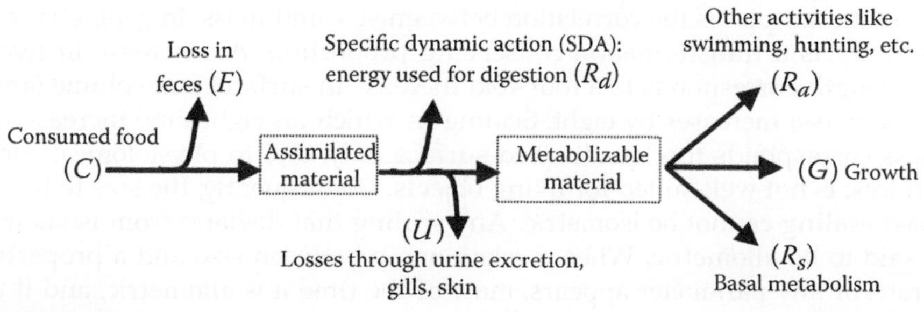

FIGURE 14.3
Schematization of metabolic energy flows in an organism.

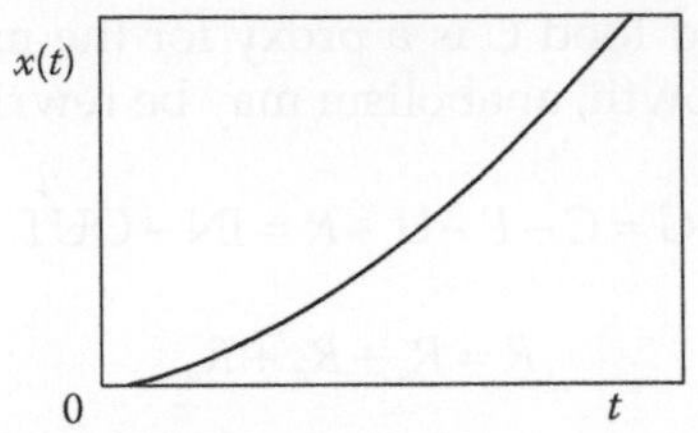

FIGURE 14.4
Solution of the general metabolic model for unlimited growth ($m < n$).

where h is a parameter incorporating all the constants of proportionality of the terms related to nutrition, n the scaling exponent for these terms, k is the coefficient of fasting catabolism, and m is the exponent of fasting catabolism. Equation 14.9 is derived starting from the well-known Ursin model for fish growth (Ursin, 1967) whose rationale is summarized in Equation 14.8. However, this equation may be employed with large generality to describe limited/unlimited growth in most living systems. If $m > n$ (limited growth), the solution of Equation 14.9 is a sigmoid function like in Figure 14.2. If $m < n$ (unlimited growth), the solution has a biological meaning only for the earlier stages of growth (Figure 14.4).

The basic assumptions underlying this model are energy conservation, mass–energy equivalence, and equal chemical composition of biomass and substrate. Further assumptions on proportionality relationships of body size with anatomical structures or physiological functions may shed some light on allometric principles entailing universal properties.

14.3 Allometric Principles

Living objects show an impressive variety of sizes, shapes, and functions. However, despite this astounding heterogeneity, notably regularities emerge suggesting the existence of an underlying general principle. One of the most evident examples is the correlation between size and mass. In geometry, an isometry is a transformation conserving proportion. An increase in two-fold length corresponds to a four-fold increase in surface, and volume (and hence mass) increases by eight. Scaling in which an eight-time increase in mass corresponds to a quadrupled surface only, due to physiological constraints, is not well-suited for living objects. Consequently, the size to body mass scaling cannot be isometric. Any scaling that deviates from isometry is said to be allometric. When a relationship between size and a property, a rate, or any parameter appears, most of the time it is allometric, and it is referred to as an *allometric principle*. Therefore, one may say that allometries are relationships between dimensions of organisms (and changes in the relative proportions of these dimensions) and changes in absolute sizes. In nature, allometries exist between individuals of the same species (intraspecific) and also among different species (interspecific). These relations may span over several orders of magnitude in organisms' size. Let C indicate a property, a rate, or a parameter (e.g., the metabolic rate), then the allometric principle may be written as

$$C = C_o\ x^n \tag{14.10}$$

where n is the scaling exponent, x the body mass, and C_o a constant.

Since the 1930s, the ecological literature reports plenty of allometric relationships and the most famous is between the metabolic rate and the weight. The parameters n and C_o in Equation 14.10 depend on the pertinent mechanisms of regulation at play (e.g., for temperature endothermy or poikilothermy). For mammals and small birds, that are all endotherms, the relation is depicted in Figure 14.5, and the exponent term n displays a value close to 3/4 or 0.75. This value can be justified on the basis of topological and physiological constraints (Brown et al., 2004). The analysis of the same topological constraints explains that if the scaling were isometric, the exponent would have been 2/3 or 0.67. Such relations are very useful as they make possible to foresee the metabolic rate of organisms by knowing its weight.

Indeed, several other allometries are known, which are not fully understood. For instance, for fishes, respiration R, food consumption

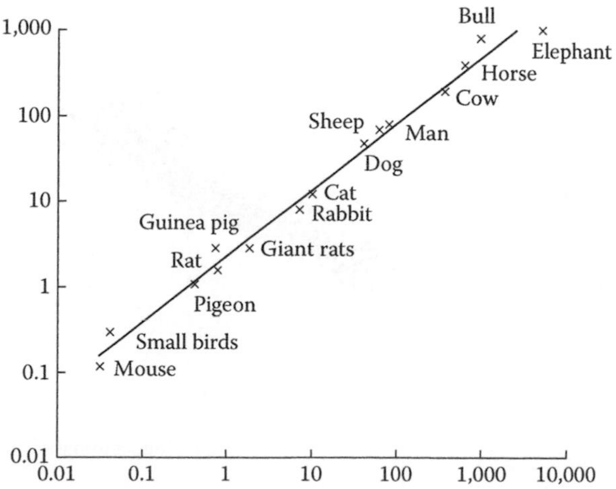

FIGURE 14.5
Logarithmic plot of metabolic rate (watts) against body mass (kg) for small birds and several mammals (Benedict, 1938).

C, and excretion are shown to be dependent on the body weight x (Jørgensen, 2008):

$$R = R_o\ x^m$$

$$C = C_o\ x^n$$

$$\text{Excretion} = k_o\ x^s$$

where the values of the scaling exponents m, n, and s are the expressions of the universal underlying constraints, and their values are the subjects of heated debate among scientists.

Generally, for a variety of different organisms, similar relations are found for the uptake of toxic substances (e.g., heavy metals like Cr or Cd) and the relations linking other specific size indicators like some organismic length, to generation time (or life expectancy).

Also, while studying food webs (see Section 19.5), this approach suggests some speculations (and occasionally well rooted explanations) that may be inferred by searches on scaling relations. For example, the relationship between consumption and biomass for 1599 consumer compartments from 56 aquatic trophic flow networks relative to water ecosystems located all over the world is shown in Figure 14.6. Despite the diverse ecosystems considered (fresh, transitional, and marine waters), exhibiting differences in latitude, depth, surface, temperature, biomass, and primary productivity, a general scaling relationship exists in aquatic trophic network models, linking

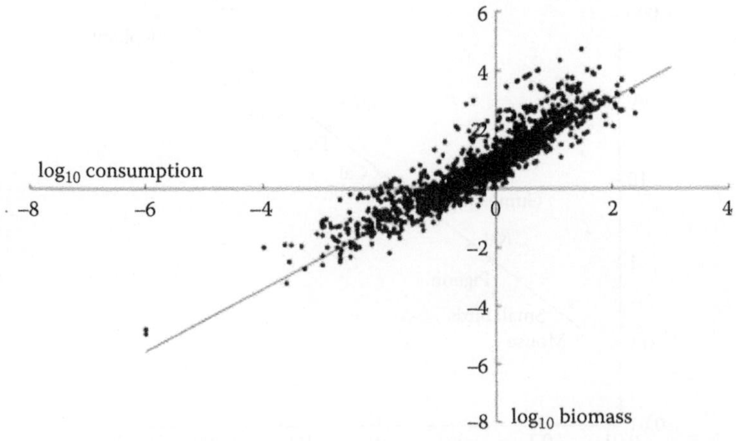

FIGURE 14.6
Logarithmic plot of the relationship between consumption and biomass for 1599 consumer compartments from 56 aquatic trophic flow networks. Data are well approximated ($R_2 = 0.82$) by a common power law and lies in a remarkably narrow window on the plot, despite clear ecological differences among the analyzed ecosystems (unpublished data).

the food consumption of a compartment (e.g., populations, taxonomic or functional groups) to its total biomass.

In conclusion, scaling appears in living systems whenever similar constraints on processes occurring at different scales are operating. Apart from the basic assumptions underlying the metabolic model that are of thermodynamic (energy conservation, mass–energy equivalence) or biochemical (e.g., equal chemical composition of biomass and substrate) nature, other constraints of topological or physiological type are patterned by allometries. Even if these allometries are not explained in terms of fundamental principles, and even if the discussion on the methodologies to employ for evaluating trend lines is still animated, estimating and studying scaling exponents is of primary interest for the manifest potential of proving general properties of living systems. Moreover, from a practical point of view, allometric relations allow the ability to foresee the value of key physiological parameters in an indirect way, which may be difficult or impossible to estimate directly.

15

Primary Production

15.1 Introduction

Primary production is the synthesis of organic carbon compounds from the inorganic carbon source represented by carbon dioxide, which is performed by organisms known as autotrophs or primary producers. In general, ecologically important autotrophs are photoautotrophs, which can fix inorganic carbon using the sunlight's energy in the process of photosynthesis (see Chapter 13), however, primary producers also include chemoautotrophs, that is, organisms capable of fixing CO_2 based on the energy of chemical reactions, such as bacteria living near hydrothermal vents on the bottom of the oceans. Algae (microalgae, macroalgae, sea grasses, etc.), plants, and some particular bacteria (photosynthetic cyanobacteria, also known as blue-green algae) are the common examples of primary producers in aquatic and terrestrial ecosystems. The term *phytoplankton* indicates the fraction of plankton (organisms in the water column that float or, anyway, cannot swim against currents) made up of primary producers, such as microalgae and cyanobacteria, which yield most oceanic primary production.

Primary producers use the organic carbon and energy they fix to fuel their metabolism (Section 14.2), that is, to support their respiration and to promote their own growth or production of biomass; indeed, in some cases, the term primary production is used to indicate such fixed energy or produced biomass. Primary production is then consumed by herbivorous organisms, and consequently carbon and energy enter the food web of predator–prey trophic interactions (Chapters 16 and 19) reaching other living organisms, which can only assimilate organic carbon and cannot fix CO_2. Hence, all the organic carbon flowing in the biosphere can be traced back through trophic interactions to the inorganic carbon fixed by primary producers, and primary production represents the ultimate source of carbon to all other living organisms. Thus, primary production can be considered as the basis of earth's food webs and ecosystems. Consequently, primary production is also a direct or indirect source of food for human beings, who exploit such resource intensely. For example, the primary production required to sustain the catches of world's fisheries directly or indirectly is about 8% of the global aquatic primary production (Pauly and Christensen, 1995). In addition, men directly exploit primary production for other uses such as obtaining wood and fibers, but can also enhance it by enriching terrestrial and aquatic ecosystems in nutrients, that is, through the application of fertilizers, which when washed out by rains and transported through groundwater lead to the eutrophication of water bodies.

Primary production can be divided into gross and net primary production (GPP and NPP, respectively). GPP is simply the total amount of carbon fixed through photosynthesis, while NPP is equal to GPP subtracted the carbon lost as CO_2 by primary producers during respiration. A common currency for GPP and NPP is $gC\ m^{-2}\ y^{-1}$, that is, a carbon flux per unit area per unit time.

15.2 Limitation

All organisms need to acquire substrates and resources to live, grow, and thrive, and primary producers make no exception. For photosynthesis, the availability of water, visible light, inorganic carbon, and other nutrients, such as phosphorus and nitrogen, is required (Chapter 13). The growth of primary producers can also be influenced by external environmental conditions (temperature, pH, irradiance levels, etc.) that affect the rate of their physiological processes. For instance, it is well known that temperature influences enzymatic reactions and metabolic rates. Given the importance of primary producers for terrestrial and aquatic ecosystems, in ecology it is common to study their growth in relation to resource availability and environmental conditions. In particular, research is focused on how these factors can limit the maximum achievable growth, which corresponds to perfect or optimal conditions for growth. The most common types of the numerous resources and environmental factors affecting the growth of autotrophs are reviewed in the next sections, with a focus on photosynthetic organisms. The growth limitation due to more than one resource (colimitation) is also discussed. Although not treated here, the role of pH has been briefly introduced in Chapter 13 concerning photosynthesis.

15.2.1 Rationale behind Michaelis–Menten

For many types of resources, such as nutrients, it makes sense to assume that their scarcity limits primary production, and that increasing resource availability would reduce such limitation without harming primary producers. Therefore, the relationship between resource availability and primary production will always be positive and, at very high resource levels, growth will tend to a maximum value, with further increases in the resource level having little or no effect on it. One common description of such a situation is represented by Michaelis–Menten-like* kinetics (Section 9.3): Equation 9.14 is used to describe the growth rate μ of primary producers as a function of resource availability C in the external environment, analogously to enzymatic reactions that are common in biology.

$$\mu = \mu_{MAX} \frac{C}{k_C + C} \tag{15.1}$$

Thus, the relationship between growth rate and resource availability is saturation-like, and only a fraction of the maximal growth rate μ_{Max} is realized depending on the value of C and of a semisaturation constant k_C, which is resource specific. C is a proper measure of resource availability such as a concentration, for example, the concentration of bioavailable nutrient forms in the environment, or a flux per unit surface (e.g., of visible light). The growth rate μ can depict the increase in the number of individuals (abundance) or total biomass in a population or a community of primary producers (e.g., phytoplankton), or even the growth in size or mass of a single organism (e.g., a tree).

15.2.2 Nutrient Limitation

The scarcity of bioavailable nutrients in waters and soils can limit the growth of primary producers, as the nutrients perform many different physiological functions. The best-known and common limiting nutrients for primary production are nitrogen and phosphorus, which are found in many key biochemical molecules and physiological apparatuses. For example, the machinery that primary producers use to acquire resources, such as chlorophyll and proteins, is high in nitrogen, while growth machinery such as ribosomal RNA is rich in both phosphorus and nitrogen (Arrigo, 2005). Nitrogen is the typical limiting nutrient in marine systems, while in freshwater systems, phosphorus is often the limiting nutrient. Although CO_2 is needed to perform photosynthesis, carbon is rarely the limiting nutrient, since CO_2 is quickly cycled in ecosystems due to its release during cellular respiration in living organisms (see Section 12.4 regarding the C-cycle). Other less-common limiting nutrients include: silicon, which can limit the growth of phytoplankton termed diatoms, characterized by silicon-made shells and generally blooming in spring; and iron, a micronutrient (as opposed to macronutrients needed in larger quantities such as N and P), limiting photosynthesis in the case of the phytoplankton of the open oceans (Lavery et al., 2010).

Nutrient limitation theory dates back to the first half of the nineteenth century and Liebig's law of the minimum, which is based on the assumption that the chemical composition of organisms is roughly constant, meaning that the key

* Although we have written about "Michaelis–Menten kinetics," the phrase "Monod equation" is commonly used when applying the famous saturation-like relationship to the description of growth processes, based on Monod's work on the growth of microbes; when the focus is on uptake processes (Section 15.4), the phrase "Michaelis-Menten equation" is used to highlight the analogy with enzymatic processes (Section 9.3) (Healey, 1980).

elements (C, H, O, N, P, Si, S, etc.) are found in the bodies of primary producers according to constant stochiometric ratios. According to Liebig's law, the limiting element is the scarcest nutrient with respect to the balanced nutrient composition ratio, and the growth is positively related to such a limiting nutrient and unrelated to other ones. For example, the chemical composition of phytoplankton is often described by a ratio of C:N:P of about 40:7:1 by weight, the *Redfield Ratio*, and thus P is limiting if $N/P > 7$ by weight, while N is limiting if $N/P < 7$. Although it is being realized that phytoplankton species (and all primary producers, of course) actually differ markedly in their stochiometric composition, reflecting different growth strategies (as discussed, for example, resource acquisition machinery is high in N), and that the Redfield Ratio appears rather to reflect the average of the current typical algal composition of aquatic environments (Arrigo, 2005), the general idea behind Liebig's law remains valid.

The simplest approach to describe the nutrient limitation is to focus upon the availability of nutrients in the external environment, for example, using phosphorus or nitrogen concentration in soil or water in Equation 15.1. When doing so, it is important to realize that nutrients cycle through different forms (see biogeochemical cycles of N and P in Sections 12.5 and 12.6) but not all nutrient forms are bioavailable; moreover, not all bioavailable forms are up taken by primary producers at the same rate. Hence, the choice of the appropriate forms of nutrients to be taken into account in nutrient limitation studies, monitoring campaigns and models should be done carefully.

An exclusive focus on the external nutrient concentration can potentially lead to misleading conclusions about limitation, since it is the internal pool or intracellular concentration of nutrients which is actually relevant for the growth of primary producers. For example, the concentration of nutrients in the environment could be low, but primary producers could have stored enough nutrients in their bodies to sustain growth efficiently over time. This situation can be described by adapting Equation 15.1 to describe the two processes that lead to limitation, that is, the nutrient uptake from the environment, limited not only by external nutrients but also by the internal pool (if high, uptake will be low) and growth, limited only by the internal pool (Jørgensen and Bendoricchio, 2001). Another example is that nutrient concentration does not necessarily reflect the rate of nutrient cycling and fluxes, as in the case of *regeneration* fluxes from non-bioavailable to bioavailable forms (e.g., from sediments to water column). If the rate of nutrient regeneration matches that of a primary producer uptake, that is, when nutrients are up taken as soon as they become bioavailable, then even an extremely-low bioavailable nutrient concentration can effectively sustain growth.

15.2.3 Temperature Dependence

Temperature affects physiological rates in all living organisms in multiple manners through its effects on chemical reactions (Chapter 9), for example, enzymatic processes, and on physical processes, such as diffusion (Section 7.2) and mass transport between phases and at interfaces (Section 7.7). Primary production is strongly temperature dependent since, for example, this forcing influences the rates of photosynthesis (Chapter 13), respiration (Section 15.3), and evapotranspiration (Section 8.4). For single species of algae and plants, the relationship between growth and temperature is an optimum curve, reflecting that increasing temperature favors primary production but too high values can damage plants, reduce water availability, deactivate enzymes, and so on. Such curve is acclimation and species dependent. Over a short range of temperature, such relationship can be approximated by linear or exponential Arrhenius models (Jørgensen and Bendoricchio, 2001) (Figure 15.1). Also, an exponential model can be a proper description of the temperature dependence of primary production in a whole community (Bowie et al., 1985), as different species are characterized by different optimal temperatures and consequently, as temperature increases, some species decrease but are replaced by other ones that thrive in warmer climates. Also, an exponential relationship between temperature and primary production reflects the process of acclimation (Chapter 13) to light (photoacclimation), in the case of phytoplankton. Warmer water bodies tend to be characterized by higher light availability, and algae become acclimated to such high light conditions by developing less chlorophyll, to avoid damages caused by absorbing too much light. In such a case, the efficiency of light exploitation per unit pigment by phytoplankton is high (due to the low chlorophyll concentration); that is, chlorophyll-specific primary production is high, hence the positive correlation between temperature and primary production (O'Malley, 2010). Indeed, it is important to remember that temperature can be correlated with light intensity when interpreting results in correlation and modeling studies.

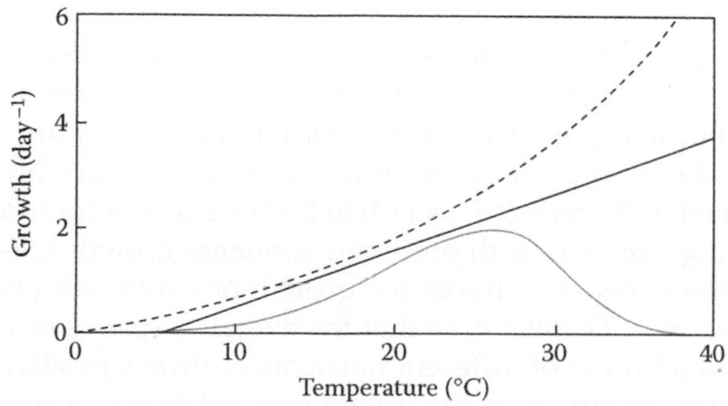

FIGURE 15.1
Examples of three common relationships between temperature and algal growth according to Jørgensen and Bendoricchio (2001): the linear, exponential, and optimum models (respectively, continuous black, dashed black, and gray line). The optimum model is generally the most realistic model for describing the behavior of a single species but also the most complex one from a mathematical point of view.

15.2.4 Light Dependence

Light availability is a fundamental requirement for photosynthesis, and the intensity of visible radiation is positively related to primary production. A typical description of this relationship is through the Michaelis–Menten formulation of Equation 15.1, where resource availability C is light intensity. However, the absorption of too much light can damage primary producers (Chapter 13) and reduce photosynthetic rates (*photoinhibition*). Therefore, when the environmental range of light intensity values is wide or if, anyway, the highest observed values lead to photoinhibition, an optimum description of the relationship between light intensity and growth of primary producers can be appropriate (Jørgensen and Bendoricchio, 2001). As discussed in Chapter 13 and Section 15.2.3, the same photosynthetic species can display different productivity levels at a given light intensity, depending on the typical past levels of light in the ecosystem, a process known as photoacclimation. For example, algae acclimated to scarcely lit environments grow faster in low light conditions, but grow more slowly in high light conditions than those algae acclimated to well-lit waters.

15.2.5 Colimitation

The growth of primary producers typically does not depend on a single resource but on the availability of more than one, a process known as resource colimitation. Actually, this term is generally used to designate two different cases: one for colimitation due to multiple resources, for example A and B, if the increased availability of *either* A *or* B augment primary production (Case 1); but colimitation due to A and B can also indicate that only the *simultaneous* increase of *both* A and B boosts productivity while the increase of either A or B has no effect on growth (Case 2). When resources A and B are both nutrients, Liebig's law of the minimum (Section 15.2.2) is not applicable. Three common cases of colimitation in marine ecosystems can be identified (Arrigo, 2005). If only the combined addition of A and B affects phytoplankton growth (Case 2), *multinutrient colimitation* is taking place. In the case of two limiting nutrients, this means that both of them are at too low levels to allow uptake. An example is an N-limited ecosystem where N_2 fixer organisms, not limited by N, reduce P as PO_4 to limiting concentrations. Instead, if increases in single resources all positively influence growth (Case 1), it could be that each single resource limits the growth of a different phytoplankton species (*community colimitation*), so that the overall response of primary productivity to the addition of different nutrients is always positive, or, it could be that one resource enhances the uptake/assimilation of a second limiting resource (*biochemical colimitation*). An example of biochemical colimitation is that due to light and iron. If an ecosystem is light-limited, higher irradiance will increase the growth of algae. Since iron is employed in the light-absorbing machinery of phytoplankton, its addition could also increase the growth, again by acting upon light limitation (Arrigo, 2005).

Illustration 15.2.1

It is often discussed in lake management which nutrient is limiting the phytoplankton growth and thereby control the eutrophication. On average, the concentration of nitrogen in phytoplankton is seven times the concentration of phosphorus.

For a considered lake, drainage water from agriculture contains 20 times nitrogen as phosphorus, while the waste water discharged to the lake has only 5 times nitrogen as phosphorus. Below which ratio of waste water to drainage water, denoted *R*, would phosphorus not be the limiting nutrient?

SOLUTION

The following R would correspond to the value above which *P* is not the limiting element for phytoplankton growth:

$$R5 + 20 = 7(R + 1) \text{ or } 2R = 13; R = 6.5$$

If the ratio is less than 6.5, *P* becomes limiting, whereas if the ratio is more than 6.5, nitrogen is the limiting element.

15.3 Respiration

Nearly all living organisms but obligate fermenters respire, and respiration is one of the distinctive traits of life. Respiration is a catabolic process (Section 14.2); that is, it represents the part of metabolism devoted to breaking down (typically organic) substrates to release the energy needed by organisms to maintain internal organization and sustain life in general. Also primary producers respire, and their respiration is particularly important, because it makes a part of the carbon and energy fixed through photosynthesis (Chapter 13) unavailable to grazers, a fraction of (gross) primary production always ending up as respiration. Remarkably, respiration in primary producers takes place all the time, even at dark, unlike photosynthesis. The difference between GPP and primary producer's respiration is termed as NPP. We remark that, in primary producers, respiration has to be distinguished from the process of photorespiration.

Respiration is a complex set of several reactions (e.g., del Giorgio and Williams, 2005; Abedon et al., 2008) taking place at the cell level, hence called *cellular respiration*. Two types of respiration can be identified, aerobic and anaerobic, depending on whether oxygen or another substance is utilized as an electron acceptor in the process. Anaerobic respiration can be observed in simpler, more primitive organisms, for example, some bacteria. Actually many bacteria are facultative anaerobes (e.g., *Escherichia coli*), which means that when anoxic conditions develop they can switch from aerobic respiration, which is much more energetically-efficient, to anaerobic respiration or to fermentation (see also Section 12.3). Although facultative and obligate anaerobes are not as abundant as strictly aerobe organisms, they have ecologically important roles since they can live and thus exploit resources in hypoxic-anoxic waters and soils; they also play a key part in several biogeochemical cycles, for example, that of nitrogen in the case of denitrification (Section 12.5), or that of sulfur. Some anaerobic bacteria can reduce SO_4 to gaseous H_2S within the anoxic sediments and waters of aquatic ecosystems, and then such gas can move vertically reaching the upper part of the water body where it can be employed as an electron donor in anoxygenic photosynthesis by green–purple sulfur oxidizing bacteria.

Aerobic respiration is by far, the most common type of respiration, particularly in more complex, higher plants and animals (although anaerobic metabolism can also be carried on for brief periods in the case of oxygen shortage, e.g., during intense physical activity). In eukaryotes, aerobic respiration is primarily performed by organelles found within most cells and called mitochondria. From the biochemical point of view (Abedon et al., 2008), aerobic respiration involves the oxidation of an electron-donor substrate, which is typically organic, for example, carbohydrates, proteins, fats, cellulose (chemoorganotrophic aerobic respiration), but can also be inorganic in the case of chemolitotrophic respiration using oxygen (e.g., see *Nitrosomonas* and *Nitrobacter* bacteria in the nitrification process, Section 12.4). It also involves the reduction of O_2, which is the final electron acceptor. Four main stages can be identified, made up by multiple chemical and enzymatic reactions: glycolysis, when glucose is transformed into pyruvate molecules; oxidation of pyruvate molecules to form acetyl coenzyme A and CO_2; the Krebs or citric acid cycle; electron transport chain and chemioosmosis. They result in a flow of electrons from the substrate to the acceptor and, concurrently, in a flow of protons, which is used to synthesize adenosine triphosphate (ATP) molecules during the process of oxidative phosphorylation. The overall effect of aerobic respiration can be summarized as

$$C_6H_{12}O_6 + 6\ O_2 \rightarrow 6\ CO_2 + 6\ H_2O + \text{Energy} \tag{15.2}$$

In this exothermic redox reaction, an energetic substrate, here exemplified by glucose (previously fixed by photosynthesis in the case of primary producers, and derived from feeding in the case of consumers), acts as an electron donor, while oxygen is consumed, acting as an electron acceptor. The substrate is broken down through oxidation, and the contained energy is then stored into ATP molecules, while carbon dioxide and water are released as waste products. ATP molecules are then used to fuel metabolic activity, for example, for the purpose of maintaining and repairing existing organic tissues and of supporting life activities in general, for example, growth processes such as biosynthesis (e.g., see Amthor [2000], regarding plants).

Equation 15.2 highlights the key role of respiration in ecosystems and in the biosphere, this process representing a strong biotic link in the biogeochemical cycles of oxygen and carbon (Sections 12.2 and 12.3). Although respiration at the level of cells and organisms, for example, biochemical reactions, gas exchanges (O_2, H_2O, CO_2) with the environment, is relatively well-understood, scientific knowledge is not as clear regarding respiration at the scale of ecosystems, even though it is a very important component of the carbon balance in primary producers (ca. 35%–70% of C fixed in plants is respired) (Amthor, 2000). A scarce understanding of respiration, which represents the largest sink of organic matter taking place in the biosphere, as a biogeochemical and ecological process appears to be a major gap in our picture of the global carbon cycle (del Giorgio and Williams, 2005). Indeed, with regard to the cycling of oxygen, carbon and energy in ecosystems, the relationship between primary production and respiration is fundamental, because the two processes have opposite effects, in the case of both consumers and primary producers: during respiration, the organic carbon originally fixed by photosynthesis is transformed back into inorganic form, while oxygen is consumed. For this reason, the interaction between primary production and respiratory flows is an important factor shaping the functioning of ecosystems, and their relative magnitude has been used as a measure of ecosystem status. For instance, an important characteristic of aquatic ecosystems is the extent of the euphotic zone, which is the depth where the rate of primary production equals that of respiration. Below this depth, which is strongly influenced by the effect of turbidity on light (see Chapter 8), the ecosystem does not receive exogenous energy and carbon input and thus, no new food inputs for heterotrophs (except for vertical transport of organic matter through settling or resuspension, or organism migrations). This is just one example clarifying why the ratio of primary production to total respiration and the closeness of the value of this ratio to one have been identified as indicators of ecosystem maturity and system proximity to a well-developed, climax status (Odum, 1969, 1983a). This is because only in a well-organized ecosystem, all primary production can be efficiently exploited and used to maintain a great amount of living biomass, which corresponds to large respiratory flows. Indeed, respiration and biomass are positively related; for example, respiration r in heterotrophs can be expressed as a power of body size m through an allometric relationship (Section 14.3):

$$r \propto m^{\alpha} \cdot e^{-E/kT} \tag{15.3}$$

where the exponent α is generally found in or close to the range 2/3–3/4, E is called activation energy, k is the Boltzmann's constant, and T is temperature (e.g., Brown et al., 2004). Equation 15.3 also highlights that respiration and temperature are positively related. For example, in the case of algal growth, this relation can be described through the exponential Arrhenius function (Jørgensen and Bendoricchio, 2001).

15.4 Nutrient and Metal Uptake

As mentioned in Section 15.2, the scarcity of resources such as nutrients is a well-known phenomenon limiting the growth of primary producers. Nutrient scarcity in the external environment is not negative for primary producers *per se*: for example (Section 15.2.2), a low environmental nutrient concentration does not necessarily correspond to a poor nutrient flux obsorbed by algae, since a vigorous growth could simply be sustained by intense regeneration fluxes of nutrients from sediments that are immediately taken up from phytoplankton, thus permanently leaving few bioavailable nutrients free in the water column. This is just one example highlighting that the rate of nutrient uptake from the external environment is an important process in the study of primary producer growth. It is evident that uptake by primary producers is affected by nutrient concentration in water or soils, but other factors come into play, such as the internal quota of nutrients found in the primary producer body. For example, in case of single phytoplankton cells, uptake rate decreases to zero as the maximum possible internal concentration is achieved. The joint effect of external nutrient concentration C and internal nutrient quota q on uptake rate U can be accounted for by multiplying U by a proper limitation function (Jørgensen and Bendoricchio, 2001):

$$f(C,q) = \frac{q_{\max} - q}{q_{\max} - q_{\min}} \cdot \frac{C}{K_C + C} \tag{15.4}$$

where $q_{\max}$ and $q_{\min}$ represent the maximum and minimum possible internal quota, respectively, and K_C is the usual semisaturation constant in the Michaelis–Menten formulation of resource limitation (Section 15.2.1). All parameters depend on both the primary producer species and the nutrient studied. Equation 15.4 shows that the relationship of external nutrient concentration with uptake rate is commonly, empirically described as a saturation-like Michaelis–Menten process, although model parameters can also vary over the time depending on cellular physiology and acclimation (Chapter 13) (Smith et al., 2009). Such a relationship can be observed in the case of both algae and terrestrial plants, and resembles the relationship between external nutrient concentration and primary producer growth described in Section 15.2.2. This similarity is justified by the obvious link between the processes of uptake and growth, and it is worth noticing the analogy with enzymatic processes, that is, how such nutrient-based kinetics resembles the relationship of substrate availability (here, nutrient availability) with the rate of substrate consumption (here, nutrient uptake) and with the rate of creation of a final product (here, producer growth) in enzymatic reactions (Section 9.3). As in the case of growth, parameters of the Michaelis–Menten relationship can take an ecological meaning and be used to predict the outcome of competition (Chapter 19) for a limiting nutrient between algal populations (e.g., Healey, 1980).

The capability of plants to uptake nutrients such as N, P, and metals (Fe, Mn, Zn, Cu, Mg, Mo, etc.), even those with no use for growth and other biological functions, has several popular applications in engineering. One is the well-known use of wetlands to remove nitrogen and phosphorus from wastewaters and surface waters (Kadlec and Knight, 1996) to prevent the onset of eutrophication (Section 20.2) in surface water bodies, such as lakes, coastal lagoons, and seas. Also, plants are commonly used to remove toxic heavy metals and radionuclides from contaminated soils and effluents in a cost-effective manner, for example, when metals are found in low environmental concentrations. This kind of application is part of a set of ecological engineering techniques known as phytoremediation, in which plants are used to remove, degrade, or control wastes and pollutants (Salt et al., 1995; McCutcheon and Jørgensen, 2008). Plants used in the phytoremediation of metals should be able to bioaccumulate high metal concentrations, tolerate the toxic effects of such metals, grow quickly, and build up a great biomass. Although not strictly related to engineering, the capability of plants to uptake (or, sequestrate) inorganic carbon from the atmosphere has also important implications for society and the environment, due to the fundamental role of this process in the biogeochemical cycle of carbon and, consequently, in climate change (Section 12.4).

Indeed, a huge number of environmental factors other than nutrient (or metal) concentration can affect the uptake process. Trivially, nutrients (metals) should be in a bioavailable (readily available) form to be taken up (see Sections 12.5 and 12.6 for two common limiting nutrients, N and P). However, it is important to notice that plants can mobilize and uptake unavailable metals bound to soil in a number of ways (Salt et al., 1995; McCutcheon and Jørgensen, 2008), for example, by changing the pH in the *rhizosphere,* which is the soil near roots where uptake takes place. Also, to be taken up, nutrients and metals should obviously reach the primary producer, so that physical, chemical, and biological processes and environmental conditions that can affect mass transport and availability (e.g., to the plant roots) all potentially come into play; for example, advection and stratification in water bodies; diffusion; adsorption and desorption; chelation; ion exchange; presence of mass/electrochemical gradients; solubility; temperature; pH and redox potential; competition for nutrients; presence of symbiotic fixing organisms, as in the case of nitrogen (Section 12.5); the availability of other chemical elements (e.g., trace metal cofactors) that can affect nutrient uptake, as in the case of biochemical colimitation (Section 15.2.5) (Arrigo, 2005); and so on.

15.5 Excretion

Excretion is the emission of metabolic wastes and other useless materials for an organism into the external environment. All living beings excrete matter: feces, urine, exudates, and gases such as CO_2 and H_2O resulting from respiration (which is a type of excretion, but is considered as a separate process in this book and dealt with in Section 15.3). The ecological role of excretion is twofold. On one hand, it represents a loss term in a mass- or energy-budget describing the growth of an organism or a population. For instance, the growth of primary producers such as phytoplankton can be described according to the general equation (Jørgensen and Bendoricchio, 2001):

$$\frac{dA}{dt} = (\mu - r - ex - s - m)A - G \tag{15.5}$$

where the derivative over time of algal biomass A (e.g., Chl-a concentration) is a function of a gross growth rate μ, respiration rate r, rate of excretion ex, also called exudation in the case of phytoplankton cells, rate of settling s, a process leading to the death of algae when they settle out of the euphotic zone (Section 15.3), a rate m representing all other natural mortality sources but predation, and a flow term G representing predatory mortality due to grazers. This model highlights how a fraction of photosynthetic production (Chapter 13) is actually channeled into excretion flows.

On the other hand, excretion flows contain organic matter and carbon, nutrients and energy, and thus can be considered as a source of food for several organisms feeding on them, such as bacteria in the case of phytoplankton excreta. Indeed, bacterial and phytoplanktonic populations are generally tightly coupled in the so-called microbial loop of matter and energy flows (Chapter 17), and their biomass or production rates positively correlate, since bacteria feed on algal extracellular exudates and remineralize nutrients that, in turn, are taken up by phytoplankton. Exudation of organic matter in phytoplankton can reach high fractions of the total primary productivity, reportedly ranging from about 5%–40% (Lignell, 1990).

Excretion fluxes reflect the diet, habitat and habits of an organism, and if it has ingested or bioaccumulated from environmental media toxic substances (Chapter 18), for example, metals, it is possible that a fraction of them is found in excreta. This is why it can be important to consider excretion in ecotoxicological models. Excretion flows take place in different manners depending on the organism. Simpler organisms, such as phytoplankton, release exudates in the form of dissolved organic matter (DOM) through their cell membranes and walls, while other organisms often display dedicated organs, such as kidneys, or lungs in the case of respiration. Similarly to what happens for phytoplanktonic cellular membranes, more complex organisms can also excrete through their coverings, which act as exchange surfaces. Indeed, we all know that our body emits sweat through the skin. As another example, the release of gases by plants to the atmosphere during processes such as photosynthesis (O_2; Chapter 13), transpiration (H_2O; Section 8.4) and respiration (CO_2, H_2O; Section 15.3) is not limited to few spatial sites but can take place through numerous micropores called stomata. Plants also produce other kinds of excreta, such as resins.

Excretion is a metabolic process and, as such, its rate generally depends on the power of body size with an exponent lower than one and is positively related to temperature. Excretion can also be affected by the numerous physiological and environmental conditions and gradients that affect mass transport, for example, those driving diffusion (Chapter 7) (as in the case of gas exchanges across stomata and exudates diffusing around phytoplankton cells), osmotic dynamics, and so on.

15.6 Fixation

One fundamental ecological characteristic of primary producers is their capability to fix inorganic carbon (by definition) and nitrogen (in some cases) into organic forms, thus literally providing the material basis for food webs, since biological fixation is the only natural pathway by which these inorganic nutrients can become available to living organisms.* Carbon is fundamental for life, since this element is a major component of organic matter. The amount of carbon fixed during the process of CO_2 fixation, also termed carbon sequestration, is determined by the interplay of two processes, photosynthesis and respiration, already described in Chapter 13 and Section 15.3, respectively, and whose effects are opposite (although the two processes are not directly related, e.g., the physiological sites where they take place are different). Indeed, the former process fix carbon dioxide into organic matter, consumes water and release oxygen, while the latter one represents the cost of metabolism and is fueled by organic matter and oxygen, while (waste) products are water and carbon dioxide. Carbon fixation is a fundamental process to understand the C cycle and climate change (Section 12.4).

Biological fixation of atmospheric N_2 into ammonia represents the major input of bioavailable nitrogen to ecosystems, if neglecting anthropic sources such as fertilizers based on industrially-synthesized ammonia (Section 12.5). As for carbon, nitrogen too is necessary to living organisms, being a key component of amino acids, RNA, DNA, and chlorophyll. Nitrogen fixation is carried out by many different types of diazotrophic or N-fixing microorganisms, based on the enzymatic complex of nitrogenase; not all of these organisms are primary producers (Rascio and La

* Actually, other minor natural pathways from inorganic to organic forms exist for nitrogen; see Section 12.5.

Rocca, 2008). Indeed, some primary producers can fix nitrogen directly, as in the case of cyanobacteria or of green–purple sulfur anaerobic bacteria belonging to the genera *Chromatium* and *Chlorobium*, which all carry out photosynthesis. Cyanobacteria often have ecologically important roles, being ubiquitously found in most waters and soils. But, some primary producers can also fix nitrogen indirectly, as in the case of plants hosting symbiotic diazotrophic bacteria. About 90% of existing legume plants can host nitrogen-fixing bacteria in root nodules, often fixing noticeable N quantities, for example, 500 kg N ha^{-1} y^{-1} in the case of broad bean *Vicia faba*, with important applications in agriculture; other important cultivated crops such as some cereals (rice, maize, wheat, sorghum) and sugar cane can host nitrogen-fixing symbiotic bacteria (Rascio and La Rocca, 2008). A famous example of a symbiotic relationship involving N-fixation is represented by lichens, which consist in a fungus and an autotroph such as a cyanobacterium or a microalga.

The nitrogenase complex is made of two different enzymes, dinitrogenase reductase and dinitrogenase, both of which are deactivated by even small quantities of oxygen. Several nitrogen fixers perform oxygenic photosynthesis (Chapter 13), that is, produce oxygen, or are aerobic, thus they have developed strategies to avoid nitrogenase being inactivated by oxygen, such as performing nitrogen fixation in spatially-segregated sites (e.g., different cells) or at night. Also, nitrogen fixation is an expensive process from an energetic point of view, and it is not generally carried out if "cheaper" nitrogen forms are bioavailable, such as nitrates, nitrites, and aminoacids.

15.7 Mortality

Mortality is a process that can potentially limit primary production, and it can be useful to examine separately the effects of different sources of mortality, as highlighted, for example, by Equation 15.5 where this process is split into three terms: predation due to herbivores, settling, and a "natural" mortality term embodying all other processes causing mortality. In the following, some common sources of mortality for primary producers are reviewed.

Grazing (see Chapter 16), or predation by herbivorous organisms, is a well-known source of mortality for primary producers in terrestrial ecosystems (e.g., cattle grazing grass; Figure 15.2) as well as aquatic ones (e.g., zooplankton feeding on phytoplankton). Grazers can sometimes exert *top down* control on primary producers, meaning that the growth and abundance of the producer is strongly affected and ultimately determined by predator abundance. One famous example is the top down control exerted on kelps by sea-urchins, if natural predators of sea-urchins such as sea-otters are killed (Jackson et al., 2001). Recently, top down control on phytoplanktonic production by zooplankton grazers has been demonstrated even in coastal oceans

FIGURE 15.2
(**See color insert.**) Grazing cows on an alpine pasture in the Asiago plateau, Italy.

(e.g., Daskalov, 2002), where abiotic conditions such as nutrients, light, and temperature used to be considered the predominant forcing of primary production.

A description of grazing mortality can be a complex task. Not only ingestion rate clearly depends on a power function of the body size of the predator and on temperature in a similar fashion to respiration (e.g., Peters, 1983; Brown et al., 2004), but all the numerous factors affecting predator abundance, its ingestion rate, and the composition of its diet should be accounted for, such as abiotic conditions, functional responses of predators to prey density (see Chapter 16), intra and interspecific competition, migrations and spatial distributions, availability of other preys, presence of carnivores preying on the herbivores and, in general, food web interactions, just to name a few, in addition of course to prey availability to predators (i.e., primary producer abundance itself; when prey abundance controls predator abundance one speaks of *bottom up* control). Due to this last reason, joint models of both primary producer and consumer abundance are often appropriated (Jørgensen and Bendoricchio, 2001).

The scarcity of those factors limiting primary production (Section 15.2) can also lead to the producer's death and thus represent a mortality source, if their scarcity is extreme. For example, plants and crops can die if they are not watered, and microalgae die if they settle out of the well-lit euphotic zone. Physical processes such as fires and winds can also be important mortality causes in terrestrial ecosystems such as forests. Primary producers themselves (e.g., invasive species) can be a source of mortality in the case of competition for resources, for example, a dense tree canopy shading the underlying soil and grasses.

Finally, primary producers can die because of other, varied causes that are typically pooled under a single "natural mortality" term in ecological models, such as senescence, illness, parasites, and so on, unless a particular mortality source is explored in more detail in a specific model.

16

Production of Upper Trophic Levels

16.1 Introduction

Biological production in ecosystems is classified in ecology according to the trophic position, or trophic level of the species studied. The trophic level of a species is a measure of the number of predator–prey interactions that separate it, on the average, from primary producers or detritus (dead organic matter). The term predator–prey, trophic or feeding interaction indicates one organism or population feeding on another to acquire energy, organic matter, and nutrients. Primary producers are assigned a trophic level of one since they represent the energetic and material basis of ecosystems (Chapters 13 and 15), and the same is done for detritus. Herbivores, which are the consumers feeding on primary producers, are assigned a trophic level equal to two since they are one trophic link away from primary production, while predators feeding on herbivores have a trophic level of three, and so on. The trophic level thus depends on an organism's diet. Most organisms display a certain degree of omnivory, which means that they feed on different prey species or trophic levels, reflecting the fact that ecosystems are not simple chains of trophic interactions, but intricate *food webs* (Section 19.5). Thus, the trophic level of an organism will not be an integer number but will be calculated from the trophic levels of its prey items by weighting them on the amount that the corresponding preys represent in the predator's diet:

$$TL_i = 1 + \sum_{j=1}^{N} DC_{j,i} TL_j \tag{16.1}$$

Thus, the trophic level TL_i of the predator i is one link away from the average of the trophic levels of its N preys, after weighting each of such trophic levels (TL_j) on the fraction $DC_{j,i}$ of the predator's diet represented by the given prey species j.

Predation is a fundamental process regulating energetic and material flows, including pollutants, from primary producers to heterotrophs (consumers, predators, etc; i.e., those organisms unable to fix carbon and which have to obtain it through predation, as opposed to autotrophs) and across the food web. The multitude of different heterotroph organisms and their different ecological functions highlight the importance of this ubiquitous process in structuring the functioning of aquatic and terrestrial ecosystems. One important ecological law dictated by thermodynamics is that the production at one trophic level has to be smaller than the production at the trophic level just below it, the famous *trophic pyramid*, since not all the production of a trophic level can be channeled to the upper trophic level through predation, but a fraction has to be dissipated through the metabolic cost of respiratory processes. Thus, a given amount of production (e.g., kg of carbon per year) at the upper trophic levels is, from an *emergy* point of view (Section 4.4), more expensive than the same amount produced at the lower trophic levels. For this reason, the exploitation of biological production at higher trophic levels (for example, for human consumption) is less sustainable. One example is the intensely fished Atlantic bluefin tuna *Thunnus thynnus*, a top predator characterized by a trophic level higher than four (typically, 4.5), which mostly feeds on medium-low trophic level small pelagic fishes, such as sardine, anchovy, horse mackerel, mackerel, and so on. Assuming a commonly observed biomass transfer efficiency between trophic levels of 10% (Pauly and Christensen, 1995), it means (roughly) that consuming only 100 g of tuna is equivalent to more than 1 kg of sardine and anchovy (whose trophic level is about three), 10 kg of zooplankton, and 100 kg of phytoplankton. This example is also useful to remark that the above-mentioned concept of trophic pyramid refers to production and is not valid in the case of biomass. For example, in some aquatic ecosystems (Table 16.1), the biomass of small pelagic fish can exceed that of zooplankton, whose production is however much higher, since plankton is composed of numerous and small-sized individuals (small size means high production rates; see, e.g., Section 14.3).

TABLE 16.1

Distribution of Living Biomass and Biomass Flows among Integer Trophic Levels from One to Four in the Pelagic Food Web of the Northern Adriatic Sea during 1965

Trophic Level	Living Biomass ($t_{\text{wet weight}}$ km^{-2})	Through-Flow ($t_{\text{wet weight}}$ km^{-2} y^{-1})
I	11.52	769
II	4.97	277
III	13.22	99
IV	3.07	17

Source: Galli et al., A trophic network model to study biomass oscillations of anchovy and sardine populations in the northern Adriatic pelagic ecosystem, unpublished data.

Note: Only flows originating from primary producers are reported. Trophic flows decrease as the trophic level increases according to the trophic pyramid concept derived from thermodynamics, while biomass is concentrated on the first and third trophic level.

16.2 Grazing

The biological production of herbivorous organisms that feed, or "graze", on autotrophs is called secondary production. Secondary production can be quantified as the amount of organic tissue built up at trophic level two per unit area per unit time, measured in a suitable currency such as carbon, or wet/dry weight. Examples of grazers include zooplankton (microcrustaceans such as copepods and cladocerans, jellyfish, etc.), herbivorous fish, filter-feeding benthos consuming microalgae and other bottom invertebrates (e.g., sea-urchins, which mainly consume algae) in aquatic ecosystems, and terrestrial herbivores, such as insects and cattle. Grazing does not necessarily lead to the death of the targeted primary producer, especially when it is a relatively large sized, multicellular organism, such as grasses, trees or macroalgae (as opposed to small-sized unicellular phytoplankton). In such case, primary producers can be consumed by grazers only partially. Based on the definition of trophic level (Equation 16.1), secondary producers are sometimes also considered to include those organisms feeding on dead particulate or dissolved organic matter; that is, decomposers, which should however be distinguished from grazers.

Grazing can be an important source of mortality for primary producers and can limit their population biomass level (top-down control) (Section 15.7). The opposite situation is that of bottom-up control by primary producers on grazers, meaning that the biomass of herbivorous organisms is limited by the availability of food. The type of trophic control pattern, top down *versus* bottom up, characterizing the grazing predator–prey interaction can depend on many factors, and general rules are difficult to identify. The notion that pelagic ecosystems are resource-limited, and fish abundance is controlled by food supply and thus, ultimately, by microalgae, dates back to the work of Hensen at the end of the nineteenth century, and was an unchallenged dogma for decades (Verity et al., 2002). Recently, however, the presence of top down control by zooplankton on phytoplankton has been demonstrated even in large marine ecosystems (Daskalov, 2002).

Trophic control patterns can be influenced by both man and the environment. In kelp forests, the human removal of apex predators (e.g., sea otters) through fishing released herbivores such as sea urchins from predatory pressure, and the grazing of kelps by sea urchins increased so much that clear shifts were observed in those ecosystems from forested areas to barren, unforested states (Jackson et al., 2001). In contrast, trophic cascades triggered by fishing in the Black Sea enhanced microalgal biomass (Figure 16.1) (Daskalov, 2002). A trophic cascade takes place when the effect of a change in predation pressure propagates across consecutive trophic levels in the food web. For example, predatory fishes were overexploited and decreased in abundance in the Black Sea, thus their zooplanktivorous fish preys were released from predation and increased in abundance, zooplankton biomass

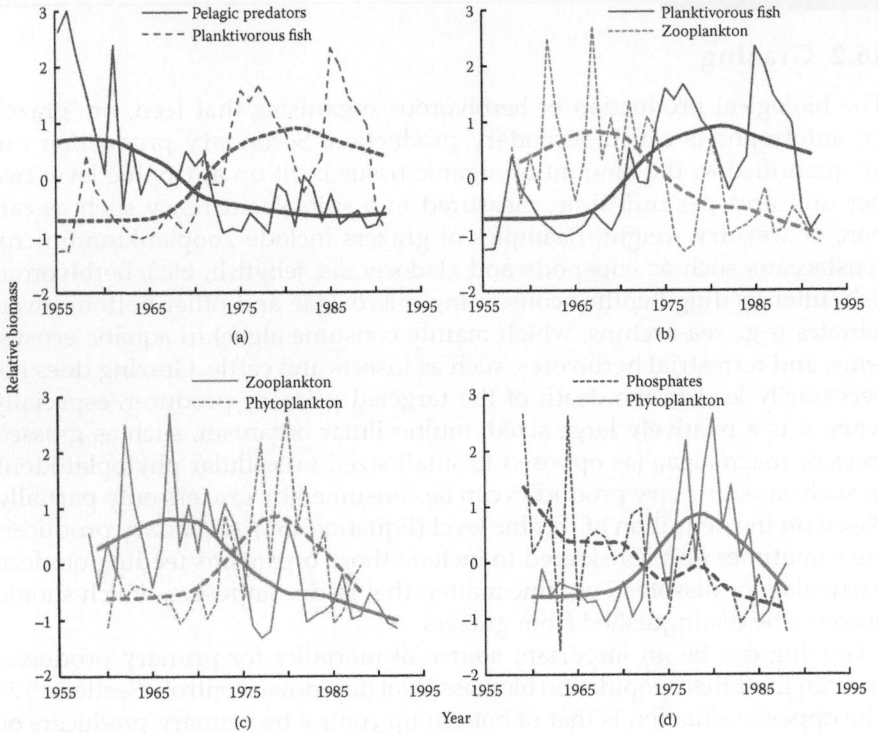

FIGURE 16.1
Release from top down control affects phytoplankton biomass in the Black Sea ecosystem. Data have kindly been provided by Georgi Daskalov and are described in Daskalov (2002). Time series of relative biomass/concentration are plotted: the lighter lines represent the original data, while thicker curves represent long-term trends as given by Loess smoothing. The continuous line indicates the organism (e.g., the predator), which impacts top-down the prey/resource which is indicated by the dashed line. The overfishing of pelagic predators led to a pervasive trophic cascade, as their planktivorous fish preys were released from predation and increased (a), zooplankton decreased due to increased predation by planktivorous fish (b), and phytoplankton increased due to release from grazing pressure (c). The increase in algal biomass even affected the concentration of phosphate in the water surface layer (d).

was consequently suppressed, and phytoplankton biomass increased following the reduction in grazing pressure (Figure 16.1). Concerning the influence of the environment on trophic control, in Africa savannas, a stable coexistence of grass and trees is only present in regions characterized by scarce mean annual precipitation, while abundant average rainfall leads to the dominance of woody cover which outgrows its grazers (Sankaran et al., 2005).

The numerous abiotic and biotic factors and processes influencing grazing are reviewed on Section 15.7. As grazing is a predation process, some thoughts reported on Section 16.3 also apply to grazers.

16.3 Predation

Predation is the act of feeding on a living organism (prey) by another living organism (predator), to obtain from the prey's tissues those nutrients and energy needed to live, grow, and reproduce. It is a special case of the consumer–resource (or trophic) interaction, which includes processes that cannot be classified as predation such as detritivory (the consumption of dead organic matter). Predation represents the basis for other biological interactions such as competition (Section 19.3), and is one of the best studied and most important ecological processes, to the point that the structure of the network of predatory interactions (the "food web," Section 19.5) in an ecosystem can be studied to gain information on the developmental status (from an ecological succession point of view) of that ecosystem (Ulanowicz, 1997), although such status is shaped by multiple ecological processes not certainly limited to trophic interactions.

Predation involves the death of the prey, which is being eaten, but this is not always true (e.g., grazing) (Section 16.2). Predators can be carnivorous (eat other animals; in the case of fish, one speaks of "predatory" fish), herbivorous (eat primary producers), and omnivorous (eat both animals and primary producers). Man can act as a predator in ecosystems, for example, when hunting or fishing, and cause trophic effects whose magnitude is comparable to those of natural predators (e.g., Pauly and Christensen, 1995).

Body size has a strong influence on predation, given that many key ecological processes and traits scale with or are related to body size (Woodward et al., 2005; Brose, 2010). For example, individual metabolism q, which is approximately proportional to food ingestion, is related to body mass m through the famous scaling relationship:

$$q \propto m^{\alpha} \cdot \mathrm{e}^{-E/kT} \tag{16.2}$$

where the exponent of m is about 0.6–0.8 and is usually higher for fish and ectotherms, E is called activation energy, k is the Boltzmann's constant, T is temperature (Peters, 1983; Brown et al., 2004; Glazier, 2005). This relationship highlights that temperature influences predation rates (here, temperature and predation are positively related, but temperature can influence many ecological processes including prey dynamics, therefore its relationship with predation can be more complex). Body size also affects the structure of predator–prey interactions. Diet changes as organisms grow, and its width typically widens (Woodward et al., 2005). In aquatic ecosystems, the saying that "big fish eat little fish" highlights that predators are usually larger than their prey. Predator mass and mean prey mass are positively correlated, and consumer-resource body mass ratios are usually found within a certain range (very small preys are difficult to spot, while very large preys are difficult to kill, e.g., Brose, 2010).

Predation involves costs for the predator in addition to benefits (energy intake), since predators consume energy and could be vulnerable to the attacks of their own predators while hunting or handling (eating, digesting, etc.) preys. These and other tradeoffs involved in predation are investigated in foraging ecology (e.g., by the *optimal foraging theory*) (Kie, 1999).

The most famous mathematical description of predation is probably represented by the Lotka–Volterra predator–prey coupled differential equations:

$$\frac{dN_1}{dt} = r_1 N_1 - p_1 N_1 N_2 \tag{16.3}$$

$$\frac{dN_2}{dt} = p_2 N_1 N_2 - d_2 N_2 \tag{16.4}$$

where N_1 is prey abundance, N_2 predator abundance, r_1 the intrinsic growth rate of the prey, d_2 the natural death rate of the predator, and p_1 and p_2 represent the effect of the predator–prey interaction. The positive term on the right-hand side of Equation 16.4, causing the growth of the predator population over time, is proportional to the number of captured preys, which is proportional through p_2 to the product of prey and predator abundance (i.e., the probability of a predator meeting a prey in the case of the assumption of random encounters in an homogeneous environment). The decrease in prey population abundance is due to the killings caused by the predator, which are proportional through p_1 to the product of prey and predator abundance as above. When solved over time, Equations 16.3 and 16.4 can lead (with a proper choice of the values of the parameters and initial conditions) to lagged oscillations of predator and prey abundance typical of top-down dynamics (Figure 16.2), since predation causes the decline of the prey as well as the growth of the predator population, which, in turn, leads to higher predation mortality; when the prey reaches a low density, the growth term for the predator in Equation 16.4 becomes lower than the natural mortality term, thus causing the decline of the predator population, which releases the prey from predation.

The Lotka–Volterra model is a simple description of the process of predation, based on several assumptions that are not necessarily realistic (e.g., the predator growth term is linearly related to prey abundance or the model does not take into account complex food web interactions, Section 19.5), and many modifications or alternative models exist ; for example, Equations 16.3 and 16.4 can be adapted to include the concepts of logistic growth and carrying capacity (Jørgensen and Bendoricchio, 2001). Particularly, the relationship between prey abundance density and predator consumption rate, called functional response (Holling, 1959; Jørgensen and Bendoricchio, 2001) can take markedly different forms, reflecting different behaviors or ecological conditions, as well as tradeoffs involved in predation. One example of such tradeoffs is that predators have to decide what amount of their time should be spent searching for preys, and what amount for handling preys. A typically-observed functional response is the type II functional response, in which the consumer food intake rate can be expressed as a Michaelis–Menten function

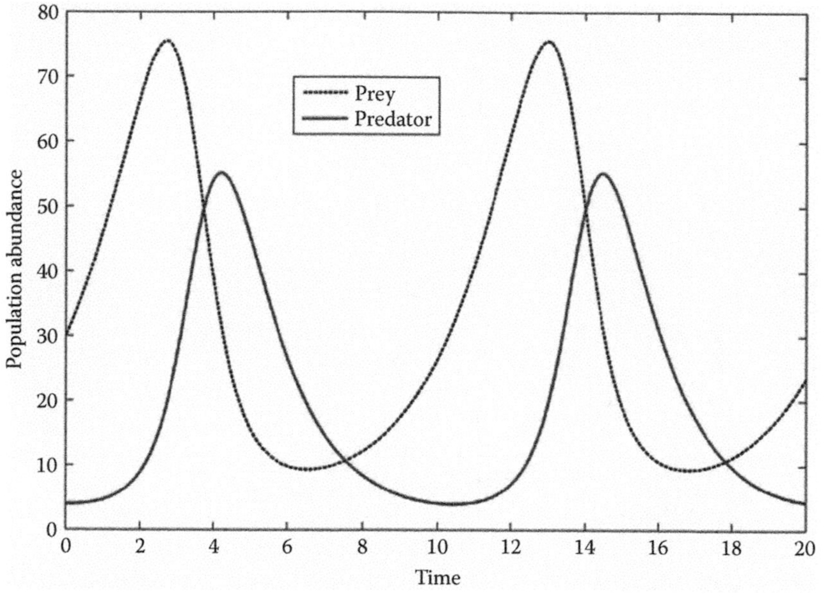

FIGURE 16.2
Oscillating Lotka–Volterra dynamics (Equations 16.3 and 16.4), showing the predator–prey cycles over time caused by the interplay of the processes of predation, intrinsic growth and natural mortality. Prey abundance is the dashed line, predator abundance is the continuous line. Note the lag between the peaks and troughs in the abundance of the prey and predator populations.

(Equation 15.1) of the prey density, thus implying that per capita food consumption by the predator cannot increase indefinitely, even if the prey is very abundant. Another example of functional response is the optimum (or humped) relationship between food intake rate and prey density, which can be observed when a large prey density has a negative effect on the predator (e.g., when preys produce toxic wastes or support each other by assuming a defensive position).

17

Microbial Processes

17.1 Introduction

This chapter broadly describes the role of microorganisms (or microbes) in some key ecological processes. As suggested by the word itself, microbes are organisms of very small size; that is, they typically are microscopic, meaning that man can see them only by using a microscope. Microbes are mostly unicellular organisms ubiquitously found in all the earth's ecosystems and include both prokaryotes and eukaryotes. Examples include bacteria, protozoa, microalgae, archaea, and fungi, therefore, both autotroph (Chapter 13) and heterotroph microorganisms exist (mixotrophic microbes also exist, which exploit both organic and inorganic carbon sources). Microorganisms even exist inside the body of other organisms (e.g., in the digestive system of man, i.e., the intestinal flora, or cows). Although viruses are generally not considered as living beings and cannot be classified as microorganisms, they are studied in the field of microbiology due to the size and the significant influence they exert on the dynamics of bacterial populations. In aquatic ecosystems, viruses can represent a major source of bacterial mortality.

Due to small size and fast generation times, microorganisms represent some of the lowest trophic levels of food webs (Chapter 16) and are a potential source of food for other organisms. For example, in marine ecosystems, bacteria represent an alternative energy pathway leading to upper trophic level organisms (fish), as opposed to the "classical" food chain represented by algae–mesozooplankton–planktivorous fish (Section 17.2). In addition to representing a carbon and energy source, microorganisms contribute to the functioning of ecosystems by promoting nutrient cycling: indeed, a very important ecological process carried out by several microorganisms is the decomposition of dead organic matter (Section 17.3) and the consequent remineralization of nutrients (N, P, etc., Chapter 12). Without microorganisms, several biogeochemical cycles would not be possible. For example, the cycle of nitrogen is strongly influenced by bacteria, which are involved in several transformations affecting this nutrient in the environment, such as remineralization of organic nitrogen, nitrification, denitrification, and atmospheric nitrogen fixation (Section 12.5). As a broad rule, microorganisms and the processes they carry out are sensitive to environmental conditions such as temperature, pH, moisture, nutrient availability, and so on, that is, bottom-up factors (Chapter 16). In aquatic ecosystems, however, low trophic level planktonic microorganisms represent the basis of food webs, channeling energy and matter to upper trophic levels, and they can undergo intense predation, which can even lead to top-down control (Chapter 16). Traditionally, these opposite situations have resulted in different descriptions of aquatic ecosystems. In models focused on low trophic levels and biogeochemical cycles (e.g., NPZ models, which stands for Nutrient–Phytoplankton–Zooplankton), planktonic microorganisms, such as zooplankton, represent the final step of the food chain. In the fisheries model, zooplankton, microalgae, or the microbial loop (Section 17.2) typically represent a "closure term," which is assumed to be constant or is only roughly simulated. Yet, the significant influence of both bottom-up and top-down processes on the dynamics of aquatic microorganisms, and the ecological importance of microbial processes, highlight that a more integrative description of the dynamics of planktonic microorganisms (e.g., end-to-end modeling or model coupling) (Rose et al., 2010) is needed to describe the behavior of aquatic systems.

Several ecological processes carried out by microorganisms are exploited by man. For example, the biotechnological process of brewing employs yeasts to carry out fermentation (Section 17.4). Biogas production is based on microbial degradation processes. In wastewater treatment, biological treatments can be exploited to decrease the concentration of organic matter, nitrogen, and phosphorus in water.

17.2 Microbial Loop

The phrase "microbial loop" was coined in 1983 by Azam et al. to describe the flow of energy and nutrients to upper trophic levels in marine food webs through a trophic pathway mainly based on microorganisms such as bacteria and protozoa. According to the paradigm that was dominating before the work of Azam and colleagues, the basis of the marine food web can be depicted as a directed chain through which energy and nutrients flow: in this chain, primary production, mainly represented by large microalgae such as diatoms and dinoflagellates, is predated by mesozooplankton (typically copepods), which in turn are consumed by small planktivorous fish. Thus, the "classical" grazing food chain is directly fuelled by primary production. Azam et al. (1983), based on a large body of scientific evidence that had accumulated over the previous years (Fenchel, 2008), showed that an alternative trophic pathway of comparable importance, which is fuelled by detritus (dead organic matter), coexists with the grazing food chain. Within this "microbial loop," the dead particulate and dissolved organic matter (DOM) found in the water column (represented by algal exudates, dead organisms including plankton, excreta and egestion, settling particles, inputs of organic matter from coasts, marine snow, and so on) are decomposed by bacteria feeding on such energy- and nutrient-rich substrates. Bacteria are then consumed by protozoa, particularly flagellate ones, which in turn are eaten by larger zooplankton such as microzooplankton, eventually coupling the microbial loop to the classical grazing food chain.

Over the past 30 years, the microbial loop was discovered to be a small but intricate food web much more complex than the above mentioned schematization (Fenchel, 2008) (Figure 17.1). For example, viruses have been found to cause an amount of bacterial mortality comparable to protozoan grazers. Unicellular photosynthetic prokaryotes and the small prochlorophytes can represent key primary producers in marine systems (Fenchel, 2008). Yet, the microbial loop concept remains fully valid: bacteria can build up a large biomass by up-taking energy, carbon, and other nutrients from DOM and particulate organic matter (POM) which are not easily bioavailable to other organisms. Thus, the microbial loop represents an important energy and element recycling pathway in marine food webs. A large amount of the carbon channeled through the microbial loop can be dissipated as microbial respiration (small-sized organisms display very high respiratory rates) so that the microbial loop does not necessarily represent a large carbon source to upper trophic levels, but it can often be a carbon "sink" (Fenchel, 2008).

Another reason why the microbial loop is an important ecological process in marine systems is that the bacterial decomposition of organic matter contributes to the recycling of the nutrients that such matter contains. When these nutrients have been mineralized by bacteria or excreted by other components of the loop such as flagellates and microzooplankton, they can be easily uptaken by phytoplankton, thus one key effect of the microbial loop is to stimulate primary production by allowing the fast recycling of the

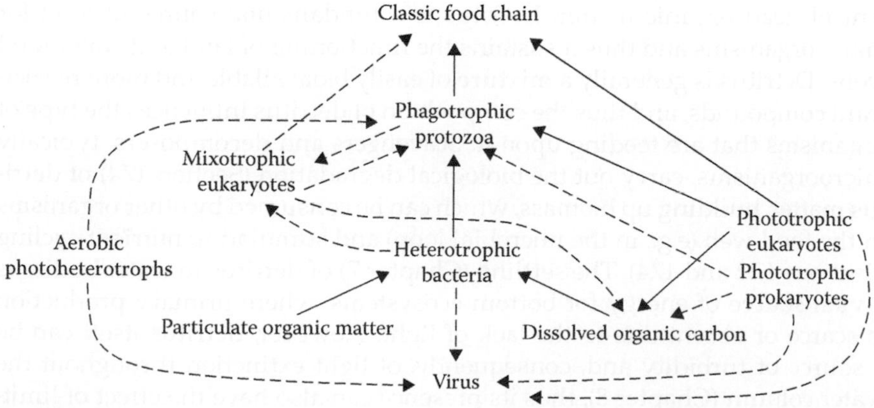

FIGURE 17.1
The microbial loop according to Azam et al. (1983) (solid arrows) and with later additions (dashed lines). (Redrawn after Fenchel, T., *J. Exp. Marine Biol. Ecol.*, 366, 99–103, 2008.)

nutrients contained in POM and DOM. Even when there is strong competition for nutrients between bacteria and phytoplankton, the microbial loop still provides nutrients to the food web, supporting the system productivity. In such case, for example in oligotrophic systems characterized by nutrient scarcity, the microbial loop is favored over the classical food chain, that is, most nutrients are channeled in the microbial loop, because bacteria have a larger surface to volume ratio and thus are favored over larger microalgae in the uptake of nutrients (uptake takes place through the cell surface). But, even if bacteria outcompete phytoplankton for nutrients, the final outcome is, again, that nutrients from DOM and POM are fed back into the marine food web.

17.3 Role of Detritus

Detritus is the pool of dead organic matter found in ecosystems, including feces and excretions by living organisms, bodies of dead animals and plants or their fragments, egested food, algal exudates, mucus, and so on. Detritus also includes the biomass of the associated microorganisms that decompose it, such as bacteria. In aquatic ecosystems, detritus is suspended in the water column as settling particulate matter (the so-called marine snow); at times, the term detritus is also meant to include all other types of dead organic matter such as dissolved organic substances and colloids. In land ecosystems, detritus typically accumulates on the soil as dead animals, fallen leaves and twigs, and so on, whose degradation lead to the formation of the stable, organic portion of the soil known as humus.

The ecological importance of detritus lies in the energy and nutrient content of dead organic matter. Detritus is a fundamental source of food for many organisms and thus it sustains the functioning of land and water food webs. Detritus is generally a mixture of easily bioavailable and more recalcitrant compounds, and thus the composition of detritus influences the type of organisms that are feeding upon it. Scavengers and decomposers, typically microorganisms, carry out the biological degradation (Section 17.4) of detritus matter, building up biomass, which can be consumed by other organisms in the food web (e.g., in the microbial loop) and stimulating nutrient cycling (Sections 17.2 and 17.4). The settling (Chapter 7) of detritus in water bodies is a vital source of energy for bottom ecosystems, where primary production is scarce or absent due to the lack of light. However, detritus itself can be a source of turbidity and, consequently, of light extinction throughout the water column (Chapter 8), thus its presence can also have the effect of limiting algal growth and system productivity. In bottom ecosystems, settling matter is not only fed upon by microbes, but also by bottom-dwelling micro- and macroorganisms (invertebrates, fish) known as benthos. A common strategy to uptake detritus in benthic ecosystems is filtering water, a process carried out by several organisms known as filter feeders, such as bivalves (clams, oysters, mussels), ascidians, sponges, and so on.

The consumption of detritus leads to the cycling of matter and energy in the food web, which otherwise would have been unavailable to the biota. Detritus-feeders such as microbes and benthos are consumed by predators, so that detritus effectively sustains ecosystem functioning. Thus, the consumption of detritus can be related to the degree of self-sufficiency of ecosystems, and even to their stability (Ulanowicz, 1997). Indeed, detritus-based trophic flows are not negligible: the carbon or energy flows originating from detritus in the food web often display a magnitude that is comparable to, if not larger than, the trophic flows originating directly from primary production. In his seminal paper on the development of ecosystems, Eugene P. Odum (1969) linked several trends that he thought had to be expected as ecosystems develop to the role of detritus in the food web. For example, during system development, there should be a shift from a grazing-based food chain (i.e., a linear trophic structure) to a detritus-based food web containing feedbacks, and detritus should become more important in nutrient regeneration. Thus, mature ecosystems should rely on detritus more heavily than early developing systems. This expected trend should be considered as a rule of thumb rather than as a strict tenet, and it should be applied following case-specific considerations. For example, linking detritus cycling and ecosystem maturity is not straightforward, because as mentioned above detritus-based trophic flows can dominate even in early developmental ecosystems, or because system maturity is not only linked to the amount of cycled matter in the food web but also to the number of circular pathways and their length (Baird et al., 1991).

17.4 Biological Decomposition

Biological decomposition or biodegradation is the breaking down of dead organic matter, such as detritus, carried out by both biotic factors and living organisms (known as decomposers, e.g., bacteria or fungi), into simpler organic and inorganic compounds. The ecological importance of this process has already been discussed in Sections 17.2 and 17.3: biodegradation leads to the recycling of energy and matter (carbon and other nutrients) in dead animals, excreta, algal exudates, which otherwise would remain unavailable to the other biotic components of the ecosystem, because they are contained, for instance, in hard-to-degrade substances, such as lignin, and other recalcitrant compounds. Biological decomposition makes available to primary producers through mineralization those key nutrients supporting algal and plant growth, such as nitrogen and phosphorus (Chapter 12). During the degradation carried out by decomposers, some chemical elements are released in the environment, and other ones are uptaken by decomposers from detritus. Consequently, decomposers grow in biomass, and their subsequent predation by other biota leads to the cycling of the previously uptaken chemical elements in the food web (Chapter 19). For all these reasons, decomposition is tightly linked to important biogeochemical cycles (Chapter 12), such as carbon, nitrogen, phosphorus, and even oxygen, as the common process of aerobic biodegradation consumes O_2. Concerning the relationship between biodegradation and the carbon cycle, it is worth noting that the carbon stored in the soil in the form of organic matter is approximately twice as much as the carbon stored in the vegetation and the atmosphere. The magnitude of such carbon reservoir and its dynamic exchanges with the environment, for example, through biological decomposition, make this process a key player in the global carbon cycle, which deserves more study if the ongoing climate change is to be fully understood (Wang and D'Odorico, 2008). Biological decomposition is also exploited in many human activities, such as those based on fermentation, or the degradation of toxic compounds (Chapter 18) in wastewater treatment plants and during the remediation of polluted sites.

The process of nutrient *mineralization* is part of the biological decomposition process, but it should be distinguished from it: biological decomposition is the metabolic breaking down of organic carbon compounds by decomposers to obtain organic carbon and energy, and causes the release of CO_2, but does not necessarily involving the mineralization of nutrients. Instead, mineralization typically refers to the biological transformation of nutrients from an organic to inorganic form, which is generally bioavailable to primary producers (Wang and D'Odorico, 2008).

Biological decomposition is a complex process that can be influenced by many factors. Indeed, even if biological decomposition is mainly due to the enzymatic (Chapter 9) chemical reactions resulting from the metabolic activity (Section 14.2) of living organisms, the most common decomposers being microorganisms such as bacteria and fungi, this process is actually the final outcome of multiple, interacting physical, chemical, and biological processes. Furthermore, biological activity itself can be influenced by physical and chemical factors; that is, by environmental variability.

The biological decomposition of dead organic matter in terrestrial and aquatic ecosystems, as distinguished from mineralization, can be divided in three phases (Wang and D'Odorico, 2008): leaching, fragmentation, and chemical alteration. During leaching (a physical process), small soluble organic compounds (e.g., sugars) and ions (e.g., K^+) dissolve in water, thus leaving the decomposing matter (for example by diffusion, Chapter 7) and becoming available to the uptake by bacteria, plant roots, and so on. Fragmentation is the breaking down of dead organic matter into smaller particles. One important effect of fragmentation is the increase in the surface-to-volume ratio of detritus, which in turn enhances its availability to decomposers (which act on the surface) such as bacteria. Fragmentation is mainly a physical and biological process, depending on the factors carrying it out. Common physical factors causing fragmentation are water currents and oscillations in temperature (e.g., freeze-thaw cycles) or in humidity, while biota (benthos, bacteria, soil fauna such as worms, etc.) fragment dead organic matter through the chemical–mechanical action of feeding. Finally, the phase of chemical alteration turns dead organic matter into simpler organic compounds or even into inorganic ones.

Mineralization takes place differently in the case of nitrogen and phosphorus (Wang and D'Odorico, 2008). The former nutrient can be found in the form of an ester link (C–O–P), thus phosphorus can be separated from the organic matter through mineralization even without decomposing the carbonaceous structure of such matter. Instead, nitrogen can form direct chemical links with carbon; thus it can be released during the phases of fragmentation and chemical alteration. During these first steps of N-mineralization, nitrogen is released as dissolved organic nitrogen (DON), which is a bioavailable form. DON can be further degraded by microorganisms to ammonium ions NH_4^+, which can be consumed by primary producers or bacteria, or can undergo additional transformations of the nitrogen cycle, such as oxidation to nitrites and nitrates (Chapter 12).

From a biochemical point of view, three distinct types of decomposition can be carried out by biota: aerobic respiration, anaerobic respiration, and fermentation (Odum, 1983a) (Chapter 15). All these processes lead to the oxidation of organic matter and to energy production. Aerobic respiration is the most common of the three process among animals and has opposite effects to those of primary production (Section 15.3), since the consumption of oxygen (and hence its presence) is required to perform this process, as oxygen represents the electron acceptor in this chemical reaction. Aerobic respiration generates water and CO_2 but it does not necessarily degrade all energy-rich organic substances present in the dead organic matter, for example, because they are recalcitrant or because all oxygen is consumed by the aerobic respiration process before complete degradation is achieved. Anaerobic respiration does not consume gaseous oxygen to take place, since it exploits other external inorganic substances as electron acceptors (Section 15.3). Anaerobic respiration is performed only by certain organisms, for example, bacteria. Finally, fermentation (Ciani et al., 2008), similarly to anaerobic respiration, does not involve gaseous oxygen, but in this case the terminal electron acceptor is an organic compound, which is often the metabolic intermediate resulting from the oxidation of another organic substance. Again, as in the case of anaerobic respiration, fermentation is limited to certain organisms, such as bacteria and yeasts.

Biological decomposition of dead organic matter and mineralization are typically modeled as first-order kinetics (Chapter 18); that is, the degraded substance is proportional to the amount of substance present. Several abiotic factors can influence the rate of biological decomposition and mineralization. The role of temperature cycles in decomposition has already been mentioned above, however temperature can affect in general the rates of numerous ecological processes including metabolic rates and, thus, decomposition (Chapter 14) (Brown et al., 2004), typically through Arrhenius or humped "optimum" relationships. Broadly speaking, warmer locations are characterized by faster biological decomposition rates. Leaching is favored by humidity, being enhanced in locations where rainfall is higher; however decomposition is suppressed by water logging in wetlands (Wang and D'Odorico, 2008), since this condition slows the interphase transport (Chapter 7) of O_2 (needed to perform aerobic degradation) to the decomposing matter and the removal of generated CO_2 from it. The influence of humidity may reverse the positive effect of temperature on biological decomposition in dry, hot terrestrial ecosystems, since high temperatures cause increased evapotranspiration (Chapter 8); that is, water loss, they can actually slow biodegradation down in case they cause excessive water scarcity.

The rates of biological decomposition and mineralization are also increased by the availability of nutrients (even in the substrates to be degraded) and of energetic, quickly assimilable substrates and, in general, by the presence of environmental conditions fit to the bacterial community of decomposers (proper pH, temperature, absence of toxic substances such as heavy metals, etc.). Sugars, fats, and proteins are typically quick to decompose, while cellulose, lignin, chitin, bones, and so on are slow to degrade. The characteristics of the existing microorganism community, such as biomass and affinity for the substrate also influence decomposition and mineralization. If decomposers are not abundant or acclimated to the substrate and to the environmental conditions, it can take time before maximum degradation rates are reached.

An important difference between biological decomposition in aquatic and terrestrial ecosystems is that, in water, an important source of detritus, such as DOM, is directly uptaken by bacteria and transformed into biomass, without being decomposed into an inorganic form as it happens in terrestrial systems (or even in water bodies in the case of POM). Thus, in the case of DOM, it may be more correct to speak of biological cycling of matter rather than of complete decomposition.

18

Ecotoxicological Processes

18.1 Introduction

Ecotoxicology is the scientific study of the effects of toxic substances, generally of human origin, upon ecosystems as well as on their living components such as communities and populations including man. This applied discipline represents the integration of toxicology with ecological knowledge. This chapter illustrates the basic processes that influence the transport and fate of toxicants in ecosystems and their distribution among different media (soil, water, air, and biota).

18.2 Biodegradation

Biodegradation is the process of degradation (or decomposition, Section 17.4) of toxic chemicals performed by living organisms in soil, water, or air either directly or indirectly. In the case of organic compounds, one can distinguish between primary biodegradation, which is any transformation that alters the chemical structure of the toxicant, and ultimate biodegradation; that is, the biologically mediated transformation of organic compounds into inorganic form (Jørgensen and Bendoricchio, 2001).

Biodegradation is a central process in ecotoxicology as it decreases the negative effects of toxic compounds on the environment and on human health without direct human effort, since the natural processes carried out by organisms and within biological communities generally decrease the mass of toxicants over time. However, it is important to notice that, in some cases, the biodegradation of toxic chemicals can lead to the creation of intermediate or final reaction products, which are toxic themselves, even to a larger degree than the original compound.

Most of the considerations made in Section 17.4 concerning the process of biological decomposition also apply to biodegradation. This section focuses on the approaches that are commonly used to describe the biodegradation process carried out by microorganisms, in nature or in human applications of biodegradation, such as wastewater treatment or environmental remediation of polluted sites. Similar to a large number of other biological and biologically mediated processes in nature, the biodegradation of toxic compounds characterized by an environmental concentration C available to decomposers can generally be approximated as a first-order kinetics process:

$$\frac{dC}{dt} = -k \cdot C \tag{18.1}$$

Thus, the reduction in toxicant concentration over the time is related and proportional to the available concentration in the ecosystem. Equation 18.1 is commonly used in ecotoxicological and sanitary engineering applications. For example, it is employed to describe the dynamics of biochemical oxygen demand (BOD) in water, that is, the aerobic biodegradation of organic wastes like sewage. When focusing on the reduction in concentration, it is important to distinguish biodegradation, which causes a reduction in toxicant mass and consequently concentration, from other processes that only reduce concentration (e.g., dispersion or diffusion) (Newell et al., 2002) (also in Chapter 7 and *washing by dilution* as discussed in Section 23.2).

If the biodegradation rate k, whose units are the inverse of time (e.g., day^{-1}), is assumed to be constant over time, the integration of Equation 18.1 yields a negative exponential function $C(t) = C(0) \cdot e^{-kt}$, highlighting that achieving the complete biodegradation of a toxicant can require a long time, because the process slows down at low concentrations. Exploiting the assumption of time-constant k, one can easily calculate an inverse relationship between k and the commonly used biological half time of a substance $\tau_{1/2}$, which is the time needed to reduce the initial concentration by 50%:

$$k = -\ln(0.5)/\tau_{1/2} = 0.693/\tau_{1/2} \tag{18.2}$$

This relationship highlights that higher biodegradation rates mean shorter degradation times. The value of k can display a huge variability and be site-specific (Table 18.1) (Trapp and Matthies, 1998), so that field measures are often required. For example, k in groundwater can be determined by fitting a solute transport model to field measurements of concentration, or by comparing the transport of an inert tracer with the transport of the contaminant that is being studied, as one major factor causing differences in the plume of the inert tracer and that of the contaminant is, often, biodegradation (Newell et al., 2002). In general, k can depend on biotic and abiotic environmental conditions; for example, it is often linked to temperature through an exponential Arrhenius function (e.g., Section 17.4).

TABLE 18.1

Observed Variability of First Order Biodegradation Rates for BTEX Hydrocarbons in Groundwater, as Observed on Field in 30–50 Studies

Contaminant	Biodegradation Rate (Day^{-1})			
	Minimum	25th-percentile	75th-percentile	Maximum
Benzene	0	0	0.007	0.085
Toluene	0.0001	0.002	0.065	4.7
Ethyl Benzene	0.0001	0.0006	0.013	0.08
m-Xylene	<0.0001	0.0011	0.017	0.13

Source: After Suarez, M.P., & Rifai, H.S. 1999. Biodegradation rates for fuel hydrocarbons and chlorinated solvents in groundwater. *Bioremediation*, 3(4): 337–362.
Note: Several orders of magnitude in the biodegradation rate value are spanned for each contaminant.

Equation 18.1 is actually derived by applying some simplifications (Jørgensen and Bendoricchio, 2001) to a slightly modified version of the commonly used Michaelis–Menten equation (Chapter 14):

$$\frac{dC}{dt} = -\frac{1}{Y}\frac{dB}{dt} = -\mu_{max} \cdot \frac{B}{Y} \cdot \frac{C}{k_C + C} \tag{18.3}$$

In this relationship, B represents the decomposer biomass, Y stands for yield, which is the decomposer biomass produced per unit of substrate (the toxic compound) consumed, μ_{max} is the maximum growth rate of the decomposer population in absence of substrate limitation (here expressed through the Michaelis–Menten formulation), and k_C is the semisaturation level of substrate concentration. Such equation can further be complicated to reflect the dependence of bacterial biomass on environmental factors, such as oxygen concentration (in the case of aerobic degradation), humidity, pH, presence of toxic compounds for them, such as heavy metals, and so on. For example, the growth of aerobic bacteria can be described as dependent on both the substrate concentration C and the oxygen concentration O_2 by multiplying two distinct saturation terms:

$$\frac{dx}{dt} = \mu_{max} \cdot x \cdot \frac{C}{k_C + C} \cdot \frac{O_2}{k_{O_2} + O_2} - d \cdot x \tag{18.4}$$

where x refers to the biomass of aerobic bacteria, k_{O_2} is the semisaturation constant for oxygen, and d is the decaying rate of microbes. Yield coefficients such as that found on Equation 18.3 can be used to couple the relationship from Equation 18.4 with the dynamics of the substrate and of oxygen.

Finally, the processes influencing the availability of the contaminant to decomposers also play a role in biodegradation, such as the diffusion

(Chapter 7) and adsorption (Chapter 11) of the soil contaminant to be degraded to sites that are unavailable to the bacteria.

To get from Equations 18.1 through 18.3, two assumptions are required. The first one is that the substrate concentration is small when compared to the semisaturation constant ($C \ll k_C$), meaning that the toxicant in the environment is found at trace or (relatively) low concentrations. This assumption simplifies Equation 18.3 to

$$\frac{dC}{dt} = -\mu_{max} \cdot \frac{B}{Y} \cdot \frac{C}{k_C} = -k_1 \cdot B \cdot C \tag{18.5}$$

where $k_1 = \mu_{max}/(Y \cdot k_C)$. This is a first-order kinetics with respect to C, but it also contains the decomposer biomass. The second assumption is that such biomass is approximately constant in the environment, being determined by external forcing factors (availability of food and nutrients, and predatory pressure), which can constrain its value. For example, in water bodies, B is strongly linked to the amount of suspended matter (Jørgensen and Bendoricchio, 2001). This assumption yields Equation 18.1 by considering $k = -k_1 \cdot B$.

18.3 Bioaccumulation, Bioconcentration, and Biomagnification

Living organisms can uptake contaminants, such as organic compounds and heavy metals, from all environmental media (water, soil, and air), and they can even absorb them from the food they eat, that is, from other organisms. Several contaminants, which are studied in ecotoxicology, tend to bioaccumulate, which means that, as a result of the above-mentioned uptake from environmental media and food, contaminants are retained into the body tissues of the biota. For bioaccumulation to take place in a given organism, the contaminant mass budget over its body must be positive; that is, the input (uptake from environmental media and food) should be equal or greater than the sum of output (excretion) and the amount of contaminant biodegraded in the organism's body. This idea can be conceptualized through a simple mass balance dynamic equation (Jørgensen and Bendoricchio, 2001) in the case of a persistent contaminant that cannot be biodegraded (e.g., a heavy metal):

$$\frac{dC_b}{dt} = U - k_e \cdot C_b \tag{18.6}$$

where C_b (g of contaminant per g of body weight) is the body concentration of the contaminant, U represent the absorbed mass of contaminant per unit time (e.g., it can be proportional to the contaminant concentration in the environment), and the negative term on the right-hand side of the equation represents the amount of excreted contaminant through a first-order kinetics.

The excretion rate can be expressed as $k_e = a \cdot m^b$, that is, proportional to the organism's body mass through coefficients *a*, *b*, based on allometric principles which have been confirmed by field observations (Chapter 14; *b* generally ranges from 2/3 to 3/4). Thus, to describe bioaccumulation, the process of uptake and that of excretion should be characterized, which are typically contaminant- and species-specific.

Uptake has already been described on Section 15.4 in the case of primary producers. In the case of consumers, bioaccumulation can be related to the amount of contaminants ingested through predation (Section 16.3) and, thus, to the structure of trophic interactions in the ecosystem (Chapter 19), and to the direct uptake from water, air, and soils. Concerning direct uptake, however, it can be difficult to develop general models. For this reason, in ecotoxicology, the focus is upon the final outcome of the process of bioaccumulation that is on how pollutants are partitioned between living organisms and other environmental media (e.g., their food). Indeed, over long time scales, a condition of equilibrium or dynamic equilibrium can be expected between the concentration in environmental media and the body concentration of the contaminant (instead, in the case of short-time scale events such as an oil spill, dynamic models may be needed). In the case of bioaccumulation caused only by the uptake of contaminant by water (e.g., contaminants diffusing across epithelia and gills in fish), one usually speaks of bioconcentration (the reader is cautioned against the different uses of the terms "bioaccumulation" and "bioconcentration" that are sometimes used interchangeably in the scientific literature).

Finally, if bioaccumulation is only due to food, one speaks of biomagnification. This last word highlights that, often, the concentration of bioaccumulating contaminants is larger ("magnified") in upper trophic level organisms (Chapter 16); that is, it increases up the food chain (Chapter 19). This happens because predators typically eat several preys during their life, and also because predators are typically larger than their preys and thus, due to allometric principles (Chapter 14), they are characterized by slower metabolic rates and live longer. For example, bluefin tuna in the Mediterranean Sea eats numerous anchovy and thus uptakes mercury from all of them. This large number of preys eaten over tuna's long lifetime, in addition to the body-specific rate of excretion, which is lower in tuna with respect to anchovy, causes mercury biomagnification in tuna. Biomagnification does not always take place, for example because the uptake from food is not necessarily the process that dominates the contaminant accumulation in the biota. Biomagnification tends to be important in upper trophic level organisms in the case of contaminants, which are lipophilic, highly persistent, scarcely soluble, easy to uptake, and difficult to excrete. When biomagnification is present, it can be difficult to spot. For instance, only some of the top predators can display high contaminant concentration because, unlike other predators, they intensely feed on preys, which consume large amounts of highly polluted compartments such as detritus in sediments. Thus, the structure of food webs can strongly influence the biomagnification patterns, and the reader is referred to the sections on predation and food webs (Chapters 16 and 19).

As the focus here is on the partition of contaminants, a large body of literature exists, which reports values or empirical/semiempirical models to calculate contaminant- and species-specific bioconcentration factors (BCF) and bioaccumulation factors (BAF), which are defined as

$$\mathrm{BCF} = \frac{C_b}{C_w} \tag{18.7}$$

$$\mathrm{BAF} = \frac{C_b}{C_{\mathrm{env}}} \tag{18.8}$$

where C_w represents the contaminant concentration in water and C_{env} its concentration in the surrounding environmental medium such as soil. Such coefficients are reported for both plants and animals.

In the case of lipophilic chemicals, there is a well-known correlation between BCF and the octanol–water partition coefficient K_{ow}, which is the ratio at equilibrium and at a given temperature of the contaminant concentration in a known volume of n-octanol to the concentration in a known water volume. Low values of this ratio characterize soluble substances; high values are typical of compounds that tend to bioconcentrate. As a broad rule, if $K_{\mathrm{ow}} < 1000$, the contaminant does not bioaccumulate. Since K_{ow} is easier to measure in the laboratory, numerous reported values exist for contaminants (e.g., Trapp and Matthies, 1998), which can be used to estimate the bioconcentration factor through empirical regressions:

$$\log \mathrm{BCF} = a \cdot \log k_{\mathrm{ow}} + \log f_{\mathrm{lipid}} w \tag{18.9}$$

where f_{lipid} is the fraction of body fat in the organism where the compound mainly tends to accumulate, and the a exponent of K_{ow} is generally close to 1 (which would lead to the intuitive relationship $\mathrm{BCF} = f_{\mathrm{lipid}} \cdot k_{\mathrm{ow}}$; that is, the contaminant accumulates only in the lipid tissue, which is assumed to behave as octanol). Similar semiempirical relationships exist linking the bioaccumulation in soils or sediments to the fraction of organic carbon in the soil/sediments, as well as to K_{ow} and f_{lipid} (Jørgensen and Bendoricchio, 2001).

Illustration 18.3.1: Toxic Substance Concentrations and Biomagnification Series

A food chain consists of four levels. For each of the four levels, the fraction of food (dry matter basis) assimilated can be considered to be 2/3 of the food intake. The amount of food intake for all the four levels per unit time are, respectively, 10, 2, 1, and 0.36, expressed in kg/24 hours. The respiration is 20% of the biomass per 24 hours. The food for the first trophic level contains 2 ppm of a toxic substance that is assimilated 90% by all four trophic levels. The toxic substance is excreted at a rate of 0.01 per unit of time for all the four trophic levels.

Calculate the concentrations of the toxic substance in the four levels at steady state and indicate the series of biomagnification.

SOLUTION

At steady state, the input is equal to the output for each tropic level. It means that the assimilated food is the food for the next tropic level minus the respired biomass (which is equal to $0.2 \times$ biomass). As we know the assimilated food and the food for the next tropic levels, we can find biomass as five times the difference between assimilated food and food for the next tropic level.

It means that biomass level $1 = 5(10 \times 2/3 - 2) = 23.5$

Biomass level $2 = 5(2 \times 2/3 - 1) = 1.7$
Biomass level $3 = 5(1 \times 2/3 - 0.36) = 1.55$
Biomass level $4 = 5 \times 0.36 \times 2/3 = 1.2$

Similar equations can be developed for the toxic substances in each level (unit applied ppm × kg = mg), that is

Assimilated toxic substance minus excreted toxic substance (which is equal to 0.01 times the toxic substance in the trophic level) minus toxic substance in the food for the next tropic level (which is equal to the food times the concentration of the toxic substance in the trophic level). The amounts of toxic substance in the tropic levels are indicated as X_1, X_2, X_3, and X_4. Concentrations of toxic substance in the four levels are the toxic substance of the level divided by the biomass.

Toxic substance level $1 = 0.9(10 \times 2) = 18$ mg $= 0.01X_1 + 2X_1/23.5 = 0.097X_1$ or $X_1 = 195$ mg. The concentration of the toxic substance in level $1 = 195/23.5 = 8.7$ ppm (mg/kg).

Toxic substance level $2 = 0.9(2 \times 8.7) = 15.7$ mg $= 0.01X_2 + 1X_2/1.7 = 0.6X_2$ or $X_2 = 26$ mg. The concentration of toxic substance in level $2 = 26/1.7 = 15$ ppm (mg/kg).

Toxic substance in level $3 = 0.9(1 \times 15) = 13.5$ mg $= 0.01X_3 + 0.25X_3/1.55 = 0.17X_3$ or $X_3 = 80$ mg. The concentration of toxic substance in level $3 = 80/1.55 = 52$ ppm (mg/kg).

The toxic level $4 = 0.9(0.25 \times 52) = 11.7$ mg $= 0.01X_4$ or $X_4 = 1170$ mg. The concentration of toxic substance in level $4 = 1170/1.2 = 975$ ppm (mg/kg).

The biomagnification series is $8.7 \rightarrow 15 \rightarrow 52 \rightarrow 975$ ppm (mg/kg).

18.4 Equilibrium between Spheres: Partition of Toxic Substances among Air, Water, Soil, and Biota

As discussed in Chapter 11, the area near the surface of the planet can be conventionally divided into four interconnected geospheres: lithosphere, hydrosphere, atmosphere, and biosphere. Toxic and xenobiotic substances and other elements of relevance for living systems are present in various forms in different spheres. The various spheres can reach an equilibrium throughout dynamic exchange phenomena. Exchanges between spheres are typically described by relationships between the equilibrium concentrations (for instance Equation 11.1). The transfer of a compound takes place until equilibrium concentrations are reached. The rate of transfer is generally proportional to the distance from equilibrium (or concentration gradients) and depends on the diffusion coefficient of the compound and on the resistance at the interface between the spheres. The resistance at the interface and the diffusion coefficient depend in turn on the temperature, the surface exposed to the atmosphere, or to the liquid volume and the flow rate (e.g., the discussion on transport at interfaces in Section 7.7).

A change, caused by mass transfer processes, in the concentrations of the various components or elements in the ecospheres can have dramatic consequences on the environmental conditions of the ecosystem considered. Moreover, the identification of trends in global changes of concentrations might be strategic, as this may cause, in the long term, serious threats for life conditions on the planet.

In the previous sections, we have introduced several of the most relevant processes governing exchanges of substances among the four spheres, which are ultimately determined by the mass transfer processes and the equilibrium concentrations. In the following, some of these processes are recalled and appropriately contextualized in terms of equilibria among ecospheres.

18.4.1 Atmosphere–Hydrosphere (Henry's Law)

As shown in Section 7.7, Henry's law describes the solubility of a gas at a given concentration in the atmosphere. In this case, the partition coefficient in (Equation 11.1) is the dimensionless Henry's constant of the substance, which determines the equilibrium distribution between the atmosphere and the hydrosphere. Figure 7.6 shows the resistance to gas transfer in liquid phase as a function of Henry's constant values. Reareation and volatilization (Sections 7.8 and 7.9) are special cases of hydrosphere–atmosphere transfer processes that may be described by the two-film theory introduced in Section 7.7.

18.4.2 Hydrosphere–Lithosphere (Adsorption)

Adsorption and ion exchange (Sections 11.2 and 11.3) are the processes responsible for the soil–water distribution dynamic equilibrium, which is usually attained quickly. The partition coefficient in Equation 11.1, in this case, is not constant. Indeed, this coefficient varies with adsorbate concentration in the dissolved phase because these processes are limited by the capacity of the adsorbent to accept dissolved elements. From experimental observations of adsorption processes in nature, a valuable description of these phenomena is provided by one of the adsorption isotherms presented in Chapter 11.

18.4.3 Hydrosphere–Biosphere, Atmosphere–Biosphere, and Lithosphere–Biosphere

Processes, such as respiration, deposition, uptake, decay (Chapter 15), and so on, all involve transfers in or out of the biosphere. The water cycle as well (Chapter 12) comprises the biosphere. Definitely water is an unavoidable player for all living systems. Plants and animals are largely made up of water. Some organisms may contain up to 90% water in their tissues. For example, the human body has approximately 60% water content. The 83% of water in human blood makes the plasma a significantly hydrophilic milieu. The total water throughput for an adult human is about 2.5 L. As stated in Chapter 11, water is the matrix through which substances of biotic relevance are conveyed. For this reason, water plays an important role in supporting the plants and animals in the biosphere. Plants absorb water from soil moisture and nutrients dissolved in water. Water carries away the waste by-products of the cellular metabolism.

Throughout processes such as uptake described in Section 15.4, besides nutrients, a large number of other elements, such as metals, toxicants, or xenobiotic substances can be transferred to the biosphere from the other spheres. Often, the biosphere represents the end point of concern in the propagation of environmentally harmful substances. The fate of these substances, in propagating through the *trophic chain* (see Chapter 19), is in the long term that of bioaccumulation and bioconcentration, as described in Section 18.3.

19

Biological Interactions

19.1 Introduction

Organisms in ecosystems are not living isolated but are continuously in interaction with other organisms either of the same species or with organisms of other species. When we build ecological models, it is crucial to describe these interactions, because the model results are strongly dependent on a close to correct description of at least the most important exchange processes among the state variables. The resources in ecosystems are limited according to the mass and energy conservation principles, which imply that the organisms are competing for the resources. Therefore, competition is an important biological interaction. Many population dynamic models have been constructed to describe this competition (e.g., Lotka–Volterra model) (Jørgensen and Fath, 2011 and Section 19.3). An equally important interaction process is, however, *cooperation*, because without cooperation no ecosystem will endure. Cooperation is necessary to cycle the resources—for instance the biologically important elements—(see Chapter 12). Without cycling, the limiting elements would be used very fast and further development of the ecosystem would stop, and due to the Second Law of Thermodynamics, the ecosystem would inevitably decompose completely and reach thermodynamic equilibrium. This means that the system is without gradients, free energy, and exergy and the entropy is at maximum. Cycling processes are described in Chapters 13 through 17, and how these processes ensure the cycling is presented in Chapter 12. Chapter 18 considers the processes of particularly toxic substances in ecosystems and their influence on other biological processes.

Ecosystems are organized as hierarchical systems and the biological and nonbiological components form networks to ensure that a cycling of the resources can take place. A detailed description of the hierarchical and network organization of ecosystems and interaction processes (e.g., competition and cooperation), which are a direct consequence of this organization, is beyond the scope of this book. Here, the focus is on the ecological processes and not on the ecosystem organization or on the interactive processes resulting from this organization, which would require a profound treatment of ecosystem theory. The direct use of systems ecology and our knowledge about the ecosystem level to build models that capture the right organization and give an approximate good description of the influence of the forcing functions (impacts) on the state variables of the ecosystems, is covered in ecological modeling textbooks (Jørgensen and Bendoricchio, 2001; Jørgensen and Fath, 2011).

Ecological process *adaptation* and *shifts in species' composition* are included in this chapter, because these processes can be described as single processes, although they are very complex. They have a clear result, namely the change of the species' biological properties, which is very important to capture in many ecological model cases. The next section is devoted to these two processes that can be described using practically the same methodology, because they are both a matter of how the biological properties change, when the forcing functions (the impacts on the ecosystem) are changed.

19.2 Adaptation and Shift in Species' Composition

If we follow the generally applied modeling procedure presented in most textbooks on ecological modeling, we will develop a model that describes the processes in the focal ecosystem, but the parameters will represent the properties of the state variables as they are in the ecosystem during the examination period. They are not valid for another period as ecosystem can regulate, modify, and change them, if needed as response to changes in the conditions (Figure 19.1), determined by the forcing functions and the interrelations between the state variables. Usually models have rigid structures and a fixed set of parameters meaning that no changes or replacements of the components are possible. We need, however, to introduce parameters (properties) that can change according to variable forcing functions and general conditions, because we know that these changes are a result of important ecological processes. The state variables (components) optimize continuously the ability of the system to move away from thermodynamic equilibrium, according to what is named the *Ecological Law of Thermodynamics* (ELT). ELT states that a system receiving more exergy (free energy) that needed to cover the maintenance of the system, the system will use the surplus free energy to move further away from the thermodynamic equilibrium. If the system is offered more possibilities (pathways) to move further away from the thermodynamic equilibrium, it will select the pathways that move the system most far from the thermodynamic equilibrium; it means the pathways that give the system most work capacity = eco-exergy (see the discussion in Illustration 4.6.1). ELT can be considered a translation of Darwin's theory to thermodynamics.

The state variables can change by adaptation or by change of the species' composition and several species are waiting to take over, if they are better survivors. Therefore, the idea is to test if a change of the most crucial parameters (properties of the species) would be able to move the system more away

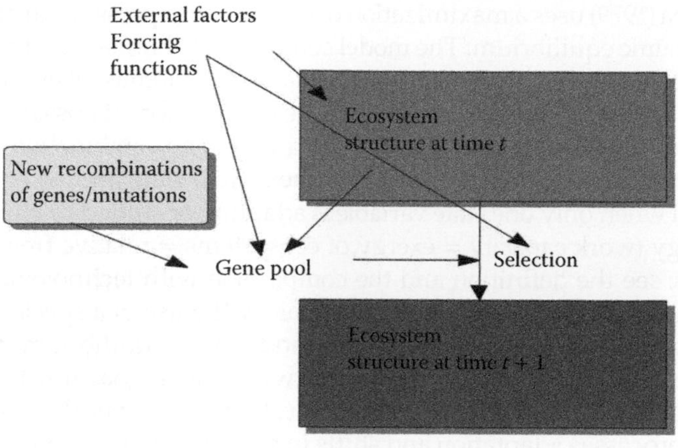

FIGURE 19.1
Conceptualization of how the external factors steadily change the species composition. The possible shifts in species composition are determined by the gene pool, which is steadily changed due to mutations and new sexual recombinations of genes. The development is, however, more complex. This is indicated by arrows from "structure" to "external factors" and "selection" to account for the possibility that the species can modify their own environment and thereby their own selection pressure; an arrow from "structure" to "gene pool" to account for the possibilities that the species can, to a certain extent, change their own gene pool.

from the thermodynamic equilibrium, and if that is the case, to use that set of parameters (properties), because it expresses the adaptation and the shifts in species composition that take place according to ELT.

The model type that can account for the change in species composition as well as for the ability of the species, that is, the biological components of our models; to change their properties, that is, to adapt to the existing conditions imposed on the species, is denoted *structurally dynamic model* (SDM), to underline that they are able to capture structural changes. They also may be called the next or fifth generation of ecological models, to notify that they are radically different from previous modeling approaches and can do more, namely describe adaptation and shifts in species composition (see also Jørgensen 1992a,b; 1994a,b; 1995a,b).

It could be argued that the ability of ecosystems to replace present species with other better fitted species could be considered by constructing models that encompass actual species for the entire period that the model attempts to cover. This approach has, however, two essential disadvantages. The model becomes first of all very complex, as it will contain many state variables for each trophic level. Therefore, the model will contain many more parameters that have to be calibrated, and this will introduce a high uncertainty to the model and will render the application of the model very case specific (Nielsen 1992a,b). In addition, the model will still be rigid and not allow the model to have continuously changing parameters due to adaptation even without changing the species composition.

Straskraba (1979) uses a maximization of biomass to express the distance from thermodynamic equilibrium. The model computes the biomass and adjusts one or more selected parameters to achieve the maximum biomass at every instance. The model has a routine, which computes the biomass for all possible combinations of parameters within a given realistic range. The combination that gives the maximum biomass is selected for the next time step, and so on. Biomass can be used when only one state variable is adapting or shifted to other species.

Eco-exergy (work capacity = exergy of ecosystems = relative free energy of ecosystems; see the definition and the comparison with technological exergy in Illustration 4.6.1) calculated for ecosystems by the use of a special reference system, such as the same system but at thermodynamic equilibrium at the same temperature and pressure, has been used widely as a goal function in ecological models for the development of SDMs, that is, to cover the two possible ecological processes adaptation and shifts in species' composition. Eco-exergy has two advantages as goal function. It is defined far from thermodynamic equilibrium and is related to the state variables, which are easily determined, modeled, or measured. For instance, this is opposite to the maximum power that is related to the flows, which are more rarely determined and more difficult to measure. Eco-exergy can also be applied when two or more species are adapting or shifted to other species, which is often the case. So, the proposed approach is holistic, as it captures that two or more species change their properties by focusing on the overall result (eco-exergy is maximized) as it was one (holistic). For instance, phytoplankton and zooplankton are often changed simultaneously when the forcing function for lakes are changed (Section 20.3).

The thermodynamic variable eco-exergy has been applied to develop SDMs in 23 cases (Zhang et al., 2010). The 23 case studies are the following:

1–8. Eight eutrophication models of six different lakes

9. A model to explain the success and failure of biomanipulation based on removal of planktivorous fish

10. A model to explain under which circumstances submerged vegetation and phytoplankton are dominant in shallow lakes

11. A model of Lake Balaton, which was used to support the intermediate disturbance hypothesis

12–15. Small population dynamic models, a eutrophication model of

16. The Lagoon of Venice and

17. The Mondego Estuary and

18. A model of Darwin's Finches

19. An ecotoxicological model focusing on the influence of copper on zooplankton growth rates

20. A model of the interaction between parasites and birds and

21. The SDM included in Pamolare 1 applied on Lake Fure in Denmark

22. Lake Chazas in Spain

23. An individual based model to show that conjugation is able to provide a better combination of parameters to obtain a higher eco-exergy level

Given in Illustration 19.2.1 is one example of structural dynamic models, namely 18 from the above list of case studies, because it is a relatively simple and easily understandable model, where the structural changes have been observed and are easy to conceive. This entails that the method to describe the ecological processes behind the structurally changes seem to work.

How SDMs are developed by determining the change of the parameters needed to describe adaptation and shifts in species composition by using eco-exergy as goal function is shown in Figure 19.2. Theoretical considerations behind the use of SDM are illustrated in Figure 19.3.

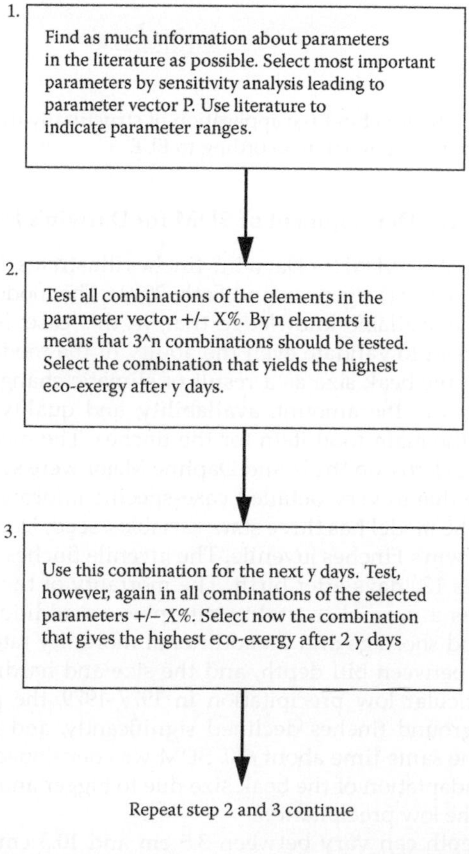
1.
Find as much information about parameters in the literature as possible. Select most important parameters by sensitivity analysis leading to parameter vector P. Use literature to indicate parameter ranges.
2.
Test all combinations of the elements in the parameter vector +/– X%. By n elements it means that 3^n combinations should be tested. Select the combination that yields the highest eco-exergy after y days.
3.
Use this combination for the next y days. Test, however, again in all combinations of the selected parameters +/– X%. Select now the combination that gives the highest eco-exergy after 2 y days
Repeat step 2 and 3 continue

FIGURE 19.2
The procedure used for the development of structurally dynamic models is shown. Structurally dynamic models have been developed successfully in 23 cases. The observed structural changes have been predicted by the model with an acceptable standard deviation, that is, general for ecological modeling. The state variables are, in addition, generally predicted with a smaller standard deviation than for the same ecological models not considering structural changes.

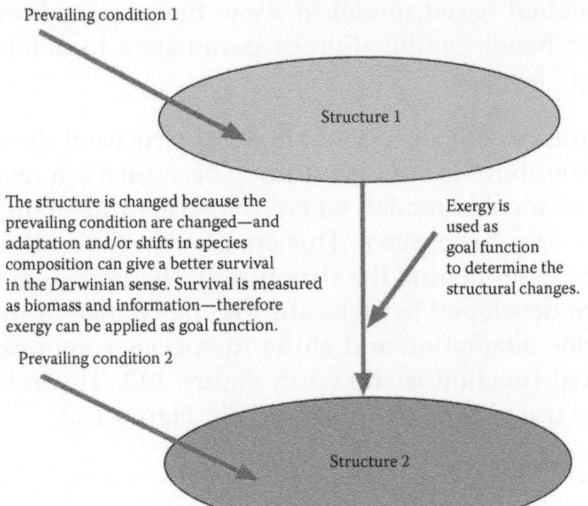
Prevailing condition 1
Structure 1
The structure is changed because the prevailing condition are changed—and adaptation and/or shifts in species composition can give a better survival in the Darwinian sense. Survival is measured as biomass and information—therefore exergy can be applied as goal function.
Exergy is used as goal function to determine the structural changes.
Prevailing condition 2
Structure 2

FIGURE 19.3
The theoretical considerations behind the application of structurally dynamic model. Exergy or eco-exergy is applied as goal function according to ELT.

Illustration 19.2.1: Development of SDM for Darwin's Finches

The development of a SDM for Darwin's finches illustrates the advantages of SDMs very clearly (Jørgensen and Fath, 2004). The model reflects—as all models—the available knowledge that, in this case, is comprehensive and sufficient to validate even the ability of the model to describe the changes in the beak size as a result of climatic changes, which are causing changes in the amount, availability, and quality of the seeds that make up the main food item for the finches. The medium ground finches, *Geospiza fortis*, on the island Daphne Major were selected for this modeling case due to very detailed case-specific information found in Grant (1986). The model has three state variables: seed, Darwin's Finches adult, and Darwin's Finches juvenile. The juvenile finches are promoted to adult finches 120 days after birth. The mortality of the adult finches is expressed as a normal mortality rate plus an additional mortality rate due to food shortage and an additional mortality rate caused by a disagreement between bill depth, and the size and hardness of seeds. Due to a particular low precipitation in 1977–1979, the population of the medium ground finches declined significantly, and the beak size increased at the same time about 6%. SDM was developed to be able to describe this adaptation of the beak size due to bigger and harder seeds as a result of the low precipitation.

The beak depth can vary between 3.5 cm and 10.3 cm according to Grant. The adapted beak size is furthermore equal to square root of $D \times H$, where D is the diameter and H the hardness of the seeds. Both D and H are dependent on the precipitation, particularly from January to April. The coordination or fitness of the beak size with D and H is a survival factor for the finches. The fitness function is based on the seed handling time and it influences the mortality, but has also an impact on the number of eggs laid and the mortality of the juveniles. The growth rate and mortality rate of the seeds is dependent on the precipitation and the temperature, which are forcing functions known as f (time). The food shortage is calculated from the food required by the finches, which is known according to Grant, and the actual available food according to the state function seed. How the food shortage influences the mortality of the adults and juveniles can be found in Grant (1986). The seed biomass and the number of finches are known as a function of time for the period 1975–1982 (Grant, 1986). The observations of the state variables from 1975–1977 were applied for calibration of the model, focusing on the following parameters:

1. Influence of the fitness function on (a) the mortality of adult finches, (b) the mortality of juvenile finches, and (c) the number of eggs laid.
2. Influence of food shortage on the mortality of adult and juvenile finches is known (Grant 1986). The influence is therefore calibrated within a narrow range of values.
3. Influence of precipitation on the seed biomass (growth and mortality).

All other parameters are known from the literature (Grant 1986).

The eco-exergy density is calculated (estimated) as 275 times the concentration of seed + 980 times the concentration of the finches. Every 15 days, it is found if a feasible change in the beak size (the generation time and the variations in the beak size is taken into consideration) will give a higher eco-exergy. If it is the case, then the beak size is changed accordingly. The modeled changes in the beak size were confirmed by the observations. The model results of the number of Darwin's finches are compared with the observations in Figure 19.4. The standard deviation between modeled and observed values was 11.6%. For the validation and the correlation coefficient, r^2 for modeled versus observed values is 0.977.

The results of a nonstructural dynamic model would not be able to predict the changes in the beak size and would therefore give too low values for the number of Darwin's Finches because their beak would not adapt to the lower precipitation yielding harder and bigger seeds. The calibrated model not using the eco-exergy optimization for the SDMs in the validation period 1977–1982 resulted in a complete extinction of the finches. A non-SDM—a normal biogeochemical model—could, therefore, *not* describe the impact of the low precipitation, while the SDM approach gave an approximately correct number of finches and could describe the increase of the beak at the same time. The results show that, at least for some models, it is indispensable to include adaptation or change of the species' composition. In these cases, it is indicated to apply SDMs.

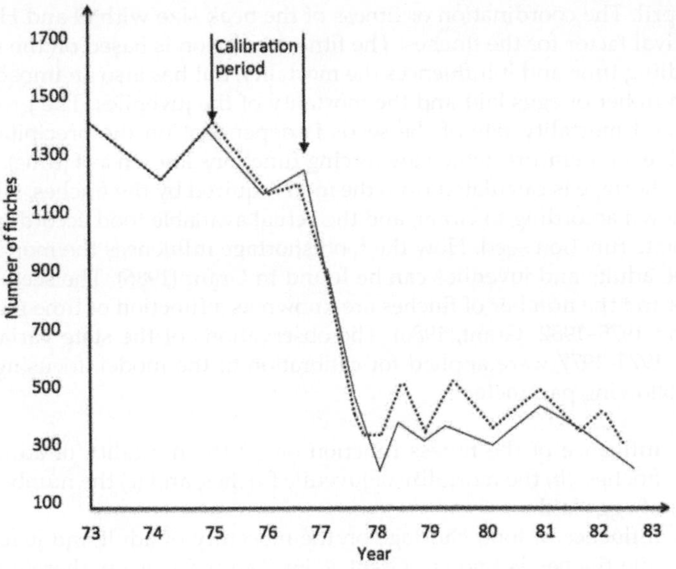

FIGURE 19.4
The observed number of finches (solid line) from 1973 to 1983, compared with the simulated result (dotted line). 75 and 76 were used for calibration and 77 and 78 for the validation.

19.3 Competition

Organisms need resources, such as food, nutrients, light (in the case of photoautotrophs) and even space to survive. In practice, however, optimal levels of resources are never present, and the growth of living organisms and populations is limited by resource availability, as discussed in Chapter 15 in the case of primary producers. *Competition* for a resource takes place when two or more organisms (or populations, functional groups, taxa, etc.) are in suboptimal conditions for growth because they are all exploiting the same resource and, because of this, their full access to such resource is limited by the other organisms. In this case, such organisms are said to be competing for that resource. According to a possible classification of biological interactions based on their effects on the interacting agents, competition is an interaction that yields negative consequences for all the organisms involved, as all of them experience resource scarcity due to the presence of the other organisms (Table 19.1). According to the *competitive exclusion principle*, or Gause's principle, the negative effect of competition can be so strong to cause, in some instances, the disappearance of one of the competing species (see the Lotka–Volterra Equations 19.1 and 19.2). On the other hand, competition can have positive effects on the integrity and diversity of ecosystems: competition by autochthonous organisms that are well adapted to the local environmental conditions can prevent the invasions of ecosystems by alien

TABLE 19.1

Possible Biological Interactions between Two Organisms (or Populations, or Group of Organisms), and Their Effect on Such Organisms

Effect of the Interaction between Species A and B		
On Species A	**On Species B**	**Name of the Interaction**
+	+	Mutualism and symbiosis
+	0	Commensalism
+	–	Predation/parasitism
0	0	Neutralism
0	–	Amensalism
–	–	Competition[a]

Source: Odum, E.P., *Basic Ecology*, CBS College Publishing, Philadelphia, PA, 1983a.

Note: + means that the corresponding species benefits from the interaction with the other one, – means that the corresponding species is disadvantaged by the interaction, 0 indicates that the interaction has no effect on the species' fitness.

[a] Competition is actually not limited to competition for resources as described in Chapter 15. A biological interaction with negative effects for all the interacting organisms can take place in other instances (Odum, 1983), for example, in the case of direct mutual interference when each species directly causes the inhibition of the others, for example, through aggression or by secreting venomous substances.

"weed" species, which would degrade the system state. Indeed, one possible mechanism favoring the blooms of alien jellyfish species in coastal seas is the overfishing of their local, natural competitors for mesozooplankton such as small pelagic fish.

As mentioned above, competition can take place between individuals, populations, groups of organisms, and so on, both in the case of consumers (e.g., predators competing for the same prey) and primary producers (e.g., microalgae competing for nutrients in the water column). Competition can be either *interspecific*, as in the case of different predator species competing for the same prey item (e.g., anchovy and sardine competing for their mesozooplankton food) or *intraspecific*, that is, among individuals of the same species. Intra- and interspecific competition is implicitly contained in the carrying capacity term of the logistic growth equation of a population, as described in Chapter 14. Such term accounts for the fact that natural resources are not infinite and thus, at high population levels, growth slows down as competition ensues between individuals of that population and/or with other populations in the ecosystem.

Competition can take place for several kinds of resources. Food is an obvious example, since energy and nutrients are needed to sustain consumer populations, and their lack can limit growth (as in the case of the situation known as bottom up control of the predator–prey interaction, or control by the resource, Chapter 16). The role of limiting factors in controlling the dynamics of primary producers has been discussed in Chapter 15: competition can ensue in the case of nutrients, which are assimilated by many different plants and algae and can thus be depleted from the environment, and for light, since the growth of primary producers can shade the environment (e.g., in the case of thick canopies or of dense microalgal populations increasing water turbidity; this effect of light extinction is described in Section 8.3). Also space can be a resource to compete for, especially in ecosystems whose spatial extension is limited. For example, benthos can compete for the space they dwell in on the seafloor, or trees can compete for the space they need to grow on the ground.

The famous Lotka–Volterra equations represent a mathematical description of this competition process. In the case of two populations of competing organisms, the logistical growth equation can be modified by adding a negative competition term, which is proportional to the product of the abundances of the two populations. This term indicates that the negative effect of competition becomes larger as both populations increase in abundance, since this increases their need for resources and their chances to meet with each other.

$$\frac{dN_1}{dt} = r_1 N_1 \left(1 - \frac{N_1}{K_1} - \frac{\alpha_{12} N_2}{K_1}\right) \tag{19.1}$$

$$\frac{dN_2}{dt} = r_2 N_2 \left(1 - \frac{N_2}{K_2} - \frac{\alpha_{21} N_1}{K_2}\right) \tag{19.2}$$

The abundances of the two competitors are represented by N_1 and N_2, r_1 and r_2 are the maximum intrinsic growth rates of the two populations, K_1 and K_2 the carrying capacities, and α_{12} and α_{21} measure, respectively, the effect of competition of population 2 on population 1, and vice versa. The study of the stability of the solutions of these coupled equations shows that they can display a very rich dynamic behavior and that the above-mentioned competitive exclusion of one species takes place in case of high values of the α coefficients, if certain conditions regarding the values of the carrying capacities are met (K's and α's ultimately determine the model stability; see, for example, Jørgensen and Bendoricchio, 2001). In particular, the two populations can stably coexist without outcompeting each other in the long term only if $K_1 < K_2/\alpha_{21}$ and if, at the same time, $K_2 < K_1/\alpha_{12}$.

19.4 Other Biological Interactions

Competition is just one of the possible interactions among biota in ecosystems. Depending on the consequence of the interaction process upon the fitness of the interacting organisms, several types of biological interactions can be identified (Table 19.1). The cooperation between nitrogen-fixing bacteria living in the nodules of legume roots, and the legume plant (see Section 12.5), is an example of *symbiosis. Rhizobium* bacteria fix nitrogen into a form that can be absorbed by the plant, which in exchange provides bacteria with carbon and energy produced during photosynthesis (e.g., sugars). *Mutualism* can be represented in Equations 19.1 and 19.2 by changing the interaction term from negative to positive:

$$\frac{dN_1}{dt} = r_1 N_1 \left(1 - \frac{N_1}{K_1} + \frac{\alpha_{12} N_2}{K_1} \right) \tag{19.3}$$

$$\frac{dN_2}{dt} = r_2 N_2 \left(1 - \frac{N_2}{K_2} + \frac{\alpha_{21} N_1}{K_2} \right) \tag{19.4}$$

In *commensalism,* one species benefits from the presence of a second one which is not affected by the presence of the former one. One classic example is the remora which attaches to the body of a shark (which is not damaged by the presence of the remora) and eats the remains of the shark's preys.

Predation has been addressed in Section 16.3. *Parasitism* differs from predation in two manners. Firstly, parasites do not kill their host (at least, not immediately) and, secondly, they are often much smaller than it, as in the case of ticks and fleas.

19.5 Trophic Networks

Predation (Section 16.3) is one of the best studied biological processes in ecology. The reason is that the outcome of predation is straightforward to observe and measure: predation means death for some organisms, and a certain amount of food obtained by the predators. All organisms are affected by predation: heterotroph consumers obtain carbon and energy from the body tissues of their preys, and although autotrophs (primary producers) do not eat other organisms, they experience predation mortality due to the action of grazers. Thus, multiple feeding (or trophic) interactions take place simultaneously and continuously in ecosystems, across all trophic levels. Traditionally, ecosystems were seen as chains of predator–prey interactions, or *food chains,* which means that trophic interactions were considered to be organized according to a linear structure: ecosystems were thought to be composed of primary producers (e.g., microalgae) eaten by secondary producers (e.g., copepods), which were eaten by tertiary producers (small fish), which were eaten by bigger predators (predatory fish), and so on. In fact, ecosystems are much more complex than a linear chain of interactions, and indeed one feature of living systems is the presence of feedback loops. Thus, in real ecosystems, feeding interactions are actually organized as a network containing ramifications, parallel pathways and cycles, rather than as straight chains. The terms "trophic network" or "food web" highlight that

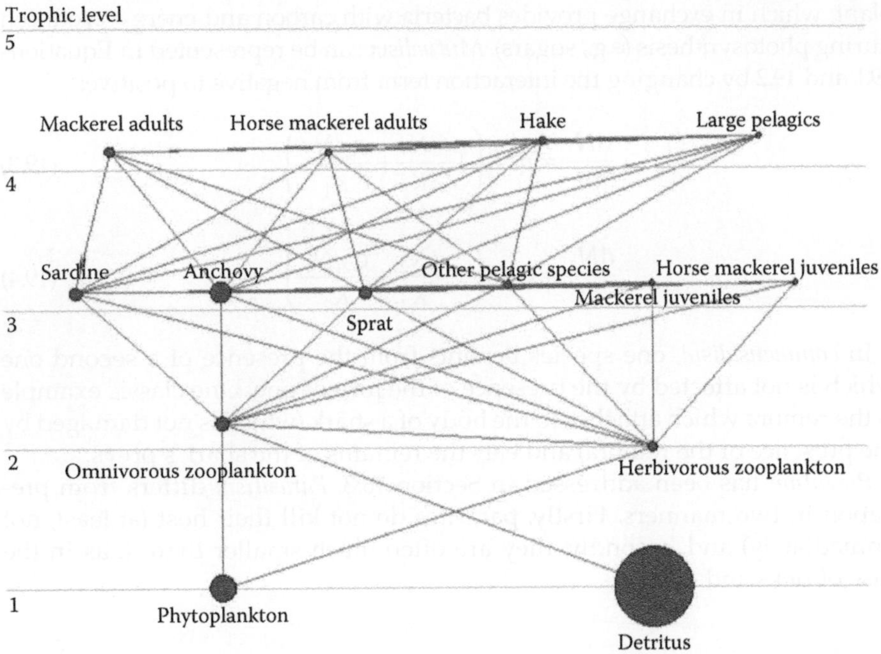

FIGURE 19.5
The structure of the pelagic trophic network of the Northern Adriatic Sea in 1965. Circles indicate groups of organisms in the food web (the area of each circle is proportional to the biomass of the corresponding group in the system), lines indicate trophic connections. The vertical position of the groups identifies their trophic level in the pelagic ecosystem (Galli, G. et al. A trophic network model to study biomass oscillations of anchovy and sardine populations in the northern Adriatic pelagic ecosystem, 2013.).

a complex structure of predator–prey interactions is in fact present in ecosystems. A trophic network model is the representation of who eats whom in the ecosystem (Figure 19.5), and, possibly, how much in the case of weighted trophic network models in which trophic flows, that is the amount of food consumed, are quantified for each trophic interaction.

The study of food webs has a long tradition in ecology, dating back to the last decades of the nineteenth century (Dunne, 2006), and represents an active research field in today's ecology, given the central influence that the structure and functioning of trophic networks can have on other ecological processes and whole ecosystem dynamics. Recently, the central influence of biological interactions on ecosystems and upon the benefits that humans derive from them has also been recognized at the policy level, leading to popular concepts such as the "ecosystem approach" and the "ecosystem approach to fisheries" (Garcia et al., 2003), which explicitly state the importance of taking the connections among the ecosystem components (which include the role of food web interactions) into account to manage the environment in a sustainable manner.

Over the past decades, in the case of fisheries science, a shift has taken place from the focus on single species toward multispecies and food-web models accounting for interspecies biological interactions such as predation.

Food webs can be viewed as intricate systems of consumers and producers, in addition to the key role played by detritus (the dead organic matter coming from both consumers and producers), which contains energy and nutrients that can be cycled back to the food web through the action of decomposers (Chapter 17). A useful concept to synthesize the structure of food webs is that of the trophic level (defined in Section 16.1) of an organism, that is, the number of predator–prey interactions that separates it from the energetic basis of the food web, as represented by the primary producers or detritus which represent, by definition, the first trophic level. The second trophic level is occupied by the grazers and decomposers, the third level by secondary producers or consumers, and so on. Noninteger trophic levels are also possible and are, indeed, common, in the case of omnivores (Chapter 16).

Traditionally, the study of food webs has developed along separate pathways. A broad research line is focused on network structure and topology, that is, the study of food web properties and regularities regarding the number of existing trophic links, for example, in relation to species richness, their species-specific distribution, the actual number of possible links that are realized (or connectance; in a food web of N species with L trophic links, connectance is equal to L/N^2), and so on; see, for example, Dunne et al. (2002) and Dunne (2006). For example, the classic work of Robert M. May in 1972–1973 challenged the paradigm that diversity and complexity necessarily lead to the increased stability of biological communities by showing that, in randomly assembled food webs, stability will be achieved only for small values of species richness and connectance (Dunne, 2006).

A second famous field of food web research focuses on the properties of quantitative, or weighted, trophic networks; that is, food webs where the magnitude of flows (predation, respiration, flows to detritus, etc.) exchanged among network compartments (representing populations, functional groups, taxa, etc.), and between compartments and the environment, is quantified in terms of the amount of flowing energy, carbon, biomass or nutrients. The study of such quantitative food webs, or flow networks, is useful to construct models of food webs and of their interactions with human activities, for example, fisheries. Indeed, one of the most famous approaches to construct quantitative trophic network models of aquatic ecosystems is the *Ecopath* software, which is strongly fishery-oriented (Pauly et al., 2000). Ecopath is based on a mass-balance approach, which means that it exploits the physical principle of mass and energy conservation to estimate the unknown model parameters (Figure 19.6).

Quantitative trophic networks have also been studied to gain information on the status, or "health," of whole ecosystems. According to the theory of ecological succession, ecosystems should display typical and recognizable patterns of change, for instance in the types of dominant ecological

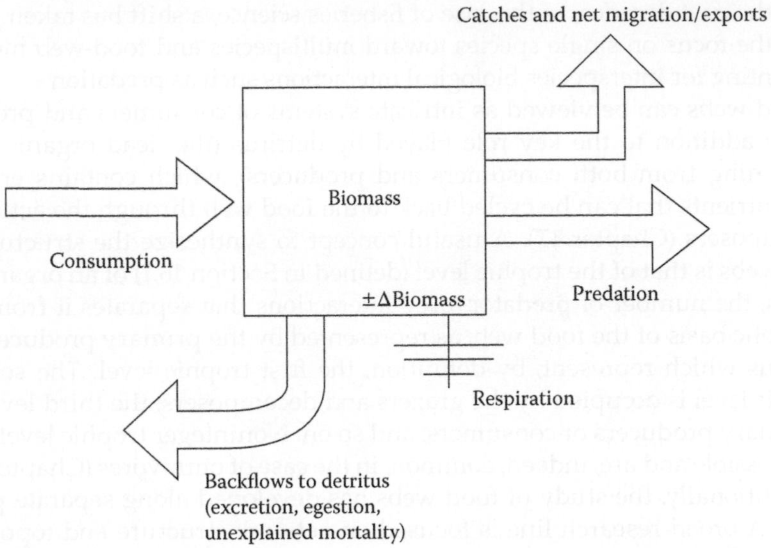

FIGURE 19.6
The mass-balance approach in *Ecopath*. For any given network node, representing a population or group of organisms, the following mass budgets must be realized over an appropriate averaging time period over which the system is in a quasistable state: (1) the production of the node should equal the different mortality sources for the node (predation, natural mortality, fishing mortality) plus the mass exported from the system (e.g., migration, other exports); (2) the flow into the node must equal the flow from the node, that is, consumption = production + respiration + flows to detritus. By writing these mass budgets for all network nodes and ensuring the mutual consistency, it is possible to estimate the values of poorly known trophic flows in the ecosystem. (From Pauly, D., Christensen, V., Walters, C., *ICES J. Mar. Sci.*, 57, 697–706, 2000.)

processes, in the complexity of ecological interactions, in the traits of the species present and in the distribution of biomass, as they grow and develop unperturbed by human actions or natural disturbances (see Chapter 4) (Odum, 1969; Ulanowicz, 1997). Since trophic flows can be easier to observe than other ecological processes, ecologists have related changes in the maturity of ecosystems to the structure of trophic flows in the food web, for example to the redundancy of trophic pathways, the amount of cycling, etc. (e.g., Ulanowciz, 1997; see also the discussion in Section 4.5). According to this approach, although ecosystem maturity is the result of many factors, which are not necessarily related to trophic processes, nevertheless the structure of matter or energy flows in food webs should reflect such maturity in a phenomenological manner. For example, a toxic chemical of human origin could kill most of the top predators in an ecosystem because of the biomagnifications process (Chapter 18), thus reducing system maturity. Although such reduction in maturity is not due to trophic processes, the structure of trophic flows in the food web would reflect it, since trophic networks characterized by scarce flows in upper trophic levels and a short average length of trophic pathways are typically observed in immature ecosystems.

The study of the dynamics of food web populations over time is a field of research in its infancy. Traditional approaches based, for example, on Lotka–Volterra equations (Chapter 16 and Section 19.3) and their many modifications are simple descriptions of reality, and typically do not take into account the key ecological factors such as the role of spatial heterogeneity in ecosystems or abiotic factors, or the dependence of many ecological processes on body size (Brown et al., 2004). As described in Chapter 16, the dynamics of food webs can be classified as top down, if it is the predators who control the abundance of its preys; as bottom up, if it is the preys and/or environmental factors that control the abundance of predators; or wasp-waist, if a dominant species determines both the abundance of its predators through bottom up control and the abundance of its preys through top down control.*

* See also the discussion on causation in the beginning of Chapter 4.

Section V

Landscape Processes

20

Aquatic Ecosystems

20.1 Introduction

Aquatic ecosystems are ubiquitous elements of natural or seminatural landscapes. Aquatic environments and the communities that inhabit them constitute the larger part of the biosphere as explained in the previous chapters. Water indeed covers about three quarters of the surface of our planet. The vast majority of aquatic ecosystems consist of salt water ecosystems, because almost all the water on the earth is found in the oceans and seas. Different zones of the ocean, each characterized by particular communities of organisms, can be distinguished as intertidal, benthic, and pelagic zones. Freshwater ecosystems, such as rivers, lakes, and ponds, represent only a limited portion of the water present on the planet. However these environments are home to a wide variety of organisms accounting for about 10% of all aquatic species.

All basic life processes started in water environments, where indeed life has begun. Water is the essential matrix for all organisms. Moreover, the history of life evolution occasionally implied some changes in water usage, storage, displacement, quality, or more generally characteristics. Definitely, describing ecological processes in aquatic environments is of utmost importance to understand the different ecological roles of the water matrix in landscape dynamics.

20.2 Eutrophication

Among the key ecological processes, primary production (e.g., growth in Chapter 14) is obviously taking place in each ecosystem. Particularly, in aquatic ecosystems, growth with limitation (Chapter 15) may result in extreme dystrophic conditions, in which the system may become highly productive due to nutrient enrichment. This altered trophic condition is known as *eutrophication* and results in extensive algal blooms induced by nutrients abundance, primarily nitrogen, and phosphorus inputs of natural or anthropogenic origin. Some of these algae itself may cause nuisance by impairing water use (e.g., producing unpleasant odors and tastes). When the excess algae die (e.g., due to light limitation by self shading) and settle, dissolved oxygen is consumed by the degradation of dead organic matter (Chapter 17) in the deeper waters. Oxygen depletion thereby affects adversely organisms of the higher trophic levels such as zooplankton and fish. This oversimplified description of the phenomenon provides a conceptual framework allowing to link nutrient cycling processes in the landscape to food-web dynamics both in seas and freshwater ecosystems. As discussed in Section 15.2.2, typically nitrogen is the growth-limiting nutrient in marine systems, while freshwater systems are phosphorus-limited.* Excesses of these two nutrients may give rise to an alteration of the trophic state. The trophic status is often employed as a basis for the definition of useful indicators of water bodies' conditions, affecting possibly water quality and usability. Trophic conditions may span from *oligotrophic* (nutrient scarcity) to *mesotrophic* or decisively *eutrophic* (nutrient excess), depending on nutrients availability and other environmental conditions such as water chemistry or hydraulics.

To appreciate trophic dynamics in aquatic ecosystems, a descriptor of feeding and growth activities may be used. Vollenweider et al. (1998) proposes a comprehensive index of eutrophication for marine ecosystems, TRIX defined as

$$\text{TRIX} = A\,[\log(\text{CL-a} \times \Delta\text{O} \times \text{DIN} \times \text{TP}) + B]$$

where A and B are scaling coefficients, CL-a is chlorophyll-a concentration, ΔO is the deviation percentage of oxygen concentration from saturation, DIN is the dissolved inorganic nitrogen, and TP is total phosphorous. This index summarizes the different factors typical of eutrophication: the productivity factors (biomass as concentration of chlorophyll-a and dissolved oxygen as absolute deviance from saturation), and the nutritional ones (concentration of DIN and total phosphorus). Numerically, the index is scaled from 0 to 10 covering a wide range of trophic conditions from oligotrophy to eutrophy, as described in Table 20.1.

The trophic state indicator for marine ecosystems TRIX, when generalized, may be considered as an evolutional criterion, as discussed in Chapter 3. The index derived from the four variables chosen by Vollenweider for marine ecosystems is generalized as (EEA, 2001)

$$X = \frac{k}{n}\sum_{i=1}^{n}\log\left(\frac{x - x_{\min}}{x_{\max} - x_{\min}}\right)_i \tag{20.1}$$

* Normally the limiting nutrient is that present in less quantity or that whose recycling time is the greater (see Section 15.2.2).

TABLE 20.1

Rescaled TRIX Values for Ecosystems' Trophic State

TRIX	Trophic State	
2–4	Good and above	Oligotrophy
4–5	Moderate	Mesotrophy
5–6	Poor	Eutrophy
6–8	Bad	

where n is the number of variables, x_i actual value of the i-th variable, x_{max} upper limit, x_{min} lower limit, and k a scaling coefficient.

The generalized trophic index (Equation 20.1) may be recast in forms similar to Equation 3.6. Here, Glansdorff–Prigogine criterion says that the variation of some x_i per unit time should display a steady behavior.

20.3 Lakes

Lakes are aquatic ecosystems in which water is standing or relatively still. This condition is common to several other aquatic ecosystems, ranging from ponds to wetlands (see next paragraph). Some authors define a lake as a surface freshwater body with depth exceeding 6 m (cfr. e.g., Mitsch and Gosselink, 2007). Three regions may be identified in lakes: the pelagic open water zone, the benthic zone that comprises the bottom and shore regions, and a deep zone (typically below the euphotic* band). These three areas can have very different biotic and abiotic conditions, hence hosting species that are specifically adapted. In general, it can be said that lakes are freshwater quasistill water bodies, in which the bottom layer (the deep zone and benthic areas) is not entirely and permanently exposed to sunlight.

As mentioned before, freshwater ecosystems are usually phosphorus limited. Because lakes are surface freshwater bodies collecting nutrient loads generated in the neighboring catchment and conveyed by surface runoff and groundwater upsurge, these ecosystems are highly subject to the risk of eutrophication. Understanding phosphorus dynamic (Section 12.6) is indeed crucial to evaluate the lake ecosystem status.

In temperate areas, lakes may become seasonally stratified, which means that a vertical inhomogeneity is established, the *thermocline,* dividing the lake in two regions, the *hypolimnion* (below) and the *epilimnion* above, with very different physical and chemical characteristics (Figure 20.1).

The epilimnion is exposed to the surface and therefore becomes turbulently mixed as a result of surface wind. Being in contact with the atmosphere, as

* See Section 7.5 for a definition of this term.

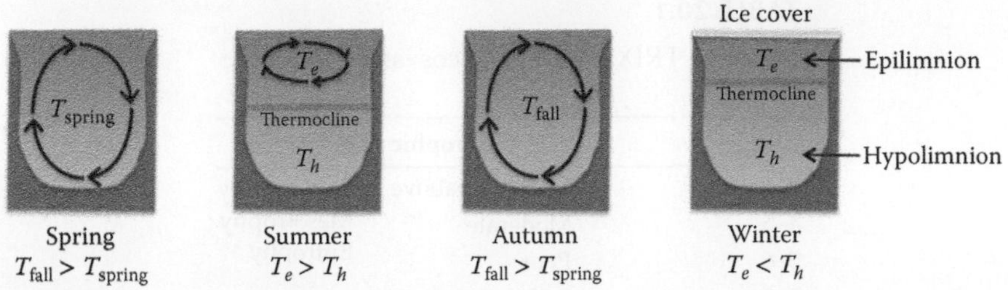

FIGURE 20.1
(See color insert.) Lake stratification and the seasonal cycle. In a thermally stratified lake, the epilimnion is the top-most layer, above the thermocline. Typically, it is warmer and has a higher pH and high dissolved oxygen concentration than the hypolimnion. Thermocline depth may change during the seasonal cycle.

described in Chapter 7, Henry's law controls exchanges of dissolved gases such as O_2 and CO_2. Primary production is dominant in this layer because it receives most part of the sunlight. As phytoplankton grows, it absorbs dissolved nutrients from the water column, and after dying it sinks to the hypolimnion. The hypolimnion is the dense, bottom layer of water in a thermally stratified lake. Typically this is the coldest layer of a lake in summer and the warmest layer during winter (Figure 20.1). The hypolimnion is isolated from surface wind mixing during summer, and usually receives insufficient irradiance (light) for photosynthesis to occur. In deep, temperate lakes, water temperature in the hypolimnion is typically close to 4°C throughout the year. The hypolimnion may be much warmer in lakes at lower latitudes.

Wind, light irradiance, depth, and temperature are the key abiotic factors that influence lake biochemistry along with nutrient loads from neighboring catchment. Biota interactions (i.e., bacteria, algae, aquatic plants, invertebrates, and fishes) may be portrayed by trophic networks (Section 19.5), which describe the dynamic of the food web allowing the assessment of lake ecosystem quality indicators and community successional patterns. Indeed, phytoplankton and zooplankton communities in lakes undergo seasonal succession in relation to nutrient availability, predation, and competition.

Toward the end of winter, increased nutrient and light availability result in rapid phytoplankton growth. The dominant species (e.g., diatoms) are small and grow quickly. Plankton is consumed by zooplankton, which becomes the dominant taxa. In early spring, as phytoplankton populations become depleted due to increased predation by zooplankton, a clear water phase occurs, whereas in summer, zooplankton abundance declines as a result of decreased prey availability (phytoplankton) and increased predation by juvenile fishes. At this stage, with increased nutrient availability and decreased predation from zooplankton, typically a diverse phytoplankton community develops and takes over. In late summer, nutrients become depleted starting from phosphorus that is readily available and quickly absorbed.

Subsequently, silicates and then nitrogen become the limiting nutrients (Chapter 15) ruling the abundance of the various phytoplankton species. Because of a minor vulnerability to fish predation, small-sized zooplankton now becomes dominant. During autumn, predation by fishes is reduced due to lower temperatures and zooplankton of all sizes increase in number. Finally in winter, cold temperatures and decreased light availability result in lower rates of primary production and decreased phytoplankton populations. Zooplankton reproduction decreases due to lower temperatures and fewer preys.

Besides eutrophication, another relevant anthropogenic impact is *acidification* (Chapter 23). Sulfur dioxide and nitrogen oxides created by the combustion of fossil fuels dissolve in atmospheric moisture and enter lake ecosystems as acid rain. Carbonate-rich bedrocks act as natural buffer, however, systems poor in carbonates are very sensitive to acid input because they have a low neutralizing capacity, resulting in pH decrease due to small inputs. In acidic conditions (pH of 5–6), algal biomass and species diversity decrease significantly, resulting in increased water transparency. At further lower pH values, all fauna becomes less diverse, and starting from the higher trophic levels, the food web is progressively disrupted.

Lake ecosystems on a global scale show latitudinal similarity patterns (e.g., both productivity and diversity* decrease with increasing latitude). This well-documented global pattern is common to other aquatic and terrestrial ecosystems, and is likely to be related with solar irradiation, climate, and weather variability.

20.4 Wetlands, Transitional Water Ecosystems

Wetlands are the most productive ecosystems of the biosphere. These important landscape elements are referred to by Mitsch and Gosselink (2007) as "the kidneys" and "ecological supermarket" of the landscape. Wetlands are very complex systems that separate and transform substances (pollutants) by physical, chemical, and biological processes that are performed simultaneously or sequentially during the staying of the water within the system. Furthermore, these are transitional water bodies connecting different portions of the landscape, and regulate the hydrologic balance for the broad system.

About 5% to 8% of land surface of the earth may be classified as wetlands (Mitsch and Gosselink, 2007) depending on the definition of wetland itself. Different types of wetlands are named differently; for example, bog, fen,

* These terms are reflected by both indicators, the emergy presented in Section 4.4 and the exergy of Section 4.6.

swamp, mangrove, marsh, moor, muskeg, pan, playa, tundra, and so on. All these systems are often grouped under the term *mire*. Several exchange processes among the different environmental matrices that have been introduced in the previous chapters, occur in wetlands. Wetland definition often depends on the particular processes that are the very scope of the survey for a wetland site. Wetlands may dry out periodically, thus entangling the attempt to give a simple definition.* Depending on the perspective, a variety of points of view may contribute to the definitions, classification, and management techniques. The wetlands inventory of the *Ramsar Convention*† is based on a very broad definition: "*Wetlands are areas of marsh, fen, peatland or water, whether natural or artificial, permanent or temporary, with water that is static or flowing, fresh, brackish or salt, including areas of marine water the depth of which at low tide does not exceed six metres*" (Article 1.1). Additionally, Ramsar sites "may incorporate riparian and coastal zones adjacent to the wetlands, and islands or bodies of marine water deeper than six metres at low tide lying within the wetlands" (Article 2.1).

Wetlands provide important ecological values. Despite humans have for centuries lived among and depended on these ecosystems, there is a legacy of fear and misunderstandings toward the role of these systems. After a period in which humans have employed much effort in wetland drainage and destruction (e.g., for land reclamation), today their many values are being recognized resulting in protection, restoration, and conservation actions in several parts of the world. Particularly, wetlands are nowadays valued highly for their elevated pollutant (e.g., nutrient) retention potential and their unique biodiversity. At present, there are an increasing number of activities aimed at restoring these sites as multifunctional landscape entities. The multiple functions of these ecosystems span in a continuum from ecological to social and several of these are often performed simultaneously. The main wetland multifunctional values are summarized in Table 20.2.

Wetland functions are characterized by the specific set of physical, chemical, and biological processes occurring in and making up the ecosystem. These processes have been described in detail in the previous chapters and include

- Water movement through the wetland into streams or the ocean
- The removal of nutrients, sediment and organic matter, and several other pollutants (e.g., heavy metals and microorganic toxicants) from water moving into the wetland
- Organic matter degradation
- Release of nitrogen, sulfur, and carbon into the atmosphere
- Growth and development of communities of organisms

* Indeed, it may occur that a wetland may be dry more often than wet.

† International treaty signed in Ramsar (Iran) in the early 1970s aiming at protecting wetland habitats all around the world.

TABLE 20.2

Multiple Functions and Values for Wetland Ecosystems

Multifunctional Wetland Values		
Ecological	Water quality improvement through the processes of assimilation and transformation of nutrients and other pollutants (self-purification).	
	Increase of the naturalistic value of the site	• Enhanced primary production by photosynthesis • Animal life production • Increase of biodiversity • Export to adjacent ecosystems of products and services
	Groundwater recharge	
Landscape Management and Social	Mitigation of flood peaks and storage of water for subsequent use (e.g., irrigation or urban water supply)	
	Uses with social value	• Landscape protection • Recreational • Commercial • Educational

The quantification of values comes usually from a very subjective estimation, of cost, merit, quality or importance of the various ecosystem services provided by wetlands. For instance, Costanza et al. (1997) presented a promising methodology for calculating the value of ecosystem services and natural capital, outlining at the same time the limits of such estimations. Wetland "values" may derive from direct outputs such as the production of food or timber and recreational uses, as well as indirectly from the various ecosystem services provided such as water quality improvement, flood control, and biodiversity or habitats conservation.

Probably the most valued function of wetlands is their self-purification capacity. The processes involved have long been qualitatively known from the theoretical point of view, but their nature and their close interconnection has to date prevented, for some of them, the acquisition of a satisfactory in situ experimental knowledge. The two main mechanisms that result in the self-purification capacity of a wetland system are ultimately the separation of the solid phase from the liquid (e.g., by adsorption), and the transformation of substances dissolved in the water column by chemical or biological degradation (see Chapter 12). Typically, wetlands are thought to be able to remove suspended solids, nitrogen, phosphorus, organic matter, pathogens microorganisms, organic pollutants, and heavy metals. Despite the impossibility of a detailed description, experimental observations suggest that overall these removal processes may be modeled by first order kinetics of common contaminants concentrations, at least for the purpose of wetland design. Kadlec and Wallace (2008) suggest that a wetland performance model can be built by combining two different approaches: the first part is a deterministic first order kinetic (e.g., Equation 9.11 with $n = 1$). This representation is used to describe the average performance or tendency of the wetland as a water purification system. Probability distributions represent the second part of the mathematical model, and are implemented as multipliers on the seasonal trend that are specific for each of the common contaminants, and the various wetland types. Given the random variation ψ, the actual concentration C can be captured through the use of a multiplier on the trend value C_{trend}:

$$C = C_{trend}(1+\psi) \tag{20.2}$$

where C_{trend} represents the seasonal concentration trend obtained through first-order kinetics by ODEs (Equation 9.11).

20.5 Rivers

Rivers are streams of water flowing in a prevalent direction (i.e., watercourses), toward another water body such as, a sea, a lake or another river. Rivers typically contain freshwater, even if in proximity to the sea, their water can be brackish or salted due to saltwater intrusion; in such cases, salinity and other sea processes (e.g., tides) can significantly influence physical, chemical, and biological processes taking place in the river ecosystem, and the composition of its biota. River ecosystems provide man with multiple services, and, for this reason, most of the ancient civilizations were born close to rivers, such as the City of Rome and the river Tiber, or the civilizations of Mesopotamia and the Tigris–Euphrates river system. For instance, rivers represent natural defenses and boundaries, provide man with food (e.g., fish) and water to drink or to be used for irrigation, and they can be navigated and used for transportation and trade. Water flow has been exploited since ancient times to produce energy through hydropower (e.g., through watermills). Also, rivers can be exploited to dispose wastes and to discharge wastewaters. Wastes are not only dilute or carried away by the water, but the self-purification capability of aquatic ecosystems is exploited to degrade them through multiple processes, including biological ones (Sections 17.4 and 18.2), into simpler and harmless forms.

Given the close relationship between man and rivers, these aquatic ecosystems are typically impacted, to a varying degree, by numerous human activities. Some examples include the construction of embankments and dams, the discharge of wastes with consequent eutrophication and/or pollution, water abstraction, fishing, introduction of nonnative species, and so on. In some heavily human-populated areas, (e.g., northeastern Italy), the anthropogenic impact is so pervasive and dates back to so many centuries ago, that it does not make sense to speak of "natural" river ecosystems (Figure 20.2).

FIGURE 20.2
(See color insert.) The Brenta river in Nothern Italy, an example of a water body that has been strongly influenced by man since centuries ago. The Brenta river used to flow into the lagoon of Venice, which it had contributed to create, but at the time of the *Serenissima* Republic of Venice its course was diverted into the Adriatic Sea because the sediment load brought by the river was filling up the lagoon.

Yet, rivers carry out fundamental ecological functions at the landscape scale, as they represent *ecological corridors* through which species are able to move and disperse, reaching ecosystems that are far from each other and that, without such ecological corridors, would be isolated (e.g., because an area occupied by intense human activities lies between them). Thus, rivers help to maintain the cohesion of fragmented ecosystems and support biodiversity by favoring species dispersal.

At the landscape scale, another important concept linked to river ecosystems is that of river catchment (or river basin, drainage basin, watershed, etc.), which is the area that drains rainfall and snowmelt into a given river, and typically includes other minor rivers, streams, and so on. Water flowing in the catchment area collects nutrients and pollutants, for example, surface runoff can remove such substances from soils contributing to the eutrophication (Section 20.2) or pollution, respectively, of the river and of the water body that the river flows into. Indeed, coastal seas receiving freshwater from large catchments are typically much productive due to the large nutrient loads they receive (de Leiva Moreno et al., 2000).

As rivers are characterized by a predominant spatial dimension, that is, their longitudinal axis is much larger than their vertical and transverse dimensions, from an ecological point of view rivers are often modeled as one-dimensional systems where resources, environmental conditions, and so on, change along the direction of the flow. A well-known monodimensional description of ecological processes in rivers is the classical *Streeter–Phelps model* (or dissolved oxygen sag curve), which depicts how the dissolved oxygen concentration in a river changes following the point discharge of a biodegradable substance and the resulting processes of aerobic biodegradation (Sections 17.4 and 18.2), which consume dissolved oxygen, and reaeration (Sections 7.7 and 7.8), that is, the exchange of oxygen between the atmosphere and the river, which tends to reestablish normal dissolved oxygen levels. This simple model is based on several assumptions such as steady state (constant river flow over time, constant discharge rate of the organic substance); uniform cross-sectional surface of the river; complete mixing of all the involved substances (oxygen, organic substance) over the cross-sectional surface, that is, uniform concentration over such surface; the biodegradation of the organic substance follows a first-order kinetics; the rate of the reaeration process is proportional to the oxygen deficit (which is the difference between dissolved oxygen concentration at saturation and the actual concentration observed, resulting from the consumption of O_2 due to biodegradation). Let the mass-balance for dissolved oxygen in the river water be

$$\frac{dD}{dt} = -k_R \cdot D(t) + k_d \cdot L(t) \tag{20.3}$$

where D is the dissolved oxygen deficit equal to $C_s(T, S)-C(t)$, that is, the difference between the dissolved oxygen concentration C_s at saturation, which is, in general, a function of temperature T and salinity S, and the actual dissolved oxygen $C(t)$ at time t; k_R is the temperature-dependent reaeration rate; k_d the temperature-dependent biodegradation rate; $L(t)$ the biochemical oxygen demand (BOD) at time t, a measure of the amount of biodegradable organic substance in water, which can be written as $L(t) = L_0 \cdot e^{-k_d \cdot t}$, L_0 being the BOD concentration at time $t = 0$ (i.e., at the discharge point).

Equation 20.3 can be solved for D to assess the environmental impact of the organic load at a given time t in a river point $x = v \cdot t$ (v being the current speed, and $x = 0$ representing the position of the point source of organic matter):

$$D = \frac{k_d \cdot L_0}{k_R - k_d} \cdot \left(e^{-k_d \cdot t} - e^{-k_R \cdot t}\right) + D_0 \cdot e^{-k_R \cdot t} \tag{20.4}$$

where D_0 is the initial oxygen deficit value, that is, at the discharge point.

In some cases, the one-dimensional approximation is not appropriate and it is advisable to take heterogeneities on the river cross-sectional surface into account, for example, when studying the dispersion of a pollutant on a relatively short time after it has been discharged into the river, and complete transverse mixing by turbulence and currents (Chapter 7) has not been achieved. Good reference texts for mixing processes in rivers are represented by the works of Fischer et al. (1979) and Rutherford (1994). Estuaries, (the semienclosed areas where rivers meet the sea), such as bays, fjords, and sounds are other examples of situations where a one-dimensional river description does not make sense. In these transitional zones, the interplay of freshwater discharge and seawater creates complex and biologically productive ecosystems, which are characterized by rapidly changing physical and environmental conditions. Such complexity and productivity result from the nutrients and sediments brought in by rivers, from the differences in density and temperature between freshwater and seawater, which affect mixing, stratification, and circulation, from the physical action of tides and of the variability in freshwater discharge (connected to processes in the watershed such as precipitation). Estuaries are important ecosystems because of their high biological productivity and the many services they can provide to the society; for example, several valuable marine fish species spawn in estuaries or lagoons (see also Section 20.4, given that estuaries are transitional ecosystems).

20.6 Open Sea

Bodies of saltwater such as seas, which are saline water bodies partly or totally landlocked, and oceans cover more than 70% of the earth's surface and are the largest world's reservoir of water. Given their large extent and volume, marine water bodies play a fundamental role in the global cycle of water (Chapter 12) and exert a strong influence on climate and weather both at a local and global scale, by storing and exchanging heat with the atmosphere. On a local scale, typically, coastal areas are characterized by a more temperate climate than inland areas due to the slower release of heat from water compared to land. The climate of terrestrial and water ecosystems is also influenced by the global system of oceanic currents driven by gradients in water density, which is in turn influenced by temperature (which is linked to depth, latitude, etc.; e.g., Sections 8.2 and 8.3) and salinity (which is influenced by freshwater inputs, evaporation fluxes; Section 8.4). Ocean currents can transport heat to cold areas, as in the case of the Gulf Stream, which makes the Northern-Western European climate warmer. Another example of an ecologically important ocean–air interaction is the process of *coastal upwelling*; that is, the wind-driven displacement of warm (and thus light) waters in the proximity of coasts, which are replaced by deeper and, thus, denser and colder waters that move vertically toward the sea surface. Deeper waters are richer in nutrients compared to nutrient-depleted coastal waters, and thus coastal upwelling contribute to enhance the primary productivity (Chapter 15) of coastal ecosystems, such as the Benguela current in Namibia and South Africa, and the Humboldt current in Peru and Chile. Consequently, coastal upwelling sustains very productive marine ecosystems and supports some of the largest and most productive pelagic fisheries of the world, such as the famous *anchoveta* (anchovy) fishery in Peru.

Vertical gradients are very important in marine ecosystems, since many environmental conditions change across the water column (also known as the pelagic environment, as opposed to the benthic or bottom environment).

Light, a key factor affecting primary productivity (Chapter 15), is found only in the upper part of the pelagial known as the euphotic zone (Section 8.3) and extinguishes as it penetrates toward the bottom of the ocean. As water pressure increases with depth, temperature typically decreases. Surface water is typically warmer and well-mixed by wind and waves (thus, its temperature and salinity are broadly homogeneous), and is separated from deeper waters, which are calmer, denser, and colder, by a relatively abrupt temperature gradient called thermocline, corresponding to a density gradient. The depth of the thermocline changes seasonally following changes in wind regime, heat fluxes, and so on. A pronounced thermocline, that is, a strong vertical stratification, can represent a powerful barrier blocking vertical mixing of nutrients, organic matter, and so on, and thus it can strongly influence ecological dynamics.

The large areas covered by the world's oceans result in a stunning variety of environmental conditions and of ecological diversity, as well as in an overall large biological productivity. Marine microalgae (phytoplankton) are important players in the global carbon cycle (Section 12.4) and consequently in the ongoing process of climate change, due to their capability of fixing large amounts of carbon dioxide into organic forms, a part of which then settle to the ocean's floor and is thus effectively removed from the atmosphere for a long time. Oceans can sequester large amounts of carbon through photosynthesis, but atmospheric CO_2 can also simply dissolve in water, according to these reactions:

$$CO_{2(air)} \rightleftarrows CO_{2(aq)} + H_2O \rightleftarrows H_2CO_3 \rightleftarrows H^+ + HCO_3^- \rightleftarrows 2H^+ + CO_3^{2-} \quad (20.5)$$

Thus, the current rise in atmospheric CO_2 levels causes a decrease in the pH of water, a process known as *ocean acidification* (Doney et al., 2009). One key ecological effect of ocean acidification is that a lower pH decreases the calcium carbonate saturation state in water, thus negatively impacting calcifying organisms that form shells or hard skeletons including some plankton and benthos (e.g., mollusks, echinoderms, corals), whose calcification and growth rates are reduced by ocean acidification. Furthermore, acidification is expected to increase the carbon fixation rate in a number of photosynthetic organisms (Doney et al., 2009). This variety of biotic responses to ocean acidification together with the important ecological role of several organisms affected by this process make it difficult to predict how the effects of ocean acidification will propagate across marine food webs, and possibly interact with the earth's climate (because some phytoplankton taxa, which are involved in the carbon cycle, will be affected by acidification).

Broadly speaking, the modeling of marine ecosystems has traditionally developed as two largely independent subareas whose integration is still far from perfect, one focusing on the biogeochemical cycles of nutrients and lower-trophic-level organisms, such as phytoplankton, bacteria, and zooplankton (e.g., the so-called NPZ or NPZD models, where N stands for nutrients, P for phytoplankton, Z for zooplankton, D for detritus), and the other one focusing on higher-trophic-level organisms (e.g., fish, sea mammals, birds), food-web interactions, and fisheries (Rose et al., 2010). Since nutrient- and plankton-microbial dynamics have been briefly described in Chapters 12, 15, and 17, we will now focus on upper trophic levels, specifically on a food-web modeling approach commonly used to study trophic interactions in fish and macro-invertebrate communities and their link with fisheries, called Ecopath with Ecosim (Christensen and Walters, 2004; see also Section 19.5). In an Ecopath with Ecosim model, the biomasses of the populations and functional groups composing a food web are simulated over time basically as the result of fluxes of matter; that is, the biomass of a given group grows following an increase in its food consumption or a net immigration flux to the system, while it decreases following the predation of other groups, biomass removal by fishing or other mortality sources. The Ecopath approach relies on a simple mass-balance approach to ensure that the many inputs needed for each model group to parameterize the food-web model (growth and consumption rate, diet, biomass, fishery catch, etc.) are accurate enough and mutually consistent. According to such mass balance, over a properly chosen period the consumption of each group must equal the sum of its production and respiration, and the change in the biomass of that group (or, its biomass accumulation) must equal the difference between its production and the sum of biomass loss terms, such as dead biomass (due to predation, fishing, and all other sources) and net migration from the system. The resulting mass balanced trophic network model (or Ecopath model) is used as the initial condition of a simulation model (Ecosim) in which the changes in the biomasses of the different food-web groups are described over time based on a system of coupled differential equations (Christensen and Walters, 2004):

$$\frac{dB_i}{dt} = g_i \cdot \sum_j Q_{ij} - \sum_j Q_{ji} + I_i - (M_i + F_i) \cdot B_i \qquad (20.6)$$

where the subscript i indicates the modeled group, while j is cycled over all model groups. B is the group biomass in the food web, t is time, g the net growth efficiency (the ratio of production to consumption for the group), Q_{ij} the biomass of group j consumed by group i, I the net immigration/import flux of biomass into the system, M the nonpredatory natural mortality rate, F the fishing mortality rate. Thus, the dynamics of the food-web groups are linked through predator–prey interactions by the two summation terms, representing food intake and losses to predators, respectively. The consumption terms Q_{ij} are calculated as a function of the biomasses of both the predator and the prey:

$$Q_{ij} = \frac{av_{ij}B_iB_j}{bv_{ij} + cB_i} \qquad (20.7)$$

Here, a, b, and c are parameters expressing the effect of an external forcing or processes, such as mediation, handling time, and so on. High values (>2) of the parameter v_{ij}, called vulnerability, indicate a top-down controlled trophic interaction, since in this case Q_{ij} becomes proportional to the product of the biomass of the prey and that of the predator, as in Lotka–Volterra dynamics (Section 16.3). Instead, small v_{ij} values indicate a bottom up driven trophic interaction (Chapters 16 and 19), since Q_{ij} becomes proportional to the biomass of the prey B_j.

20.7 Fishery

Fisheries represent a major human pressure on aquatic ecosystems worldwide. Although little human effort is needed to obtain food from aquatic ecosystems (sea do not need to be sown, unlike agricultural fields), their sustainable management has generally proved complicated, and excessive exploitation has often led to the decline in abundance of fishing resources (e.g., see Pauly et al., 1998, regarding marine systems). The huge ecological impact of fisheries is clearly depicted by the fact that 8% of worldwide primary production (Chapter 15) is needed to sustain global fishery catches, and that such percentage is typically much higher in several coastal systems (Pauly and Christensen, 1995).

Human exploitation of aquatic organisms for consumption can have many direct and indirect ecological effects in water bodies. The simplest, most obvious ecological effect of fishing is direct mortality for the fished species, which is caused by the biomass removal of that species from the ecosystem by fisheries. Fisheries do not catch only target species but also bycatch, that is, unintentionally caught organisms, which can also undergo fishing mortality. The term *catch* indicates all the species captured by the fishing gear (e.g., the net) of a ship, including bycatch, and it comprises the sum of *landings* (the fish and seafood landed and consumed as food or in other human activities) and *discards,* which are the organisms thrown back to the sea because they are of no commercial interest. Although not all discarded organisms are already dead or die because of stress or physiological damages caused by nets or exposure to air when thrown back to water, discard can be a large and typically underrated source of fishing mortality in coastal marine ecosystems. The amount of discarded organisms can be of comparable magnitude to that of landings, or even greater than them in the case of some fisheries, for example, those targeting shallow and productive systems.

Numerous models exist to study the effect of fishing mortality on a given population (Martell, 2008; Shertzer et al., 2008). The typical approach is to calculate changes in population abundance or biomass based on somatic growth rate, birth (or recruitment) rate and mortality rate (which can be due to fishing or to natural causes such as predation, illness, or aging). Surplus-production models represent a simple approach to describe changes in population biomass B. One famous example is the Schaefer model, which is a modified logistic equation with the addition of a fishing mortality term:

$$\frac{dB}{dt} = r \cdot B \cdot \left(1 - \frac{B}{K}\right) - Y \tag{20.8}$$

Here, r is the intrinsic growth rate, K the carrying capacity, and Y is the yield, or catch, expressed as $Y = F \cdot B$, F being the fishing mortality rate. Using this model, one can find the maximum sustainable yield, which is $Y_{max} = r \cdot K/4$ (corresponding to $B = K/2$) by taking $dB/dt = 0$ and $dY/dB = 0$. Such model assumes that the fished population is homogenous and can be described by lumped parameters such as r and K. Since many life history traits actually depend on size and age and individuals in a population display variability in body size, this is not always a good assumption, and size-structured or age-structured models can be used, describing separately individuals with different sizes or belonging to different life stages (Martell, 2008).

One key problem in fishery science is the description of the relationship between stock (here, this word is taken as a synonym of population) biomass and recruitment R, that is, the number of young individuals entering the population. The number of recruits can be strongly variable and be influenced by environmental factors (temperature, salinity, food availability, and so on) as well as by the spawning stock biomass S. Two common models to depict the link between R and S are the Ricker model, in which $R = a \cdot S \cdot e^{-b \cdot S}$, and the Beverton–Holt model, in which $R = a \cdot S/(b + S)$. In both cases, a and b are parameters.

The study of the effect of fishing on aquatic ecosystems is a discipline with a long applied and theoretical tradition (Jennings and Kaiser, 1998). For example, the predator–prey equations described on Section 16.3 were derived from the Italian mathematician Volterra in 1926 (but Lotka also derived them independently) based on data from fisheries in the Adriatic Sea. Fisheries science has only recently shifted from the classical focus on the management of single species, that is, from the effect of direct fishing mortality on a single population, to multispecies and food-web approaches (Garcia et al., 2003), taking into account that the dynamics of a single population targeted by fisheries is linked to the dynamics of other populations through predation mortality (Section 16.3), which can lead to potentially complex food-web interactions (Section 19.5; see also marine food-web modeling on Section 20.6). Also, it is becoming increasingly clear that fishing can have strong effects on nontarget species (e.g., through bycatch and discard mortality) and on aquatic habitats. For example, fishing gears such as hydraulic dredges and bottom trawls can scrape and scour the bottom of water bodies, collecting bycatch, killing bottom fauna, altering or destructing habitats (e.g., algae and corals) and resuspending sediments, thus changing properties of the water column such as its turbidity, nutrient concentration, and so on. Following direct effects of fishing such as fishing mortality and habitat alteration, as well as indirect effects as mediated by trophic interactions, the whole structure of local fish and invertebrate communities can radically change and, sometimes, even ecosystem functioning at large spatial scales and key system properties such as primary productivity can be profoundly altered (as in the case of trophic cascades driven by overfishing in the Black Sea) (see Figure 16.1).

Not all organisms respond to fisheries, directly or indirectly, in the same manner. In general, fishing pressure tends to be detrimental to large, slow-growing, and late-maturing species (growth rate and age at maturity are typically correlated with body size, similarly to many other biological processes, Section 14.3), which are vulnerable to intense exploitation due to such intrinsic characteristics of their life histories. Instead, small, fast-growing and fecund organisms are more resilient to fishing pressure and, in some cases, they can even benefit from it, for example, when fishing mortality reduces the abundance of their (typically larger, and thus less resilient) predators. On the contrary, if a fishery reduces the abundance of the prey of one species, then such species can be negatively, indirectly affected by fishing. A review of the multiple, complex effects of fishing in marine ecosystems can be found in Jennings and Kaiser (1998).

Catches do not only represent an ecological impact on aquatic ecosystems. They are also an invaluable source of information for assessing changes in the biomass of fishing resources over time (see, for example, Pauly et al., 1998). Indeed, fishery catches can be thought of as a sort of nonstandardized sampling, but nearly continuous over time and space. Despite the many problems with using catches to infer about changes in population abundance (e.g., catches can reflect changes in fishing effort and gear/technology, in market demands, discard practices, areas targeted by fisheries, reporting practices, etc.), landing records usually span much longer periods with respect to scientific surveys. Many of the abovementioned issues can be solved by rescaling catches (or, more typically, landings, since this is the information which is most commonly recorded) over a standardized measure of fishing effort (e.g., number of vessels, vessel total tonnage, total engine power of vessels times the number of fishing days per year) to obtain an index of relative abundance for the fished species, known as *catch-per-unit-effort* (whose common acronym is CPUE).

21

Terrestrial Ecosystems

21.1 Introduction

In landscape ecology, principal terrestrial ecosystems (*biomes*) are classified into six broad categories: *tundra, desert, taiga, temperate deciduous forest, tropical (or subtropical) rainforest,* and *grassland.* Differently from aquatic ecosystems, in terrestrial ones, the availability of water is critical, and water may become a limiting factor. Moreover, terrestrial ecosystems present less stable conditions than aquatic ones. For example, climatic characteristics, such as temperature, rainfall, wind, and airborne humidity, may change dramatically during the year, the seasonal cycle, or even in shorter time spans (days/hours).

Several terrestrial ecosystems are endangered by anthropic activities or natural effects (e.g., deforestation, nutrient or toxic substances leaching, and erosion). Common to all these systems is the role of soil, acting as a recycling device for detritus. Soil dynamic is indeed central in determining the fate of nutrients and xenobiotic toxicants. Moreover, changes in soil characteristics may lead to local vegetation/environment modifications that possibly trigger a cascade of amplifying effects, propagating from local to large scale and leading eventually to critical large-scale ecosystem transitions.

21.2 Decomposition of Organic Matter in Soil

The decomposition* rates of different fractions of organic matter in soil differ several orders of magnitudes, dependent on the composition. Dead organisms, trees overturned by the wind, autumn leaves, other plant residues, dead animals, and so on, are decomposed slowly mainly because they have a low specific surface. They are, however, decomposed to organic compounds that are in turn decomposed relatively rapidly. The organic matters

* See also biological decomposition, Section 17.4.

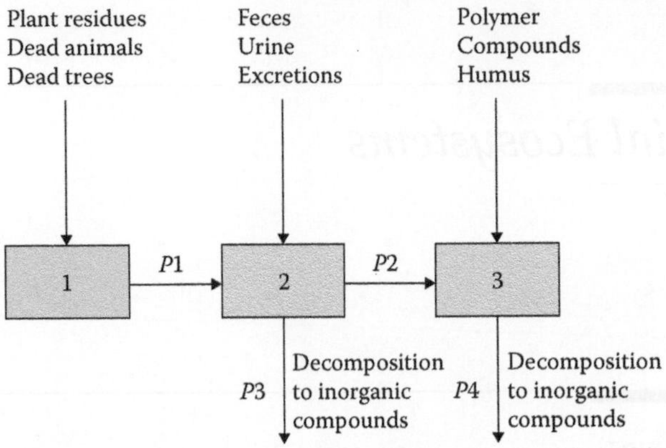

FIGURE 21.1
An approximately acceptable description of the decomposition of organic matter in soil is shown. Three fractions are considered. The specific rates of the processes are very different and with good approximation as follows: *P*1 is 0.1–1.0 × 1/year, *P*2 is 0.005–0.1 × 1/24 h, *P*3 is 0.05–0.5 × 1/24 h and *P*4 is 0.01–0.2 × 1/year.

in urine, feces, and other excretions are decomposed rapidly, while polymer organic matters, such as humus acid compounds, are decomposed slowly. An approximate acceptable description is shown in Figure 21.1, where the organic matter in soil is divided into three fractions, which are decomposed by different rates. In accordance to Lal et al. (2001) it seems that the description shown in Figure 21.1 comes close to a realistic, quantitative description of the fate of organic matters in soil.

As the number of different organic compounds in all types of soil is very high, the decomposition is very complex and the three fractions shown in Figure 21.1 should therefore be considered an attempt to capture an approximate, holistic picture of the results of the decomposition processes. Organic compounds having different decomposition rate in the soil water and on solid particles of soil add further complication to the process. Moreover, different organic compounds have different solubility in the soil water, and the soil composition (the main components being sand, clay, and humus compounds) has a major influence on the distribution between soil water and soil.

21.3 Toxic Substances in Soil

In higher aquatic plants, the absorption of waterborne pollutants by stems and leaves is often more important than the absorption from the sediment or soil by the roots (Eriksson and Mortimer, 1975). The absorption by plants from water (for instance, soil water) can with a good approximation be described by the application of a concentration factor, as it is also the case for animals (see Section 18.3):

$$\frac{C_b}{C_w} = BCF \tag{21.1}$$

where

C_b = the biotic concentration of the pollutant (g kg^{-1})
C_w = the concentration in water (g l^{-1})
BCF = a concentration factor

There is generally a correlation between BCF and K_{ow} (see Chapter 11), which is dependent on the size of the organism according to the allometric principles (see e.g., Section 14.3).

Equation 21.1 may be modified to account for the lipid phase in the organisms including plants. Generally, we can state (see, for instance, Connell, 1997) (see also Equation 18.9) that

$$BCF = f_{\text{lipid}} \times K_{ow}b, \text{ corresponding to } \log BCF = \log f_{\text{lipid}} + b \log K_{ow}, \tag{21.2}$$

where f_{lipid} is the lipid fraction in the considered organism (plant). b is usually close to 1 (often indicated to be 1.03). If C_L is the concentration of the lipophilic organic compound in the fat tissue, we have

$$C_L = \frac{C_b}{f_{\text{lipid}}} \tag{21.3}$$

$C_L/C_w = K_{ow}$, (defined as the ratio solubility in octanol and solubility in water) provided that we can consider the solubility in the fat tissue to be close to the solubility in octanol, we get

$$\log BCF = \log K_{ow} + \log f_{\text{lipid}} \tag{21.4}$$

which is the same as Equation 21.2 with b ≈ 1.0.

The bioaccumulation factor BCF_{sb} for the relationship between soil solid and biota is C_b/C_s completely parallel to Equation 21.1. If the concentration in the pore water is denoted C_w, we obtain the following expression:

$$BCF_{sb} = \left(\frac{C_b}{C_s}\frac{C_w}{C_w}\right) = BCF_{\text{org-water}} \times K_D \tag{21.5}$$

K_D is the ratio of the concentration in solid soil and in soil water. This ratio is furthermore proportional to $f \times K_{ow}$ in the exponent 0.94 based on a number of empirical results. F is the fraction of carbon in the soil. By using Equation 21.4 and the correlation between K_D and K_{ow} we get

$$BCF = f_{\text{lipid}} \frac{K_{ow} b}{x} \times f \times K_{ow} a \tag{21.6}$$

where x is a proportionality constant, f is as mentioned the fraction of organic carbon in the soil, and a is as indicated 0.94.

By using the above mentioned b value of 1.03 and use $x = 1.00$, which several experiments have shown is close to a correct value, the following expression for the BCF for the bioaccumulation factor soil–organism is obtained:

$$BCF = \frac{1.00 \times f_{\text{lipid}}}{f} \times K_{ow} 0.09 \tag{21.7}$$

This implies that BCF soil-organism has only a small dependence on K_{ow} and general properties of the soil. It depends, however, on the properties of the soil and the biota, particularly the ratio of lipid in the biota to the organic carbon content of the soil.

21.4 Erosion and Leaching

Erosion is the process by which soil, rocks, and other deposits are removed from the earth's surface by the action of water, ice, wind, and so on. A number of natural and man-controlled processes are causing the erosion: climate, including precipitation and wind force; deforestation; agricultural practices, including agricultural tillage; lack of appropriate landscape planning; construction works; and so on.

The most commonly used model for predicting soil loss from water erosion is Universal Soil Loss Equation (USLE), which estimates the average annual soil loss A as

$$A = R \times K \times L \times S \times C \times P \tag{21.8}$$

where R is the rainfall factor, which is replaced by the wind speed factor, W, when wind erosion is considered, K is the soil erodibility factor, L and S are topographic factors representing length and slope, and C and P are cropping management factors. R and W are generally proportional to the velocity energy of the water and air, respectively. It means that the rate of water flow and the wind speed in the exponent two will be included in the expression.

As expressed clearly in Equation 21.8, erosion is a complex process and cannot be calculated based on a simple relation, but it requires a very complex model to predict erosion.

Generally, terrestrial ecosystem models are more complex than aquatic ecosystem models, because soil is more heterogeneous than water. The composition of the soil can change radically within a few meters or even centimeters, which implies that the hydrological resistance is changed significantly, and even several times within the area of only one hectare, for instance.

Leaching refers to the loss of water-soluble plant components, including nutrients, from the soil, due to rain and irrigation. Soil structure, soil heterogeneity, crop planting, plant type, and application rates of manure and fertilizers are the factors that can influence the leaching process. Leaching, particularly, is an environmental concern when it contributes to groundwater contamination. As water from rain, flooding, or other sources seeps into the ground, it can dissolve chemicals and carry them into the underground water supply and thereby contaminate the ground water (see also biogeochemical cycles in Chapter 12). Of particular concern are, of course, hazardous waste dumps and landfills; and in agriculture, excess fertilizer, improperly stored animal manure, and biocides (e.g., pesticides, fungicides, insecticides, and herbicides). Leaching—like erosion—cannot be quantified by one or a couple of equations, but requires the development of rather complex models in most of the cases.

21.5 Ecosystem Transitions

The prevailing conditions for a considered area determine the type of ecosystem on a long-term basis. The type of ecosystem that gives the highest work capacity (exergy, eco-exergy) (see Section 4.6) under the prevailing conditions will be the answer to the question: which type of ecosystem should be expected in this area? Through this answer, it is presumed that Mother Nature has had sufficient time to develop the ecosystem, which may take a long time, for instance, for a forest ecosystem. It implies that there may be a long transition time, where one type of ecosystem replaces another one, because it should better fit to the prevailing conditions created partially by the previous ecosystem type. It has been observed many times that forest ecosystems are replaced by savanna and later by grassland due to climatic changes—in that case, toward a drier climate, and the climate is highly a major factor for changes from one type of ecosystem to the other. The influence of man on nature is yet another predominating factor. Indeed, agricultural practices have changed fertile land to desert, and the deforestation is another very drastic and man-controlled change of ecosystem type.

It will require a rather complex model to describe the transition of one type of ecosystem to the other due to climatic changes. It is, however, possible in the ecological modeling literature to find models that describe which ecosystem changes can be expected as a consequence of the increased emission of greenhouse gases into the atmosphere. It is possible to use the method presented for structurally dynamic model in Chapter 19 to determine whether it should be expected for given climatic changes (for instance in precipitation ad temperature) when ecosystem Type A is replaced by Type B. The work capacity (exergy, eco-exergy) is calculated for the two types of ecosystems by the creation of ecological models for the two types. The ecosystem type that gives the highest work capacity will be the type that wins.

Under all circumstances, it can be concluded that the description and prediction of the transition from one type of ecosystem to the other or even a third type later on, will require the development of complex ecological models, where many different processes are incorporated.

22

Landscapes and Urban Ecosystems

22.1 Introduction

The perspective of landscape ecology, with its large-area and long-term focus, provides a useful tool in land designing and planning in order to enhance the ecological function and sustainability of an area. Human population growth has induced deep modifications of the pristine landscapes, and the appearance of cities, towns, and urban strips constructed by humans has generated *urban ecosystems.* Although they cover a relatively small area of the world, cities are fast expanding and clumping. It is estimated that more than 60% (4.9 billion) of the estimated world population (8.1 billion) will live in urban areas by 2030. As a consequence, the growth of urban population and the supporting built infrastructure will increasingly affect both urban environments and areas that surround them, including semiurban environments peripheral to cities as well as agricultural and natural landscapes. Understanding landscape processes is thus crucial in order to quantify and measure the effects of urbanization on human and environmental health. Urban development fragments isolates, and degrades natural habitats by simplification and homogenization of species composition, disruption of hydrological systems, modifications of energy flows, and nutrient cycling.

When urban areas are considered as part of a broader ecological system, it is also possible to investigate how urban landscapes function and how they affect other landscapes with which they interact. By employing the central concept of population, which characterizes a landscape by acting as a footprint proxy for most of its transfer functions, a richer understanding of landscape dynamics is gained. Growing population aggregates in urban areas, guzzle natural resources, and produce solid and liquid wastes that interrelate to environmental processes in urban ecosystems. Hence, urban areas hoard a large share of Earth's carrying capacity from other regions in terms of resources and waste sinks.

22.2 Landscape Processes

A landscape consists of several ecosystems that are more or less connected. As ecosystems are open systems, transfer from one ecosystem to the other takes place. The transfer rate can be described as a gradient divided with a resistance, as it is applied for many other environmental processes, including diffusion. The gradient is in this case a difference in concentration for chemical compounds or a difference in density for organisms. It means that the rate of transfer can be found from a simplified model.

$$\text{Transfer rate} = \frac{(C_1 - C_2)}{R} \tag{22.1}$$

where C_1 is the concentration or density in the ecosystem denoted number 1 and C_2 is the concentration or density in the ecosystem denoted number 2, while R is the resistance. The concentrations or densities are often known by observations or by model calculations, while the resistance is usually not known and must be found by calibration of the model implementing the transfer processes.

An important environmental management problem is associated with the cut of these transfer possibilities among ecosystems, for instance due to the building of a motorway or another major construction work. An elimination of important transfer possibilities for an exchange of organisms between two or more populations in ecosystems may have a significant impact on the possibilities for the populations to survive in one or more of the concerned ecosystems. A regional set of local populations, that are occupying isolated habitat patches but are interconnected by dispersal movements, are denoted by metapopulations; see an example in Figure 22.1. All the local populations have a finite possibility of becoming extinct. Even if the local population is fairly large, extinction may still occur, for instance, through

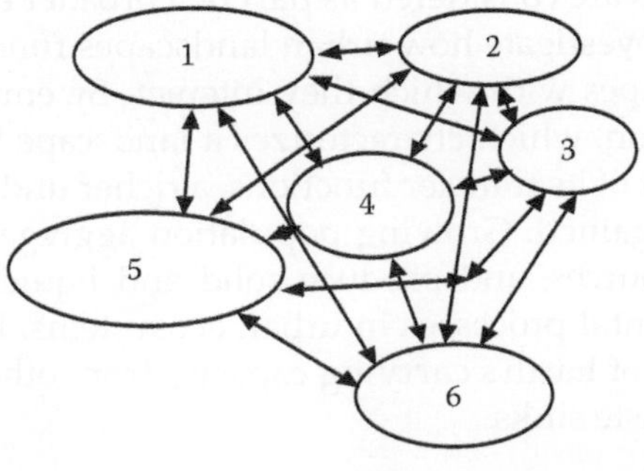

FIGURE 22.1
Schematic outline of a metapopulation. The populations occupy isolated habitats that are connected by patches.

catastrophic events. The dispersal movements or transfer processes are essential to reestablish populations that have faded or crashed to zero numbers. Species that are widely distributed in many local connected populations have reduced likelihood to be extinct regionally.

An increasing *fragmentation of landscapes* has taken place due to the human expansion. Populations that were formerly continuously distributed have become broken into separate localized groupings. Dispersal may even be inhibited by hazards in traversing the human transformed areas separating suitable habitats. Metapopulation models assess the risks of species extinctions as a consequence of such fragmentations, and identify how actions such as providing dispersal corridors can reduce the risks. Equations have been developed for these processes to deal with these problems.

The concept of metapopulation was formulated by Levin (1969) and further modified by Hanski (1994 and 1999).

$$\frac{dP}{dt} = CP(1-P) - EP \tag{22.2}$$

where P is the proportion of sites occupied by populations, E is the extinction rate of these populations, and C is the colonization rate of vacant sites by migrants from occupied patches. The change over time in the proportion of patches, dP/dt, occupied is a matter of balance between colonization and extinction.

The equilibrium proportion P_{eq} is given by

$$P_{eq} = 1 - \frac{E}{C} \tag{22.3}$$

As seen from Equation 22.3, the patch occupancy will become zero if the rate of extinction exceeds the rate of colonization.

Metapopulation models can be used to assess whether a fragmentation will have unacceptable consequences, and whether it would be necessary to establish corridors to reduce the consequences of fragmentation.

22.3 Population Growth

Quantification of population growth is important for urban planning and for the protection of wild life populations. A population is a currently changing entity, and we are interested in its size and growth = change in size.* If N represents the number of organisms and t the time, then dN/dt = the rate of

* See also Chapter 14.

change in the number of organisms per unit time at a particular instant (t) and $dN/(Ndt)$ = the rate of change in the number of organisms per unit time per individual at a particular instant (t). If the population is plotted against time a straight line tangential to the curve at any point, that is, slope of the curve, represents the growth rate. We may distinguish between absolute natality meaning the new individuals appearing per unit of time: $B_a = dN/dt$ and relative natality, which is the absolute natality divided by the number of individuals of the population: $B_s = dN/(N\,dt)$. N indicates that only the new individuals are accounted for. Mortality refers to the death of individuals in the population. The absolute mortality rate, M_a, is defined as the change of the population due to mortality: dN/dt and the relative mortality, which is the absolute mortality relative to the population, that is, $M_s = dN/(N\,dt)$. N indicates that only dead individuals are accounted for. The growth of population is the difference between absolute natality and absolute mortality.

$$\begin{aligned}\text{Growth} &= \frac{dN}{dt} = \text{Natality} - \text{Mortality} \\ &= \frac{dN}{dt}(\text{due to natality}) - \frac{dN}{dt}(\text{due to mortality})\end{aligned} \tag{22.4}$$

Correspondingly, the relative growth = relative natality – relative mortality

$$= \frac{dN}{N\ dt} = \frac{d\,N_n}{N\ dt} - \frac{dN_m}{N\ dt} \tag{22.5}$$

where N_n indicates increase in population due to natality and N_m indicates decrease in population due to mortality.

The simplest growth model considers only one population and is the basic quantifying equation in population dynamics. The interactions with other populations are in this case taken into consideration by the natality and the mortality, which might be considered dependent on the magnitude of the considered population but independent of other populations. In other words, we consider only one population as state variable.

The simplest growth model assumes *unlimited resources and exponential population growth*. A simple differential equation can be applied:

$$\frac{dN}{dt} = B_s \times N - M_s \times N = r \times N \tag{22.6}$$

where B_s is the relative natality or the instantaneous birth rate per individual, M_s the relative mortality or the instantaneous death rate, $r = B_s - M_s$, N is here representing the number of individuals in the population or the biomass of the population and t the time. Equation 22.6 represents first-order kinetics. If r is constant, we get after integration:

$$N_t = N_0 \cdot e^{rt} \tag{22.7}$$

where N_t is the population density at time t and N_0 the population density at time 0.

Figure 22.2 represents a logarithmic graph of this equation. It is a straight line and r indicates the slope. If the logarithm of the number of individuals or the biomass versus the time is a straight line with a standard deviation that is in accordance with the uncertainty, it can be concluded that the growth is following the first-order growth equation, and that the slope of the line is equal to the growth rate.

The net reproductive rate, R_0 is defined as the average number of age class zero offspring produced by an average newborn organism during its entire lifetime. The intrinsic rate of natural increase, r, is dependent on the age distribution, and it is only constant when the age distribution is stable. When R_0 is as high as possible, that is, under optimal conditions and with a stable age distribution, the maximal rate of natural increase is realized and designated r_{max}. Among various animals it ranges over several orders of magnitude.

Exponential growth is a simplification, which can only be valid over a certain time interval. Sooner or later every population must encounter the limitation of food, water, air or space, as the world is finite. To account for this, we apply the concept of density dependence; that is, vital rates, like r, depend on population size, N (while we now ignore differences caused by age). Let the carrying capacity, K, be defined as the density of organisms at which r is zero. At zero density R_0 is maximal and r becomes r_{max}. The logistic growth equation considers the density dependence by the use of a carrying capacity. The logistic equation is formulated mathematically as follows:*

$$\frac{dN}{dt} = r \times N \times \left(1 - \frac{N}{K}\right) \tag{22.8}$$

where K is the carrying capacity.

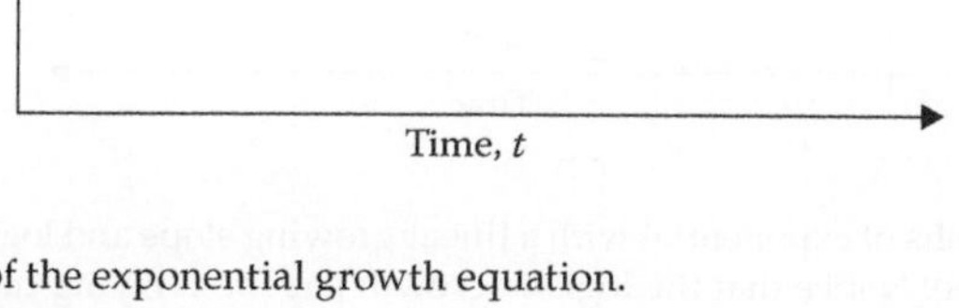

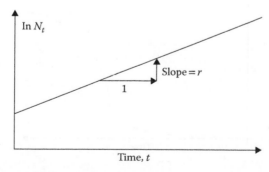

FIGURE 22.2
Logarithmic graph of the exponential growth equation.

* See also Chapter 14, Equation 14.4.

The application of the logistic growth equation requires three assumptions:

1. That all individuals are equivalent.
2. That K and r are immutable constants independent of time, age distribution, and so on.
3. That there is no time lag in the response to the actual rate of increase per individual to changes in N.

All three assumptions are unrealistic and can be strongly criticized. Nevertheless several population phenomena can be nicely illustrated by the use of the logistic growth equation.

The exponential growth equation and the logistic growth equations are compared graphically in Figure 22.3 (see also Figure 14.1).

Illustration 22.2

An algal culture shows a carrying capacity due to the self-shading effect. In spite of "unlimited" nutrients, the maximum concentration of algae in a chemostat experiment was measured to be 100 g/m^3. At time 0, 0.1 g/m^3 of algae was introduced and 4 days after a concentration of 2 g/m^3 was observed. Set up a logistic growth equation for these observations.

Solution

During the first 5 days, we were far from the carrying capacity and we had good approximations:

$$\ln 20 = r_{max} \times 4$$
$$r_{max} = 1.0 \text{ day}^{-1}$$

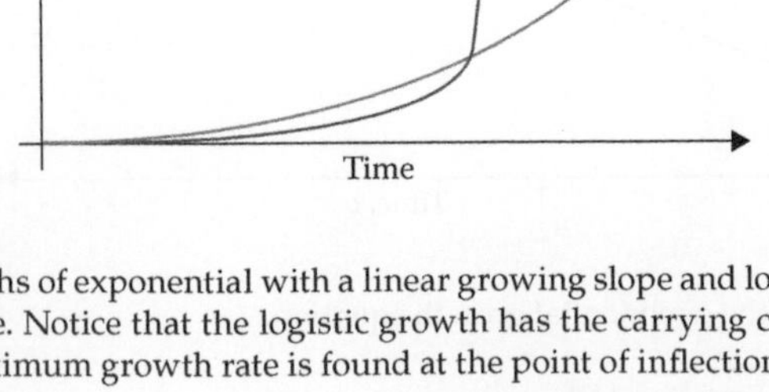

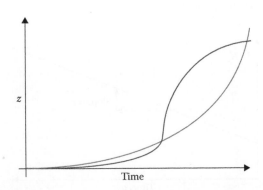

FIGURE 22.3
Comparisons of graphs of exponential with a linear growing slope and logistic growth with its characteristic s-shape. Notice that the logistic growth has the carrying capacity as an asymptote and that the maximum growth rate is found at the point of inflection.

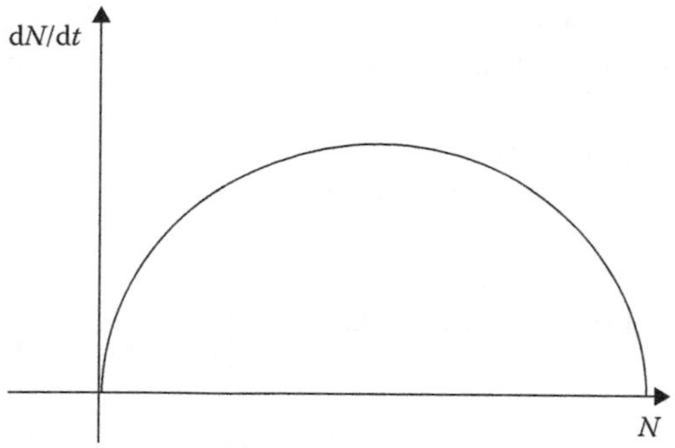

FIGURE 22.4
dN/dt is plotted versus N for the logistic growth equation.

and since the carrying capacity is 100 g/m^3, we have: (C = algae concentration)

$$\frac{dC}{dt} = 1.0 \times C \times \left(\frac{(100 - C)}{100} \right)$$

When dN/dt is plotted versus N, we obtain the graph shown in Figure 22.4. dN/dt has a maximum. The dN/dt versus N plot is applied in management of natural resources, for instance, fishery and forest models.

22.4 Environmental Processes and Urban Ecosystems

Municipalities have to provide drinking water, collect and treat garbage (solid waste) and waste water. All these activities involve environmental processes and have an unavoidable impact on urban ecosystems. Water consumption is roughly equal to the amount of waste water. The amount of solid wastes is proportional to the number of inhabitants. The water consumption and the amount of garbage per capita vary, however, from country to country. The industrialized countries use on average about 200–500 L of water per capita per day, while developing countries usually have a much lower consumption, that is, on average half as much or 100–250 L per capita per day. Similarly, the amount of garbage per capita in industrialized countries is about 500–700 kg per year, while the amount in developing countries is 250–350 kg per capita per year.

Water and wastewater treatment use a number processes, for example, chemical precipitation, settling, filtration and biological decomposition. These processes have all been treated in Sections II–IV of the book. Chemical precipitation in water infrastructures or wastewater treatment plant is following the chemical equations and rules presented in Section III, and settling and filtration are physical processes not dissimilar from the description given in Section II.

Biological treatment is an important process in almost all wastewater treatment plants. It is most often quantified by the following equation:

$$\text{Decomposition rate} = \mu_{max} \times C \times \min\left\{\frac{C}{(C + K_1)}, \frac{O_2}{(O_2 + K_2)}\right\} \tag{22.9}$$

where μ_{max} is the maximum decomposition rate (a characteristic parameter) C is the substrate concentration; that is, the concentration of biodegradable organic matter, K_1 and K_2 are Michaelis–Menten's half saturation constants (see Section 15.2.1) and O_2 is the oxygen concentration. The equation expresses that either the biodegradable organic matter or the oxygen may be the limiting factor; see Liebig's minimum law in Section 15.2.2. The first C in Equation 22.9 can be replaced by B/Y, where B is the biomass of the microorganisms carrying out the decomposition and Y is the yield of biomass per unit of C.

The sludge obtained by settling—either directly or after biological treatment—is usually decomposed anaerobic (but an aerobic decomposition may also be applied; see Section 12.3), Biogas is produced by the anaerobic decomposition. The decomposition of the sludge is usually described by the use of first-order kinetics.

The biodegradation in waste treatment plants is strongly dependent on the composition of the organic matter and on the presence of toxic substances that may slow down or even prevent the biological decomposition process. The theoretical biological oxygen demand (BOD), (Section 12.2) may be used as a suitable reference. Most often is used, however, the five-day BOD as percentage of the theoretical BOD. It may also be indicated as the BOD5-fraction. For instance, a BOD5-fraction of 0.7 means that BOD5 corresponds to 70% of the theoretical BOD = the oxygen demand for a complete decomposition of all the organic matter. C is Equation 22.9 can be replaced by BOD, and if the decomposition for 5 days is found by integration of the BOD5 will be the result. The biodegradability of the organic matter can be expressed by the BOD5 fraction. The biodegradation is, however, in some cases also very dependent on the concentration of microorganisms as made clear if C is replaced by B/Y.

In the microbiological decomposition (Section 17.4) of xenobiotic compounds an acclimatization period from a few days to 1–2 months should be foreseen before the optimum biodegradation rate can be achieved. We distinguish between primary and ultimate biodegradation. Primary, biodegradation is any biologically induced transformation that changes the molecular integrity. Ultimate biodegradation is the biologically mediated conversion of organic compounds to inorganic compounds and products associated with complete and normal metabolic decomposition.

The rate of biodegradation is expressed by application of a wide range of units:

1. A first-order rate constant (1/24 h)—it means by μ_{max}
2. Half-life time (days or hours)
3. mg organic matter for instance expressed as BOD per g sludge per 24 h (mg/(g 24 h))
4. mg organic matter (BOD) per g bacteria per 24 h (mg/[g 24 h])
5. mL of substrate per bacterial cell per 24 h (mL/[24 h cells])
6. mg COD (chemical oxygen demand determined by the amount of chromate that is needed for a complete decomposition) per g biomass per 24 h (mg/[g 24 h])
7. mL of substrate per gram of volatile solids inclusive microorganisms (mL/[g 24 h])
8. BODx/BOD∞, that is, the BOD in *x* days compared with complete degradation (∞), named the BODx-coefficient
9. BODx/COD, that is, the BOD in *x* days compared with complete degradation, expressed by means of COD

The rate biodegradation in water or soil (or sludge) is difficult to estimate, because the number of microorganisms varies several orders of magnitudes from one type of aquatic ecosystem to the next and from one type of soil to the next. Artificial intelligence has been used as a promising tool to estimate the rate of biodegradation or biodegradability. However, a (very) rough, first estimation can be made on the basis of a relationship between the molecular structure and the biodegradability. This method gives very quick estimations and provides with good approximation an idea about the biodegradability when the chemical composition is known. The biodegradability of proteins, carbohydrates, and lipids is known to be quite fast, but urban waste consists often of so many different chemical compounds with very different biodegradability, that it becomes important to estimate the biodegradability on the basis of waste composition. The following rules based on a point (zero dimensional) system can be used to establish these estimations:

1. Polymer compounds are generally less biodegradable than monomer compounds. 1 point for a molecular weight > 500 and ≤ 1000, 2 points for a molecular weight > 1000.
2. Aliphatic compounds are more biodegradable than aromatic compounds. 1 point for each aromatic ring.
3. Substitutions, especially with halogens and nitro groups, will decrease the biodegradability. 0.5 points for each substitution, although 1 point if it is a halogen or a nitro group.

4. Introduction of double or triple bond will generally mean an increase in the biodegradability (double bonds in aromatic rings are of course not included in this rule)—1 point for each double or triple bond.
5. Oxygen and nitrogen bridges (- O - and - N - (or =)) in a molecule will decrease the biodegradability. 1 point for each oxygen or nitrogen bridge.
6. Branches (secondary or tertiary compounds) are generally less biodegradable than the corresponding primary compounds. 0.5 point for each branch.

Find the number of points and use the following classification:

≤ 1.5 points: The compound is readily biodegraded. More than 90% will be biodegraded in a biological treatment plant during 24 h or less. It corresponds to the biodegradability of proteins, carbohydrates, and lipids.

2.0–3.0 points: The compound is biodegradable. Probably about 10%–90% will be removed in a biological treatment plant. BOD5 is 0.1–0.9 of the theoretical oxygen demand.

3.5–4.5 points: The compound is slowly biodegradable. Less than 10% will be remved in a biological treatment plant. BOD10 ≤ 0.1 of the theoretical oxygen demand.

5.0–5.5 points: The compound is very slowly biodegradable. It will hardly be removed in a biological treatment plant and a 90% biodegradation in water or soil will take ≥ 6 months.

≥ 6.0 points: The compound is refractory. The half-life time in soil or water is counted in years. The structure of DDT corresponds to about 7 points and the biological half-life of DDT in soil is about 14 years.

22.5 Urban Solid Waste Management

Urban solid waste management involves collection, treatment, and disposal of solid wastes generated by urban population. All these activities have significant impacts on urban ecosystems and on the other ecosystems with which these systems interact. The growing solid waste production results from increased consumption of materials, such as minerals, wood products, and food. The sustainability of the whole process is compromised by the increased use of nonrenewable materials, such as metals and fossil fuel-derived products. As a first significant impact, it should be recognized that extracting, collecting, processing, transporting, and disposing of these materials result in unbalanced greenhouse gases emissions. Differently from renewable products such as those provided by agriculture and forestry,* *life cycle assessment* (LCA) of these materials include not only greenhouse gases generation but also the large amounts of energy required for their production and disposal. Managers and policy-makers ought to understand and compare the LCAs, the energy implications and the polluting potential of the different materials to set up appropriate management strategies, such as *recycling, source reduction, landfilling, energy recovery by combustion* and *composting* (see biodegradation discussed earlier). Source reduction, reuse, and recycling are the chief management options that guarantee sustainability in an environmental friendly and socially satisfactory context. On the contrary, being more easy and immediate, the most commonly applied strategies are today's composting, waste combustion (incineration), and landfilling. Yet, these listed strategies are in growing measure of unsustainability.

* These productions generate greenhouse gases as well, however these flows are balanced by CO_2 sequestration (carbon fixation) (see also Section 12.4).

23

The Atmosphere

23.1 Introduction

Of the four ecospheres, the atmosphere has a crucial role in protecting and keeping stable conditions in the biosphere. As discussed in Chapters 11 and 12 (but see also Section 18.4), the atmosphere is involved in all life cycles, acting both as a sink and reservoir for pollutants and other compounds of biochemical relevance. In particular, the impacts of pollutants in the atmosphere depend on atmospheric dispersion processes. Among the negative impacts, of prominent importance are the ozone layer depletion, causing harmful biota exposure to ultraviolet B (UVB), and wet and dry atmospheric deposition that transfer pollutants to terrestrial and aquatic ecosystems.

The atmosphere is made up of a variable composition of the liquid and gaseous matrices. Therefore, for a more in-depth comprehension of the mechanism underlying air quality, processes of exchanges between these two phases must be considered. As usual, these processes may be described by means of Henry's law.

23.2 Atmospheric Pollution by Gases and Particles

The atmosphere is polluted by gases and particles. The gases are reacting with the atmospheric components, mainly oxygen. For instance, methane is oxidized to carbon dioxide and water. This process, as many other processes, can be described with first order kinetics, because oxygen has a constant concentration corresponding to 21% v/v. The half-life time in the atmosphere of methane is 7 years corresponding to a decomposition rate of 0.1 1/year. For other more complex chemical reactions, the simple first order reaction scheme cannot be applied. Obviously, rates are also dependent on atmospheric flows and on the flow pattern.

When considering particulate pollution, the source should be categorized with regard to contaminant type. Inert particulates are distinctly different from active solids in the nature and type of their potentially harmful human health effects. Inert particulates comprise solid airborne materials, which do not react readily with the environment and do not exhibit any morphological changes as a result of combustion or any other process. Active solid matter is defined as particulate material, which can be further oxidized or react chemically with the environment or the receptor. Any solid material in this category can, depending on its composition and size, be considerably more harmful than inert matter of similar size.

A closely related group of emissions are from aerosols of liquid droplets, generally below 5 μm. They can be either oil or other liquid pollutants (e.g., freon) or may be formed by condensation in the atmosphere. Fumes are condensed metals, metal oxides, or metal halides, formed by industrial activities, predominantly as a result of pyro-metallurgical processes; melting, casting, or extruding operations. Products of incomplete combustion are often emitted in the form of particulate matter. The most harmful components in this group are particulate polycyclic organic matter (PPOM), which are mainly derivatives of benzo-a-pyrene.

Natural sources of particulate pollution are sandstorms, forest fires, and volcanic activity. The major sources in towns are vehicles, combustion of fossil fuel for heating and production of electricity, and industrial activity.

The total global emission of particulate matter is in the order of 10^7 tons per year. Deposition of particles may occur mainly by three processes:

1. Sedimentation (Section 7.5, Stokes law may be applied, particles >20 μm)
2. Adsorption (Section 11.2) or inertial impaction (determined by differences in concentrations by use of Fick's law, particles between 5 and 20 μm)
3. Diffusion (Section 7.2, particles <5 μm)

Particles <20 μm are named *suspended particulate* matter. Particles >20 μm may be denoted *dust* that will be deposited close to the source due to the high sedimentation rate. *Dry deposition* consists of gases or dry particles. *Wet deposition* is raindrops containing gases and particles. Particles may consist of minor concentrations of dissolved salts in water drops, crystals, or a combination of the two.

Particulate pollution is an important health factor. The toxicity and the size distribution are the most crucial factors. Many particles are highly toxic, such as asbestos and those of metals, such as beryllium, lead, chromium, mercury, nickel, and manganese. In addition, particulate matter is able to absorb gases, which enhance the effects of these components. In this context, the particle size distribution is of particular importance, as particles greater than 10 μm are trapped in the human upper respiratory passage and the specific surface (expressed as m^2 per g of particulate matter) increases

TABLE 23.1

Typical Particle Size Ranges

	μm
Tobacco smoke	0.01–1
Oil smoke	0.05–1
Ash	1–500
Ammonium chloride smoke	0.1–4
Powdered activated carbon	3–700
Sulfuric acid aerosols	0.5–5

with l/d, where d is the particle size. The adsorption capacity (Section 11.2) of particulate matter, expressed as g adsorbed per g of particulate matter, will generally be proportional to the surface area. Table 23.1 lists some typical particle size ranges. However, size as well as shape and density must be considered. Furthermore, the size of a particle is determined by two parameters: the mass median diameter, which is the size that divides the particulate sample into two groups of equal mass, that is, the 50% point on a cumulative frequency versus particle size plot and the geometric standard deviation.

Particulate pollutants have the ability to adsorb gases including sulfur dioxide, nitrogen oxides, carbon monoxide, and so on. The inhalation of these toxic gases is frequently associated with this adsorption, as the gases otherwise would be dissolved in the mouthwash and spittle before entering the lungs. Particulate pollution may be controlled by modifying the distribution pattern. This method is described in detail in Section 23.3. It represents in principle an obsolete philosophy of pollution control and abatement, namely *washing by dilution*, but it is still widely used to reduce the concentration of pollutants at ground level and thereby minimize the effects of air pollution.

23.3 Modifying the Distribution Patterns

Although emissions, gaseous or particulate, may be controlled by various sorption processes or mechanical collection, the effluent from the control device must still be dispersed into the atmosphere. Atmospheric dispersion depends primarily on horizontal and vertical transport. The horizontal transport depends on the turbulent structure of the wind field. As the wind velocity increases so does the degree of dispersion, and there is a corresponding decrease in the ground level concentration of the contaminant at the receptor site. The emissions are mixed into larger volume, of air and the diluted emission is carried out into essentially unoccupied terrain away from any receptors. Depending on the wind direction, the diluted effluent may be funneled down a river valley or between mountain ranges. Horizontal transport is sometimes prevented by surrounding hills forming a natural pocket for locally generated pollutants. This particular topographical situation occurs for instance in the Los Angeles area, which suffers heavily from air pollution.

The vertical transport depends on the rate of change of ambient temperature with altitude. The dry adiabatic lapse rate is defined as a decrease in air temperature of 1°C per 100 m. This is the rate at which, under natural conditions, a rising parcel of unpolluted air will decrease in temperature with elevation into the troposphere up to approximately 10,000 m. Under isothermal conditions, the temperature does not change with elevation. Vertical transport can be hindered under stable atmospheric conditions, which occur when the actual environmental lapse rate is less than the dry adiabatic lapse rate. A negative lapse rate is an increase in air temperature with latitude. This effectively prevents vertical mixing and is known as inversion.

The dispersion from a point source—a chimney for instance—may be calculated from the Gaussian plume model (see Reible, 1998; Ganev, 2011). However, the dispersion is of course also dependent on the atmospheric flows and flow pattern.

These different atmospheric conditions are illustrated in Figure 23.1, where stack gas behavior under the various conditions is shown. Further explanations are given in Table 23.2. The distribution of particulate material is more effective than higher the stack. The maximum concentration, C_{max}, at ground level can be shown to be roughly proportional to the emission and to follow approximately this expression:

$$C_{max} = \frac{kQ}{H^2} \tag{23.1}$$

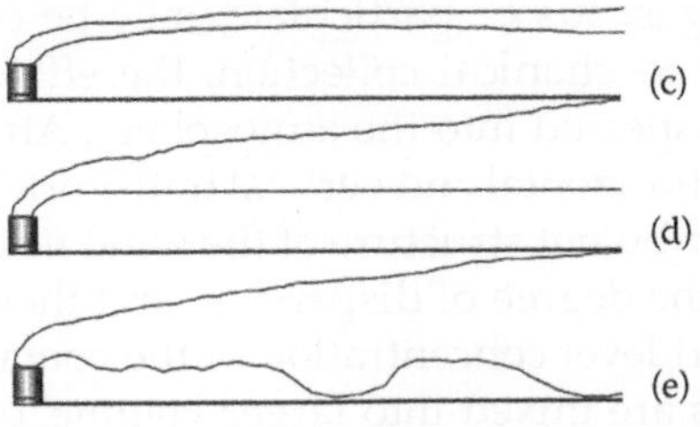

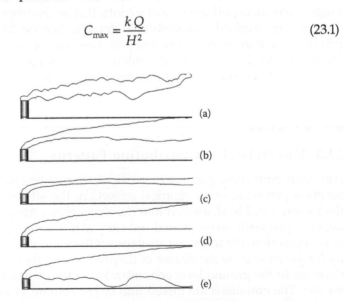

FIGURE 23.1
Stack gas behavior under various conditions. (a) Strong lapse (looping), (b) Weak lapse (coning), (c) Inversion (fanning), (d) Inversion below, lapse aloft (lofting), (e) Lapse below, inversion aloft (fumigation).

TABLE 23.2

Various Atmospheric Conditions

Strong lapse (looping)	Environmental lapse rate > adiabatic lapse rate
Weak lapse (coning)	Environmental lapse rate < adiabatic lapse rate
Inversion (fanning)	Increasing temperature with height
Inversion below	Increasing temperature below
Lapse aloft (lofting)	Approximately adiabatic lapse rate aloft
Lapse below, inversion	Approximately adiabatic lapse rate below
Aloft (fumigation)	Increasing temperature aloft

where Q is the emission (expressed as g particulate matter per unit of time), H is the effective stack height, and k is a constant. The effective height is slightly higher than the physical height and can be calculated from information about the temperature, the stack exit velocity and the chimney inside diameter. These equations explain why a lower ground-level concentration is obtained when many small stacks are replaced by one very high stack. In addition to this effect, it is always easier to reduce and control one large emission than many small emissions, and it is more feasible to install and apply the necessary environmental technology in one big installation.

The effective stack height is however a little higher than the physical stack height because the flow through the chimney brings the smoke h meters higher than the physical stack height. h can be found from:

$$h = 0.28 \times V \times D \left\{ 1.5 + \left(2.7 \times (T - 273) \times \frac{D}{T} \right) \right\} \tag{23.2}$$

where V is the stack exit velocity m/s, D is the stack exit inside diameter, T is the absolute temperature.

23.4 Ozone Depletion

There has been a decline of about 4% per decade in the total volume of ozone in Earth's stratosphere (the *ozone layer*) since 1970. Furthermore, there has been a much larger springtime decrease in stratospheric ozone over Earth's polar regions. The latter phenomenon is referred to as the *ozone hole*. The details of processes behind the ozone depletion are catalytic destructions of ozone by atomic halogens. The main source of these halogen atoms in the stratosphere is photodissociation (see Chapter 10) of man-made halocarbon refrigerants (chlorofluorocarbons [CFCs], freons, halons). These compounds are transported into the stratosphere after being emitted at the surface. CFCs and other contributory substances are referred to as ozone-depleting substances (ODS). Since the ozone layer prevents most harmful UVB wavelengths (280–315 nm) of UV light from passing through Earth's atmosphere, observed and projected decreases in ozone have generated worldwide concern leading to the adoption of the Montreal Protocol that bans the production of CFCs, halons, and other ozone-depleting chemicals such as carbon tetrachloride and trichloroethane. It is suspected that a variety of biological consequences, such as increases in skin cancer, cataracts, damage to plants, and reduction of plankton populations in the ocean's photic zone, may result from the increased UV exposure due to ozone depletion.

Three forms of oxygen are involved in the ozone–oxygen cycle (see Section 12.2): oxygen atoms (O or atomic oxygen), oxygen gas (O_2 or diatomic oxygen), and ozone gas (O_3 or tri-atomic oxygen). Ozone is formed in the stratosphere when oxygen molecules photodissociate after absorbing a UV photon whose wavelength is shorter than 240 nm. This converts a single O_2 into two atomic oxygen radicals. The atomic oxygen radicals then combine with separate O_2 molecules to create two O_3 molecules. These ozone molecules absorb by this process UV light between 310 and 200 nm, following which ozone splits into a molecule of O_2 and an oxygen atom. The oxygen atom then joins up with an oxygen molecule to regenerate ozone. This is a continuing process that terminates when an oxygen atom recombines with an ozone molecule to make two O_2 molecules:

$$O + O_3 \rightarrow 2O_2$$

The overall amount of ozone in the stratosphere is determined by a balance between photochemical production and recombination. Ozone can be destroyed by a number of free radical catalysts, the most important of which are the hydroxyl radical (OH^-), the nitric oxide radical (NO^-), atomic chlorine ion (Cl) and atomic bromine ion (Br). The dot indicates an unpaired electron and these components are thus extremely reactive. All of these have both natural and man-made sources; at the present time, most of the OH^- and NO^- in the stratosphere are of natural origin, but human activity has dramatically increased the levels of chlorine and bromine. These elements are found in certain stable organic compounds, especially CFCs, which may find their way to the stratosphere without being destroyed in the troposphere due to their low reactivity. Once in the stratosphere, the Cl and Br atoms are liberated from the parent compounds by the action of UV light, for example,

$$CFCl_3 + UV - Energy \rightarrow CFCl_2 + Cl$$

The Cl and Br atoms can then destroy ozone molecules. In the simplest example, a chlorine atom reacts with an ozone molecule, forming ClO, and the chlorine monoxide can react with a second molecule of ozone (O_3) to yield another chlorine atom. The processes are

$$Cl + O_3 \rightarrow ClO + O_2$$

$$ClO + O_3 \rightarrow Cl + 2O_2$$

The ClO from the previous reaction destroys thereby a second ozone molecule and recreates the original chlorine atom, which can repeat the first reaction and continue to destroy ozone.

The processes show that one molecule of Cl or Br can by this closed loop catalytic reactions continue very long time to break down the ozone. More complicated mechanisms have been discovered that lead to ozone destruction in the lower stratosphere as well. Due to the complexity of the processes, it is not possible to explain by a simple first order reaction or second order reaction, the ozone depletion.

23.5 Henry's Law

An increase or decrease in the concentration of components or elements in ecosystems are of vital interest, however, the observation of trends in global changes of concentrations might be even more important as they may cause changes in the life conditions on Earth.

The concentrations in the four spheres, the atmosphere, the lithosphere, the hydrosphere, and the biosphere, are in this context of importance. They are determined by the transfer processes and the equilibrium concentrations among the four spheres (see Section 18.4). The solubility of a gas at a given concentration in the atmosphere can be expressed by means of Henry's law that determines the distribution between the atmosphere and the hydrosphere (see also Section 7.7 and Chapter 11):*

$$p = H \cdot x \tag{23.3}$$

where

p = the partial pressure (atm)
H = Henry's constant (atm)
x = molar fraction in solution (adimensional)

Henry's law has been mentioned as Equation 7.34 but in the exchange between the two spheres—the hydrosphere and the atmosphere—there are good reasons to go into more details. H is dependent on temperature (see Table 23.3) and is usually expressed in atmospheres. It may be converted into Pascals, as 1 atmosphere = 101,400 Pa. A dimensionless Henry's constant may also be applied. As $p = RT\,n/v = RTc_a$ and $x = c_h/(c_h + c_w)$, where c_a

* Notice that Henry's law as presented here is expressed in terms of molar fractions (adimensional) in solution, while in Section 7.7 a slightly different form was adopted that uses concentrations, therefore He is in (atm m^3)/moles.

TABLE 23.3
Henry's Constant (atm) for Gases as a Function of Temperature

	Temperature (°C)						
Gas	**0**	**5**	**10**	**15**	**20**	**25**	**30**
Acetylene	0.72	0.84	0.96	1.08	1.21	1.33	1.46
Air (atm)	0.43	0.49	0.55	0.60	0.66	0.71	0.77
Carbon dioxide	73	88	104	122	142	163	186
Carbon monoxide	0.35	0.40	0.44	0.49	0.54	0.58	0.62
Hydrogen	0.58	0.61	0.64	0.66	0.68	0.70	0.73
Ethane	0.13	0.16	0.19	0.23	0.26	0.30	0.34
Hydrogen sulfide	26.80	31.50	36.70	42.30	48.30	54.50	60.90
Methane	0.22	0.26	0.30	0.34	0.38	0.41	0.45
Nitrous oxide	0.17	0.19	0.22	0.24	0.26	0.29	0.30
Nitrogen	0.53	0.60	0.67	0.74	0.80	0.87	0.92
Nitric oxide	—	1.17	1.41	1.66	1.98	2.25	2.59
Oxygen	0.25	0.29	0.33	0.36	0.40	0.44	0.48

Note: The values in the table are Henry's constant ($\times 10^{-5}$).

is the molar concentration in the atmosphere of component h, expressed in mol/l and c_h is the concentration in the hydrosphere expressed also in mol/l and c_w is the mol/l of water (and other possible components). If we consider only two components in the hydrosphere, h and water, and that $c_h << c_w$, we can replace ($c_h + c_w$) with the concentration of water in water = 1000/18 = 55.56 mol/L. We obtain according to these approximations the following equation:

$$\frac{c_a}{c_h} = \frac{H}{(R \times T \times 55.56)} \tag{23.4}$$

where $H/(R \times T \times 55.56)$ is the dimensionless Henry's constant.

In aquatic environmental chemistry, we often know the partial pressure in atm, p_a, and want to calculate the concentration in water. In this case, we often use the following expression:

$$C_h = K_H p_a \tag{23.5}$$

K_H is a constant = 55.56/H. If we use the values in Table 23.3 for carbon dioxide, for instance, we get that K_H at 20°C is $55.56/1.42\ 10^5 = 3.91.\ 10^{-2} = 10^{-1.41}$. For example in 2003, the partial pressure of carbon dioxide in the atmosphere is close to 0.0004 atm, which corresponds at 20°C to a concentration of $3.91.\ 10^{-2} \times 0.0004 = 0.0000156$ M in aquatic ecosystem in equilibrium with the atmosphere.

Illustration 23.1

The solubility of oxygen in fresh water is 11.3 mg/l at 10°C. The values show that it corresponds to the Henry's constant found in Table 23.3 for oxygen.

SOLUTION

Henry law is applied to find the molar fraction in water, x (the partial pressure of oxygen is 0.21 atm, corresponding to 21% oxygen in the atmosphere):

$$x = \frac{0.21}{33000} = 6.36 \times 10^{-6}$$

This is translated into mg/L

$$\text{Dissolved oxygen mg/L} = 6.36 \times 10^{-6} \times 55.56 \times 32 \times 1000 = 11.31 \text{ mg/L}$$

Reaeration is treated in Section 7.8, and it covers of course the transfer of oxygen from the atmosphere to the hydrosphere. Many empirical equations have been developed to yield the reaeration coefficient for oxygen. If other components are considered than oxygen, then the driving force is the difference in concentration and K_L according to the following equation:

$$\text{Evaporation} = b * K_L{}'A * (To - T *) \tag{23.6}$$

where A is the area, b is a coefficient that adjusts the aeration coefficient K_L to other compounds than oxygen. b can be found—with particularly good approximation when the liquid film causes the highest resistance—as the ration between the molecular diffusion coefficient, D of the considered component and that of oxygen:

$$b = \frac{D - \text{Compounds}}{D - \text{Oxygen}} \tag{23.7}$$

23.6 Acidification

As anticipated in Section 20.3, current emissions of SO_2 and NO_x of anthropogenic origin may cause regional scale acidification (pH decreasing) of the atmosphere and of sensitive ecosystems (e.g., lake ecosystems). This effect is propagated from the atmosphere to aquatic and terrestrial ecosystems through *acid rains* and *atmospheric deposition*. Pollutants in the form of gases and particulates released to the atmosphere by human activities (e.g., nitrogen, sulfur, metal compounds, as well as several other toxicants such as organic micropollutants) may eventually settle to the ground as dust (dry deposition) or fall to Earth with moisture, rain, and snow (wet deposition).

In the next years, the industrialized countries will unlikely experience significant increase in pollutant emissions into the atmosphere. However, environmental effects will still continue, and in some cases, will get worse due to the delayed acidification effects on aquatic and terrestrial ecosystems. On the contrary, over the next several decades, developing countries will probably experience major increases in emissions to the atmosphere. Consistently with the exponential population growth rates in these areas and the potential for industrial expansion, future emissions of pollutants could greatly exceed current emissions. Some of the consequences to the environment of these increased emissions are well understood from our current knowledge. However, given the magnitude of the potential change in emissions, other unpredicted effects are foreseeable.

Ocean acidification, which was first introduced in Section 12.4, is caused by unbalanced CO_2 emissions of anthropogenic origin into the atmosphere (see Section 22.5) that leave a residual net carbon dioxide inflow into the oceans of about 2 gigatonnes. This phenomenon leads to ocean acidification. Growing dissolved CO_2 in seawater increases the hydrogen ion (H^+) concentration in the ocean, and thus decreases ocean pH, according to the following reaction:

$$CO_{2(aq)} + H_2O \leftrightarrow H_2CO_3 \leftrightarrow HCO_3^- + H^+ \leftrightarrow CO_3^{2-} + 2\,H^+$$

The final effect is an increase in acidity of seawater. A decrease in pH will have negative consequences on all forms of life, yet, primarily for oceanic calcifying organisms. Indeed, as ocean pH falls, the concentration of carbonates required for saturation to occur increases, and when carbonate becomes undersaturated,* structures made of calcium carbonate begin to dissolve. Therefore, even if the rate of calcification does not change, the rate of dissolution of calcareous material increases, causing severe threats to biological communities.

* See also Section 9.6.

Appendix: The Double Logarithmic Representation

A.1 Double Logarithmic Diagrams Applied on Acid–Base Reactions

The chemistry of the aquatic environment involves many simultaneous reactions, which can be difficult to overview unless we use computer programs or double logarithmic diagrams that allow us quickly to assess the approximate concentrations. It includes a determination of components that we do not need to consider because they have relatively small concentrations. The double logarithmic diagram is applied in this handbook because it is relatively easy to construct and is very illustrative. The double logarithmic diagram for acid–base reactions plots the logarithmic of the concentration of the various species versus pH.

If we consider an acid–base reaction: $HA \Leftrightarrow A^- + H^+$ the logarithmic form of the equilibrium constant expression is

$$pK = -\log\{A^-\} - \log\{H^+\} + \log\{HA\} = -\log\{A^-\} + pH + \log\{HA\} \quad (A.1)$$

or in the form of the Henderson–Hasselbach's equation:

$$pH = pK + \frac{\log\{A^-\}}{\{HA\}} \quad (A.2)$$

Notice that "p" is a general abbreviation for "–log". K is the equilibrium constant (see Chapter 11).

From Equation A.2, it is clear that when $pH \ll pK$ is $\{HA\} \gg \{A^-\}$. As the acid–base system only has the two forms HA and A^-, $\{HA\} \approx [HA] \approx C$, where C is the total concentration of the acid–base system. Log $\{A^-\}$. at low pH values can be derived from Equation A.1: $\log\{A^-\}. = pH - pK + C. \approx \log [A^-]$. This will correspond to a straight line with the slope + pH in a logarithmic diagram. For pH = pK, the line will take the value C. A straight line through the point (pK, C) with a slope of + 1, represents, therefore, log $[A^-]$ in a diagram where log c_i (c_i symbolizing various species) versus pH.

At $pH \gg pK$, $\{A^-\}. \approx [A^-]. \approx C$. $\log\{HA\} = pK - pH + C. \approx \log [HA]$. This implies that log [HA] in a double logarithmic diagram for high pH values is represented by a straight line with the slope –1 and going through the point (pK, C).

TABLE A.1
Rationale beyond the Double Logarithmic Representation

	Log [HA]	Log [A⁻]
pH << pK	log C	pH – pK + C
pH = pK	log (C/2)	log (C/2)
pH >> pK	pK – pH + C	log C

For pH = pK, we know from Equation A.2 that {HA} = {A⁻} = C/2 ≈ [HA] = [A⁻]. Table A.1 summarizes these results.

Figure A.1 shows a double logarithmic diagram for a 0.01 M acid (log C = –2) with pK = 6.0. The diagram is drawn by the use of the straight lines for log [HA] and log [A⁻] at low and high pH, and the point (pK, C/2), which is valid for both log [HA] and log [A⁻]. The gap between low pH and pH = pK can easily be drawn and correspondingly for the gap from pH = pK and high pH.

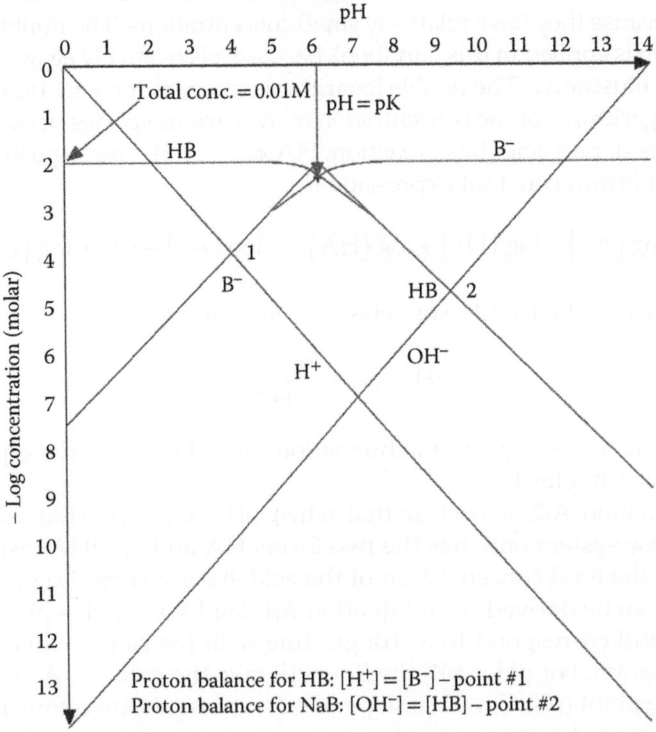

FIGURE A.1
(See color insert.) A double logarithmic diagram for an acid–base system with a total concentration of 0.01 m and pK = 6.0. The proton balance for HB and NaB is shown. The pH for the two cases are found as point 1 (pH = 4.0) and point 2 (pH = 9.0). Notice that the total composition can be red at the diagram. B⁻ at point 1 is 0.0001 and HB at point 2 is 10^{-9}.

The double logarithmic diagram represents two equations: the mass equation expression and the information that the total concentration is 0.01 M. We have, however, four unknown: [HA], $[A^-]$, $[H^+]$, and $[OH^-]$. We need therefore two more equations. Water's ion product $[H^+]\,[OH^-] = 10^{-14}$ or on logarithmic form: $\log\,[H^+] + \log\,[OH^-] = -14 = pK_w$ can be used as the third equation. The fourth and last equation is the information that a solution always will be uncharged – that the sum of the concentrations of positive charged ions times their charge = the sum of the concentrations of the negative charged ions times their charge. In most cases it is, however, more beneficial to use that the concentrations resulting from dissociation of hydrogen ions is equal to the concentrations of ions that have taken up hydrogen ions. It is called the proton balance. To use this fourth equation to assess the composition of a solution included pH, it is recommendable to write up the components that are in the solution before any reaction.

For instance, if an acid HB in water is considered, the components before any reaction is HB and H_2O. Therefore, these two components cannot result from dissociation or uptake of hydrogen ions. Dissociation of hydrogen ions from these two components can only result in formation of B^- and OH^-, and uptake of hydrogen ions is only possible for water forming the oxonium ions, which is often just written as H^+ only. Therefore, the proton balance yields the following equation:

$$[B^-]+[OH^-]=[H^+] \tag{A.3}$$

As we have an acidic solution, pH is relatively low, and it is therefore assumed—at least in the first hand—that $+ [OH^-] \approx 0$. Therefore, the proton balance gives

$$[B^-]=[H^+] \tag{A.4}$$

This equation corresponds to the point 1 in Figure A.1. This point represents the composition of the 0.01 M HB solution. It is of course necessary to check the assumption, that the hydroxide ions are negligible. As seen $[OH^-] = 10^{-9}$ at point 1. Therefore, it was fully acceptable to consider $[OH^-]$ as negligible compared with $[B^-]$ and $[H^+]$.

With the same argument can be found that the proton balance for a solution of NaB yields $[OH^-] = [HB]$ corresponding to point 2 in Figure A.1.

Figure A.2 shows a double logarithmic diagram of the system H_2B, HB^-, and B^{2-}. The total concentration is 0.01 M. In this case, the acid is able to dissociate two hydrogen ions, and has therefore two pK values: pK_1 and pK_2. The H_2B curve will at $pH \geq pK_2$ have the slope −2, as seen, because the actual process is $H_2B \Leftrightarrow B^{2-} + 2H^+$. B^{2-} will correspondingly have the slope +2 at $pH \geq pK_1$.

The proton balance for the ampholyte NaHB will with good approximations yield the equation: $[H_2B] = [B^{2-}]$ as $[OH^-]$ and $[H^+]$ both $<< [H_2B] = [B^{2-}]$. The point 3 in Figure A.2 corresponds to the composition of the various species in

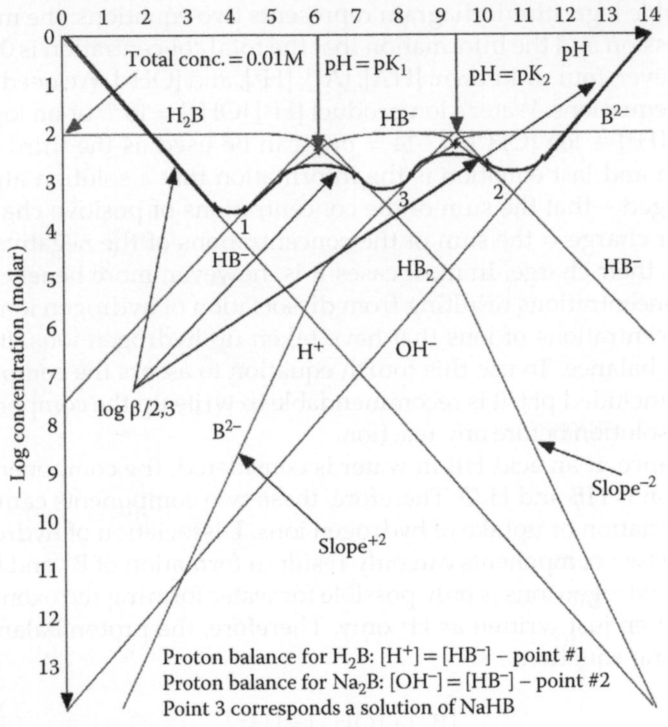

FIGURE A.2
(See color insert.) A double logarithmic diagram of the system H_2B, HB^-, and B^{2-} total concentration = 0.01 M. Notice that the slopes of the curves for H_2B and B^{2-} are +2 above pK_2 and below pK_1, respectively. The proton balance after approximations were considered and are shown for the two cases 0.01 M H_2B and 0.01M Na_2B. The composition of the two solutions can be read from the figure points 1 and 2, respectively. A 0.01 M solution of NaHB has the composition corresponding to point 3.

a 0.01 M NaHB solution with the pK values, as shown in Figure A.2, that is, 6.1 and 9.5. Point 1 corresponds to a 0.01 M H_2B solution, where the concentrations of B^{2-} and hydroxide ions are considered negligible in the proton balance. The approximated proton balance is shown in Figure A.2. Point 2 in the Figure A.2 corresponds to a 0.01 M B^2 solution. The proton balance is also shown in Figure A.2. H_2B and hydrogen ions are in this case considered negligible.

Phosphoric acid is a medium strong acid and has therefore a low pK_1 value, namely 2.12. It implies that point 1 corresponding to the composition of a 0.4 M phosphoric acid solution yields an undissociated phosphoric acid concentration different from the total concentration—compare with Figures A.1 and A.2 where a weak acid solution clearly gives an undissociated acid close to the total concentration. In the case of phosphoric acid, we have to read the concentration of phosphoric acid in a 0.4 M solution, and it is advantageous to read the concentration of dihydrogen phosphate and deduct it from the total concentration. The concentration of dihydrogen phosphate is found from the diagram to be antilog (–1.4) corresponding to the concentration of about 0.04 M. pH is 1.4. Therefore, the concentration of phosphoric acid becomes 0.4 – 0.04 = 0.36 M.

A.2 Molar Fraction, Alkalinity, and Buffer Capacity

Introduction of molar fractions makes it possible to set up equations to compute the composition of acid–base solutions. As mentioned above, the application of a double logarithmic diagram corresponds to find x unknown concentrations from x equations. It is therefore of course possible to find a composition of a complex acid–base mixture by solving the equations. However, it is in most cases faster and sufficiently accurate to use a double logarithmic diagram. Molar fractions may also be used in the double logarithmic diagram instead of concentrations. The concentrations for a number of different cases (different total concentrations) of the same components can easily be found by multiplying the molar fractions found on the diagram with the total concentration.

The molar fractions are shown in Figure A.3 for the system H_2B, HB^-, and B^{2-}. The three shown equations are found from the two mass equations and from the equations that express that the sum of the three molar fractions is one.

The buffer capacity, β, is defined as

$$\beta = \frac{dC}{dpH} \tag{A.5}$$

Acids and bases: important equations

$$\text{Molar fraction } H_2B = \frac{1}{\left(1+K_1[H^+]+K_1K_2[H^+]^2\right)}$$

$$\text{Molar fraction } HB^- = \frac{1}{1+[H^+]/[(K_1+K_2)/H^+]}$$

$$\text{Molar fraction } B^{2-} = \frac{1}{1+[H^+]/\left[\left(K_2+[H^+]^2\right)/K_1K_2\right]}$$

$$\beta = \text{buffer capacity} = 2.3\frac{([H^+]+[OH^-]+[HA][A^-])}{([HA]+[A^-])}$$

$$[\text{Alk}] = [HCO_3^-]+2[CO_3^{2-}]-[H^+]+[OH^-]+\Sigma\text{other base ions}$$

NB!! Alkalinity is conserved

FIGURE A.3
Important definitions: molar fraction, buffer capacity, and alkalinity.

where dC is strong acid or base added to the considered solution and dpH is the corresponding change of pH. When β is high, relatively much acid or base is needed to change pH, while a low β value indicates that pH is changed by addition of a minor amount of acid or base. It can be shown by differentiation according to the definition given in Equation A.5, that

$$\beta = 2.3\frac{([H^+]+[OH^-]+[HA][A^-])}{([HA]+[A^-])} \tag{A.6}$$

β can be found directly from this expression. The various concentrations can be found by calculations or from the double logarithmic diagram. It is also possible to find and draw on the double logarithmic diagram the equation log β/2.3 = log ($[H^+]$ +$[OH^-]$ + $[HA][A^-]/([HA] + [A^-])$). At very low pH, $[H^+]$ is dominating the expression (Equation A.6) and a log (β/2.3) line in the double logarithmic diagram will, therefore, follow the line for log $[H^+]$. At slightly higher pH, where $[H^+] = [A^-]$ and $[HA] \approx C = [HA] + [A^-]$ (see Table A.1), log β/2.3 = log (2 $[H^+]$) = log (2 $[A^-]$) = 0.3 + log ($[H^+]$) = 0.3 + log $[A^-]$). Where the two lines for $[H^+]$ and log ($[A^-]$) intersect, log (β/2.3) value will therefore be 0.3 above the intersection. At pH = pK, the β-expression is dominated by $[HA][A^-]/[HA] + [A^-] = C/2.C/2/C = C/4$. It means that the log (β/2.3) value will be 0.3 below the intersection of [HA] and $[A^-]$ at pH = pK. At high pH, the expression dominated by $[OH^-]$ and log [β/2.3) will therefore follow the line of log $[OH^-]$. Where log $[OH^-]$ = log [HA], log (β/2.3) = 0.3 + log $[OH^-]$ = 0.3 + log [HA].

Figure A.4 shows how a log (β/2.3) plot is found for the acid–base system in Figure A.2. The double logarithmic diagram is applied to find which concentrations we have to consider in Equation A.6 and which concentrations we can eliminate because they are negligible.

The two most important acidic components in the oceans are the carbon dioxide system and boric acid. As pH in the oceans is 8.1, the buffer capacity of the water in the oceans can easily be found. The buffer capacity of the oceans is, however, much higher than the value obtained from the buffer capacities of the ions due their content of suspended clay minerals, which are able to buffer due to the following reaction:

$$\begin{aligned} &3Al_2Si_2O_5(OH)_2 + 4SiO_2 + 2K^+ + 2Ca^{2+} + 12H_2 \\ &O \rightarrow 2KCaAl_3Si_5O_{16}(H_2O)_6 + 6H^+ \end{aligned} \tag{A.7}$$

The pH dependence is indicated by the corresponding equilibrium expression in logarithmic form:

$$\log K = 6\log(H^+) - 2\log K^+ - 2\log Ca^{2+} \tag{A.8}$$

Sillen (1961) estimated the buffering capacity of these silicates to be about 1 mole per liter or approximately 2000 times the buffering capacity of carbonates.

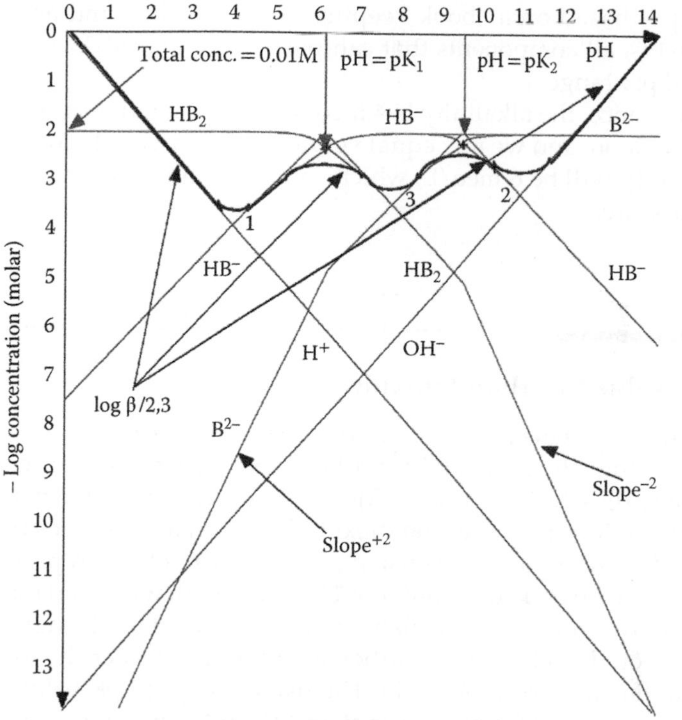

FIGURE A.4
(See color insert.) The log (β/2.3) line for Figure A.2 is shown. At points 1, 2, and 3, the line is 0.3 units above the intersections, and at pK_1 and pK_2, the log (β/2.3) line is 0.3 below the intersection.

Alkalinity is the sum of all alkaline components minus the sum of all acidic components. In aquatic chemistry, we are usually particularly interested in the amount of hydrogen ions that we have to add to a considered aquatic solution to obtain a pH corresponding to an aquatic solution of carbon dioxide. This is the alkalinity defined in Figure A.3.

Alkalinity of many natural aquatic systems is often (but not always) with good approximation equal to the hydrogen carbonate concentration. Hydrogen carbonate is at pH between 5 and 8.5, the dominating component yielding alkalinity. Other ions such as chloride and sulfate do not contribute to the alkalinity.

Notice that we cannot find the pH for a mixture of two solutions with known pH, as the weighted average of the two pH values, because the resulting number of free hydrogen ions depends on the composition of the two solutions: which ions would be able to react with the free hydrogen and hydroxide ions? Therefore, we can only find pH for a mixture when we know the concentrations of alkaline and acidic components in the two solutions and can calculate the possible neutralization reactions. The alkalinity can, however, be found easily for a mixture of two solutions, because alkalinity is based on a "book keeping" of *all* alkaline and acidic components, that is, all components that can participate in acid–base reactions in the actual pH range.

If, for instance, the alkalinity is 4 meq/L for one solution and 8 meq/L for another solution, and we mix equal volumes of the two solutions, the resulting alkalinity will be 6 meq/L, which can be used to find the resulting pH for the mixture.

A.3 Dissolved Carbon Dioxide

Open aquatic systems have in addition to the acid–base reactions of the dissolved components, an equilibrium between carbon dioxide in the atmosphere and dissolved in water. Equation A.3 can be applied to find the concentration in water of carbon dioxide. This equation implies that the carbon dioxide concentration is constant, independent of pH. A part of the dissolved carbon dioxide (in the order of 1%) reacts with water and forms carbon acid. It is however in most calculations convenient not to distinguish the dissolved carbon dioxide and the carbon acid but to consider the total amount of carbon dioxide and carbon acid. The usually applied pK values are based on this assumption. Henry's constant and the pK values for carbon acid and hydrogen carbonate are dependent on the temperature; see Table A.2.

Figure A.5 shows a double logarithmic diagram for an open aquatic system in equilibrium with the carbon dioxide in the atmosphere. The figure illustrates the system at year 1999–2002 where the partial pressure of carbon dioxide was about 0.000385 atm corresponding to a concentration of 385 ppm on volume/volume basis. In year 2012/2013, the concentration is expected to be approximately 400 ppm. Room temperature is presumed in the diagram.

Therefore, in a double logarithmic diagram, total carbonic acid = carbonic acid + dissolved carbon dioxide, denoted C_T will correspond to a horizontal line at $\log C = 0.000385.\ 10^{-1.41}$. The line representing hydrogen carbonate has a slope of +1 and intersects the carbon acid line at pH = pK_1. The carbonate

TABLE A.2

Equilibrium Constants of the Various Carbon Dioxide–Carbonate Equilibria at $I = 0$

Type of Constant	5°C	10°C	15°C	20°C	25°C	40°C	100°C
Solubility product of $CaCO_3$	8.35	8.36	8.37	8.39	8.42	8.53	
pK_1 for H_2CO_3	6.52	6.46	6.42	6.38	6.35	6.35	6.45
pK_2 for H_2CO_3	10.56	10.49	10.43	10.38	10.33	10.22	10.16
pK_w	14.73	14.53	14.34	14.16	14.00	13.53	27
K_H	1.20	1.27	1.34	1.41	1.47	1.64	1.99

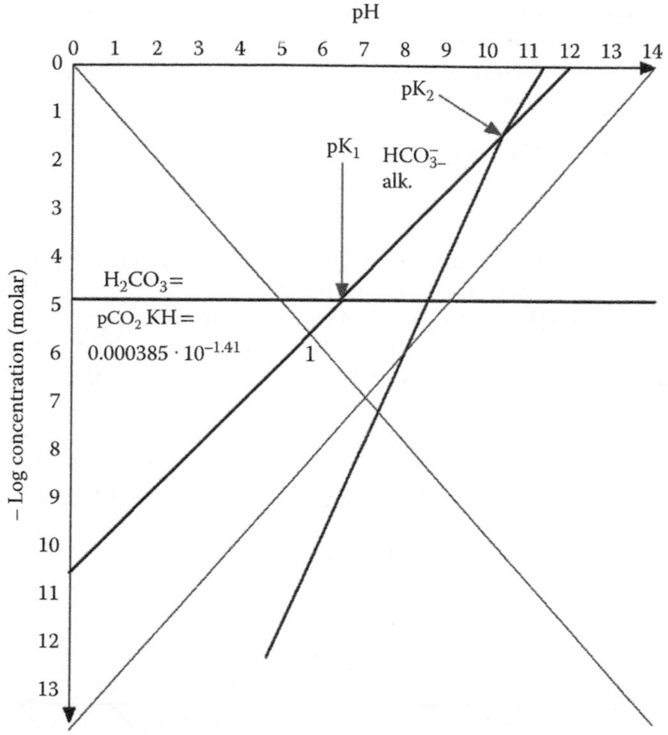

FIGURE A.5
(See color insert.) A double logarithmic diagram for an open aquatic system in equilibrium with the carbon dioxide in the atmosphere. A concentration of 385 ppm is presumed. It corresponds to the carbon dioxide concentration during the year 1999–2000.

line will correspondingly have a slope of +2 and intersect the hydrogen carbonate line at $pH = pK_2$.

The composition of an open aquatic system can easily be found by the application of the diagram Figure A.5, provided that the alkalinity is known, and other alkaline components can be omitted. If, for instance, the alkalinity is found to be 0.001 M, and it is assumed that pH is below 9, the alkalinity will, with good approximation, be equal to the hydrogen carbonate concentration. It is seen in Figure A.5 that a hydrogen carbonate concentration of 0.001 corresponds to a pH of 8.0; the assumption that pH was below 9.0 was therefore correct. The concentrations of carbon acid and carbonate can easily be read from the figure in this case.

Illustration A.1

Construct a double logarithmic diagram for carbonic acid–hydrogen carbonate–carbonate in a freshwater system at 10°C in equilibrium with the atmosphere containing 450 ppm carbon dioxide.

1. Mark on the diagram a stream with alkalinity 0.0025 eq/L and indicate the corresponding pH.
2. Mark on the diagram a stream with alkalinity 0.0005 eq/L and indicate the corresponding pH value.
3. What is the resulting pH in a lake receiving a mixture of equal volumes of the two streams?

SOLUTION

At moderate pH, the alkalinity $[HCO_3^-]$. On the diagram (see Figure A.6), is shown where $[HCO_3^-] = 0.0025$ M (point 1) corresponding to pH = 8.5. Point 2 indicates $[HCO_3^-] = 0.0005$ M corresponding to pH = 7.8.

The alkalinity is conserved when two streams are mixed. Therefore the alkalinity in the lake $\approx [HCO_3^-] = 0.0015$ M at pH = 8.3 (point 3).

In addition to the equilibrium between carbon dioxide in the atmosphere and carbon acid in water, solid carbonate may be present as suspended matter and/or in sediment and be in equilibrium with the calcium and carbonate ions in the water according the solubility product:

$$[Ca^2+][CO_3^{2-}] = K_S = 10^{-8.4} \tag{A.9}$$

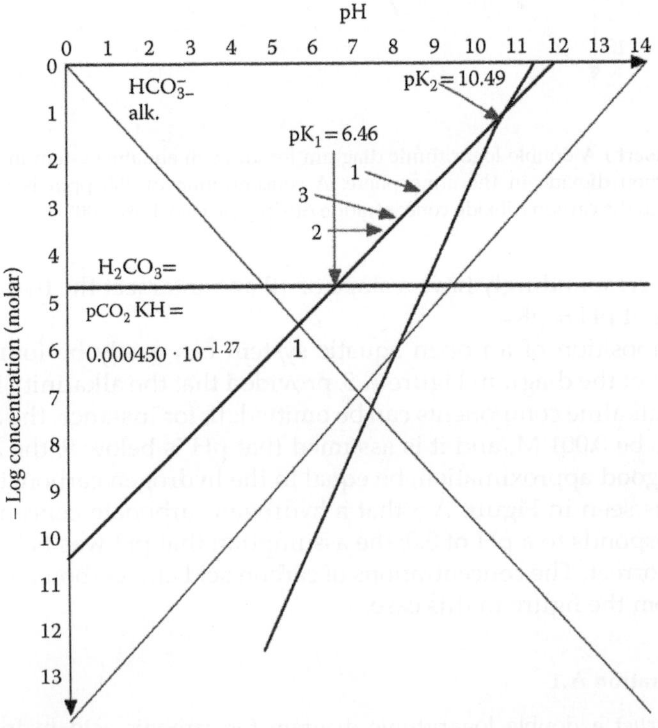

FIGURE A.6
(See color insert.) Example 12.1. See the solution.

It is possible to include also this equation in the double logarithmic diagram as shown in Figure A.7. The calcium ion line gets a slope of +2 and intersects the carbonate line at log C = −4.2. The composition under these circumstances is found by a charge balance. The sum of the negative and positive ions must balance. In most aquatic systems at pH < 9.0 the dominant cations will be the calcium ions and the dominant anions will be hydrogen carbonate ions. It implies that the following equation is valid:

$$2\left[Ca^2+\right]=\left[HCO_3^-\right] \qquad (A.10)$$

This corresponds in Figure A.7 to pH = 8.4. The carbonate ion concentration is negligible at this pH value. Notice that in this case where two equilibria are imposed on the system (equilibrium with carbon dioxide in the atmosphere and equilibrium with solid calcium carbonate), the composition of the water is given and no information of alkalinity is needed. The concentration of the carbonate ions must be included in the charge balance at higher pH, which is possible by an iteration.

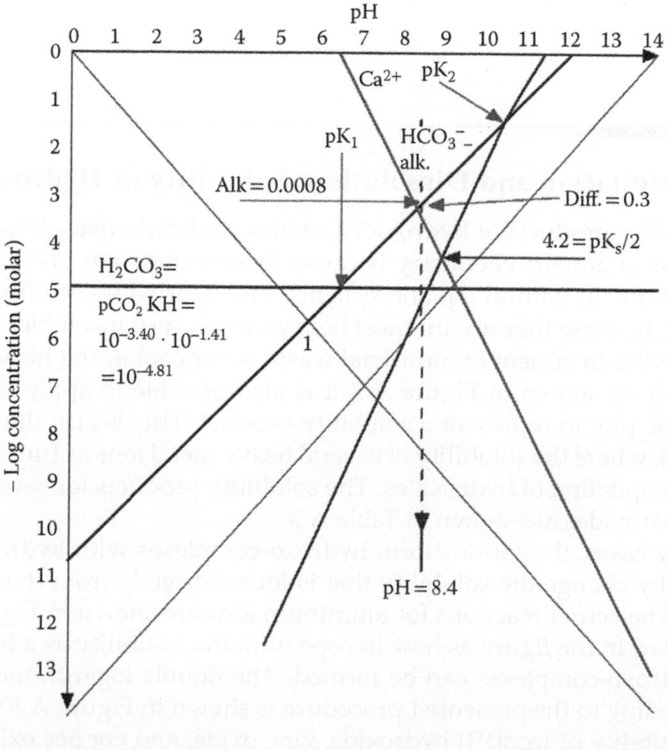

FIGURE A.7
(See color insert.) A double diagram for an aquatic system in simultaneously equilibrium with carbon dioxide in the atmosphere and solid calcium carbonate.

Illustration A.2

What is the composition when an open fresh water system is in equilibrium with solid calcium carbonate and carbon dioxide at $10^{-3.41}$ atm?

SOLUTION

The diagram Figure A.7 can be applied.
The process determining the dissolution of calcium carbonate:

$$CaCO_3(s) + CO_2 + H_2O \rightarrow Ca^{2+} + 2HCO_3^-$$

The charge balance:

$$2[Ca^{2+}] + [H^+] = [HCO_3^-] + 2[CO_3^{2-}] + [OH^-]$$

With good approximation:

$$2[Ca^{2+}] = [HCO_3^-]$$

$$pH \approx 8.4; [HCO_3^-] \approx 10^{-3} M; [H_2CO_2^*] \approx 10^{-5} M; [Ca^{2+}] \approx 5 \times 10^{-4} M; [CO_3^{2-}] \approx 10^{-5}\ M$$

A.4 Precipitation and Dissolution: Solubility of Hydroxides

The solubility products of hydroxides, oxides, and carbonates have particular interest in aquatic chemistry because these anions are present in high concentrations in natural aquatic systems. The hydroxides are furthermore of interest, because they are for most heavy metals very insoluble, which are utilized in the treatment of industrial waste water containing heavy metals. As it has been shown in Figure A.7, it is also possible to apply the double logarithmic plot to represent a solubility product. This is also illustrated in Figure A.8, where the solubility of several heavy metal ions as function of pH due to precipitation of hydroxides. The solubility products for several heavy metals hydroxides are shown in Table A.3.

In many cases, the cations form hydroxo-complexes with hydroxide ions and thereby change the solubility that is found directly from the solubility product. The actual reactions for aluminum ions are shown in Figure A.9. It is illustrated in the figure as how to cope with the solubility as a function of pH if hydroxo-complexes can be formed. The double logarithmic diagram corresponding to the presented procedure is shown in Figure A.10.

The solubility of iron(III) hydroxide, zinc oxide, and copper oxide can be found by the same method. The diagram for iron(III) hydroxide is shown in Figure A.11.

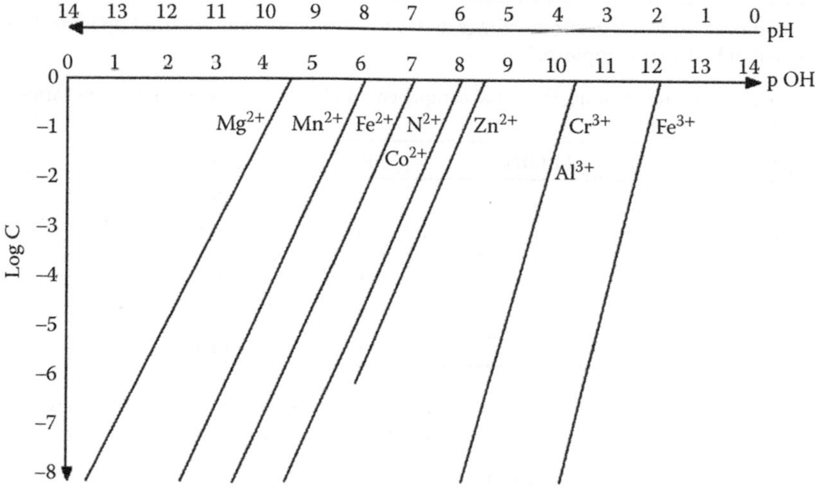

FIGURE A.8
The solubility of several heavy metal cations as function of pH.

TABLE A.3
pK_s Values at Room Temperature for Metal Hydroxides $pK_s = -\log K_s$, where $K_s = [Me^{z+}]\,[OH^-]_z$

Hydroxide	z = Charge of Metal Ions	pK_s
AgOH(1/2 Ag_2O)	1	7.7
$Cu(OH)_2$	2	20
$Zn(OH)_2$	2	17
$Ni(OH)_2$	2	15
$Co(OH)_2$	2	15
$Fe(OH)_2$	2	15
$Mn(OH)_2$	2	13
$Cd(OH)_2$	2	14
$Mg(OH)_2$	2	11
$Ca(OH)_2$	2	5.4
$Al(OH)_3$	3	32
$Cr(OH)_3$	3	32

A.5 Solubility of Carbonates in Open Systems

The solubility of calcium carbonate in open aquatic systems has already been treated in Section A.3. Figure A.12 gives for strontium carbonate,

How to make a double logarithmic diagram for the solubility as influenced by hydrolysis/formation of hydroxo-complexes?

1. Set up a diagram indicating the initial components and how they can react to form various species. For example:

Species	$Al(OH)_3$	H^+	log K*
Al^{3+}	1	3	8.5
$Al(OH)_2^+$	1	2	3.53
$Al(OH)_2^+$	1	1	−0.8
$Al(OH)_3$	1	0	−6.5
$Al(OH)_4^-$	1	−1	−14.5

2. Find the equilibrium constants (log *K_1, log *K_2, log *K_3, log *K_4) for the reactions between aluminum hydroxide and one or more hydrogen ions, here exemplified by

$$\mathbf{Al(OH)_3 + 3H^+ <-> Al^{3+} + 3H_2O.}$$

*K_1 can be found from the solubility product, $K_s = [Al^{3+}][OH^-]^3$:

$$K_s = \frac{[Al^{3+}][OH^-]^3[H^+]^3}{[H^+]^3} = K_1K_w^3$$

where K_w is the ion product of water = 10^{-14} at room temperature.

3. Plot a double logarithmic diagram log (species) versus pH using the equilibrium constants for the formation of the considered species. The solubility of Al(III) is the sum of all the soluble species, Al^{3+}, $Al(OH)^{2+}$, etc.

FIGURE A.9
It is shown how to utilize a double logarithmic diagram, when hydroxo-complexes may be formed.

iron(II) carbonate, cadmium carbonate, and zinc carbonate the same diagram as shown in Figure A.7 for calcium carbonate.

A.6 Solubility of Complexes

Complex formation is covered later in this chapter, but it is shown here how a simple complex formation under simultaneous reaction with a compound with very little solubility can enhance the solubility considerably. The calculations are made by adding two suitable reactions and determining the equilibrium constant for the total process as the product of the equilibrium constants of the two added processes. If the equilibrium constant for the total process is high, the solubility will with very good approximation be

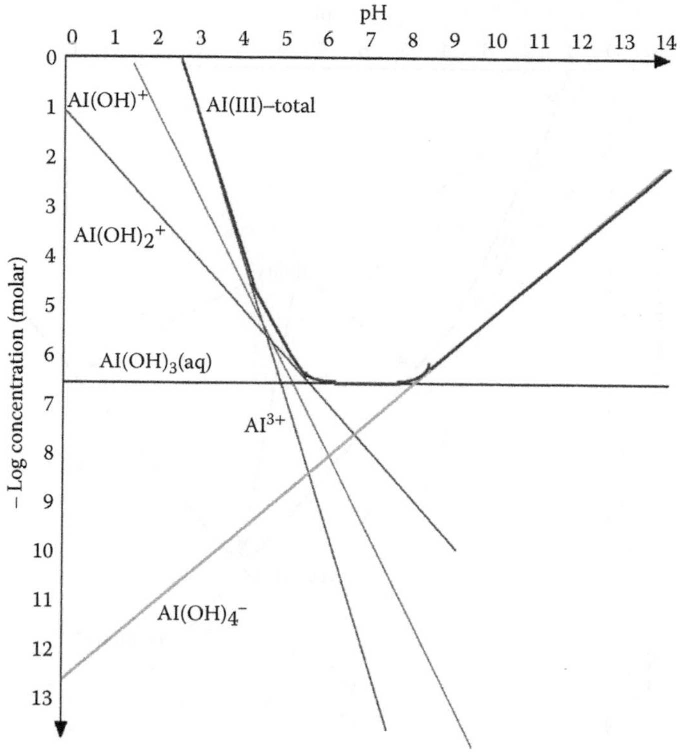

FIGURE A.10
(See color insert.) The solubility of aluminum(III) considering formation of several hydroxo-complexes as function of pH.

very pronounced. If the equilibrium constant is not very high, the solubility is made the unknown in a suitable equation.

The method is shown by use of an example; see Illustration A.3.

Illustration A.3

Find the solubility of iron(III) in a fresh water lake with and without presence of 10 μmol citrate/L. The ion strength of the lake water is O and OH = 9.0. The solubility product of iron(III) hydroxide is 10^{-24} mol²/L². The formation constant for the complex between citrate and iron(III) by pH = 9.0, where citric acid has dissociated all hydrogen ions with good approximation, is 10^{28}.

SOLUTION

From the solubility product, it is found that the solubility at pH = 9.0 (the hydroxide concentration is 10^{-5} M) is 10^{-9} M. Solubility of iron(III), when citrate is present is determined by the reaction

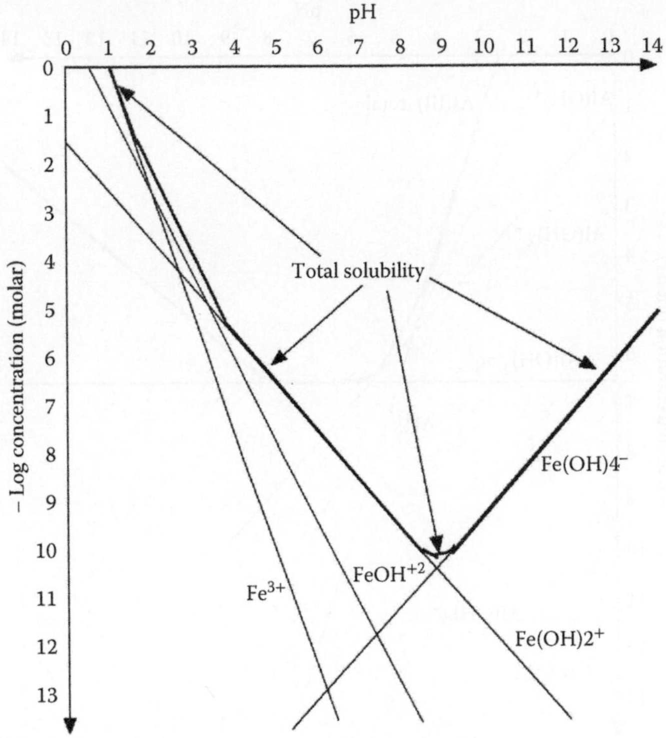

FIGURE A.11
The solubility of amorphous iron(III) hydroxide as function of pH is shown. The construction of the diagram is as explained for aluminum(III) hydroxide. The solubility of iron(III) hydroxide is determined by formation of iron(III) ions, formation of complexes with one, two, or four hydroxide ions.

$$Fe(OH)_3 + \text{citrat}^{3-} <-> Fe(III)\text{citrat} - +3OH^-$$

and the mass equation yields

$$\frac{[\text{Fe citrat} -][OH-]^3}{[\text{citrat}^{3-}]} = 10^{+4} = \frac{X * 10^{-15}}{(0.00001 - X)}$$

$$\frac{X}{(0.00001 - X)} = 10^{19}$$

X must be 10 μM/L with good approximation, which of course also corresponds to the concentrations of citrate as the equilibrium constant (10^{19}) determines how much iron(III) that will dissolve.

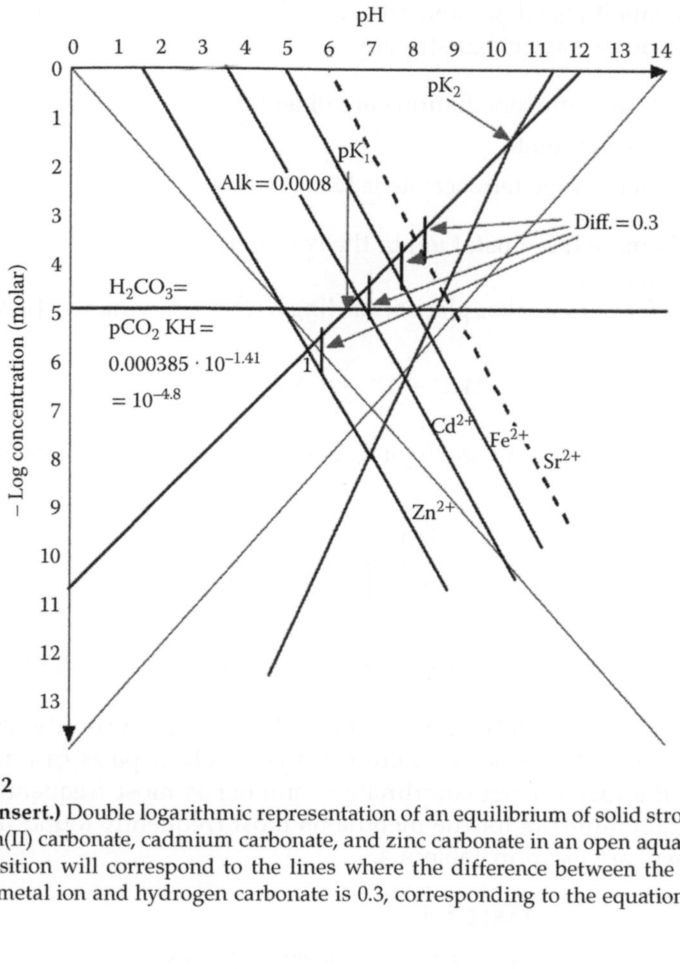

FIGURE A.12
(See color insert.) Double logarithmic representation of an equilibrium of solid strontium carbonate, iron(II) carbonate, cadmium carbonate, and zinc carbonate in an open aquatic system. The composition will correspond to the lines where the difference between the concentration of the metal ion and hydrogen carbonate is 0.3, corresponding to the equation $2[Me^{2+}] = [HCO_3^-]$.

A.7 Complex Formation

Any combination of cations with molecules or anions containing free pairs of electrons is named coordination formation. The metal ion is denoted the central atom, and the anions or the molecules are called the ligands. The atom responsible for the coordination is called the ligand atom. If a ligand contains more than one ligand atom and thereby occupies more than one coordination position in the complex it is referred to as multidentate. Ligands occupying one, two, three, four, and so on are named unidentate, bidentate, tridentate, tetradentate, and so on. Complex formation with multidentate ligands is called chelation, and the complexes are called chelates. Typical examples from aquatic chemistry are

- Oxalate and ethylenediamine are bidentate.
- Citrate is tridentate.
- Ethylenediamine tetraacetate is hexadentate.

If there is more than metal ion in the complex, we are talking about polynuclear complexes.

Complex formation takes place according to the following reaction scheme:

$$Me^{n+} + L^{m-} -> MeL^{(n-m)+} \tag{A.11}$$

where Me is a metal, and L is a ligand. The mass equation gives the following expression:

$$\frac{\left[MeL^{(n-m)+}\right]}{\left[Me^{n+}\right]\left[L^{m-}\right]} = K \tag{A.12}$$

where K is named the stability constant, complexity constant, or formation constant.

The coordination number for a metal is the number of coordination position on the metal, positions where the free electron pairs can attach the ligand to the metal. Even coordination number is most frequently found. Coordination numbers for the metal ions most frequently found in aquatic environment are listed in Table A.4.

TABLE A.4

Coordination Numbers for Metals of Interest in Aquatic Chemistry

Metal	Coordination Number
Cu^{+}	2
Ag^{+}	2
Hg_2^{2+}	2, 4
Li^{+}	4
Be^{2+}	4, 6
Al^{3+}	4, 6
Fe^{3+}	4, 6
Cu^{2+}	4, 6
Ni^{2+}	4, 6
Hg^{2+}	2, 4
Fe^{2+}	6
Mn^{2+}	4, 6

A.8 Environmental Importance of Complex Formation

Complex formation has a great influence on the effects of metal ions on aquatic organisms and ecosystems. Complex formation has the following effects:

1. The solubility can be increased:

$$\mathrm{MeY(S)+L=MeL+Y} \tag{A.13}$$

   Me represents a metal, Y a sediment, and L a ligand that can react with the metal ion.

2. The oxidation stage of the metal ion may be changed. The mass equation constants for the following two processes may differ:

$$\mathrm{Me}^{n+}+\mathrm{me}^- > \mathrm{Me}^{(n-m)+} \tag{A.14}$$

$$\mathrm{MeL}^{n+}+\ \mathrm{me}^- -> \ \mathrm{MeL}^{(n-m)+} \tag{A.15}$$

   if the stability constant (mass equation) for the two complexes in Equation A.15 are different.

3. The metal toxicity may be changed, because the complexes have another bioavailability than the metal ions.
4. The ion exchange and adsorption processes of the metal ions may be changed. The adsorption isotherms and ion exchange equilibrium constant will in far most cases be different for the complexes and the metal ions due to the bigger size of the complexes among other properties.

In the triangle diagram, Figure A.13 is shown that forms of metal ions that is usually found in fresh water and salt water. The difference is due to the high salinity (chloride concentration) and ion strength in salt water, the presence of organic ligands, and higher concentrations generally of suspended matter in fresh water.

A.9 Conditional Constant

The complex formation in aquatic systems is often very complicated because a number of side reactions (acid–base reactions, precipitations, redox reactions) are possible in addition to the main reaction, the complex formation. Reactions between metal ions and ligands are often determining the release

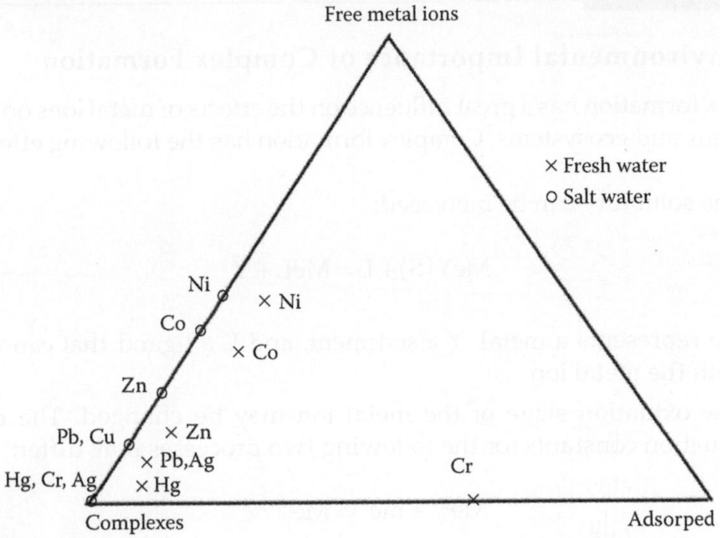

FIGURE A.13
Triangle diagram for the form of heavy metal ions in fresh water and salt water. Presence of ligands in aquatic systems will generally enhance the transfer of metal ions from sediment, soil, and suspended matter to water; but the toxicity will generally be reduced.

of metal ions from sediment, pH is often determining the strength of a complex, and the bioavailability of toxic substances is highly dependent on the form of the compound–complex bound or not, base or acid form, and so on. It is therefore of importance to be able to find the concentrations of the various forms in aquatic systems under given conditions (Alloway and Ayres, 1993).

It is possible to consider the side reactions by the application of what is called conditional constants. Let us consider the reaction:

$$Me^{n+} + L^{m-} <-> MeL^{(n-m)+} \tag{A.16}$$

It is assumed that the following side reactions are possible.

A precipitation:

$$p\,Me^{n+} + n\,Y^{p-} <-> Me_pY_n(s) \tag{A.17}$$

An acid–base reaction:

$$L^{m-} + H^+ <-> HL^{(m-1)-};\ HL^{(m-1)-} + H^+ <-> H_2L^{(m-2)-} \tag{A.18}$$

f_{Me}, defined as the ratio between the metal ion concentration and the metal concentration, including other forms determined by the side reactions, can be found from the solubility product, K_s, if $[Y^{p-}]$ e is known. If concentrations are used, we get

$$[Y^{p-}]^n[Me^{n+}]^p = K_S \tag{A.19}$$

$$\frac{[Me^{n+}]}{[Me']} = f_{Me} = \frac{K_S^{1/p}}{[Y^{p-}]^{n/p}} * \frac{1}{[Me']} \tag{A.20}$$

where the symbol Me′ represents all the metal forms except the complex, formed by what is considered the main reaction Equation A.16.

f_L, defined as the ratio between L^{m-} and the ligand concentration, including other forms determined by side reactions, can be found in a similar manner:

$$\frac{[L']}{[L^{m-}]} = \frac{1}{f_L} = 1 + \frac{[H^+]^m}{K_1 * K_2 \ldots K_m} + \frac{[H^+]^{m-1}}{K_2 * K_3 \ldots K_m} + \frac{[H^+]^{m-2}}{K_3 * K_4 \ldots K_m} + \ldots + \frac{[H^+]}{K_m} \tag{A.21}$$

where L′ symbolizes all ligand forms except the complex formed by the main reaction (Equation 9.6). K_1 to K_m are the dissociation constants for the stepwise dissociation of hydrogen ions for the acid H_mL.

A reformulation takes place:

$$\frac{[MeL^{(n-m)+}]}{[Me^{n+}][L^{m-}]} = K = \frac{MeL^{(n-m)+}]}{[Me'] * f_{Me} * [L'] * f_L}$$

$$\frac{[MeL^{(n-m)+}]}{[Me'[L']} = K * f_{Me} f_L = K_{cond} \tag{A.22}$$

K_{cond} is denoted the conditional equilibrium constant. If K_{cond} is known, it is possible to find the concentrations of free metal ions and metal complexes. By the use of the Equations A.21 and A.22 K_{cond}, is of course only valid under the conditions given by pH and $[Y^{p-}]$. If however these two concentrations are known—the conditions are known—then the usual mass constant calculations can be carried out using K_{cond} instead of K.

The most important equations for calculations of complex formation are summarized in Figure A.14. Notice that K is used for the mass equation constant where one ligand is attached to the metal ion at the time, while β_n is used for the reaction between the metal ion and n ligand.

Illustration A.4

The biological concentration factor (BCF) ($BCF = C_{Alge}/C_{vand}$) for cadmium in brown algae is 890 L/kg, while the BCFs for cadmium chloro complexes are negligible with good approximation. In the Little Belt in the mid eighties a concentration of cadmium on 32 μg/L was found in the most contaminated areas. The salinity is 1.8% and can with good approximation be considered to be entirely sodium chloride. Cadmium

Complexation constant, stability constant, formation constant.

$$K_1 = \frac{[ML]}{[M][L]}; \quad K_2 = \frac{[ML_2]}{[ML][L]}; \quad K_3 = \frac{[ML_3]}{[ML_2][L]}, \text{etc.}$$

$$\beta_1 = K_1; \quad \beta_2 = K_1K_2; \quad \beta_3 = K_1K_2\,K_3, \quad \text{etc.}$$

K* and ß* symbolize the constants where HL is the reactant (a ligand with only one H is considered in this example).

If f_L is the fraction of the L form at a given pH of the total concentration, that is, **L + HL = L′** The conditional constant

$$K_{1cond} = \frac{[ML]}{[M][L']} = f_L K_1;$$

$$K_{2cond} = \frac{[ML_2]}{[ML][L']} = f_L K_2$$

$$K_{3cond} = \frac{[ML_3]}{[ML_2][L']} = f_L K_2, \text{etc.}$$

Side reactions by M is accounted for in a similar manner.

FIGURE A.14
Important equations by equilibrium calculations of complex formations.

forms monochloro-, dichloro-, trichloro-, and tetrachloro complexes. The β-values for formation of these complexes are 25 M^{-1}, 50 M^{-2}, 32 M^{-3}, and 11 M^{-4}, respectively, by infinite dilution. The temperature is presumed 25°C and M_{NaCl} = 58.44 g/mol.

1. Find the β-values by the application of Davies equation.
2. What is the concentration of cadmium in the brown algae?

SOLUTION

$$M_w(\text{NaCl}) = 58.45 \text{ g/mol}$$

$$[Na^+] = [Cl^-] = \frac{m(\text{NaCl})}{M_w(\text{NaCl})} = \frac{18 \text{ g/L}}{58.45 \text{ g/mol}} = 0.3080 \text{ M}$$

$$I = 0.5\ (0.3080 \text{ M} * 1^2 + 0.3080 \text{ M} * 1^2) = 0.3080 \text{eq/L}$$

$$z = 1 \Rightarrow f_1 = 10^{-0.1477}$$

$$z = 2 \Rightarrow f_2 = 10^{-0.5606}$$

$$\beta_1 =$$

$$\beta_2 =$$

$$\beta_3 =$$

$$\beta_4 =$$

$$[Cd^{2+}] =$$

$$Cd^- \text{ in brown algae} = BCF\,[Cd^{2+}] = 890$$

Illustration A.5

The total concentration of mercury in a lake at pH = 6.5 is determined to be 0.05 mM. The equilibrium constants for formation of $Hg(OH)^+$ and $Hg(OH)_2^0$ by hydrolysis are 10^{-37} M, and 10^{-26} M. $M_{Hg} = 200.59$ g/mol.

1. At which concentrations can the various forms of mercury(II) be found under these circumstances?
   The LC_{50} value for Hg^{2+} ions is 1.2 mg/L for daphnia.
2. Is the LC_{50} value exceeded, when it is assumed that the hydroxo-complexes are 100 times less toxic than Hg^{2+} ions?

SOLUTION

1. The two reactions forming hydroxo-complexes:

$$Hg^{2+} + H_2O \rightarrow Hg(OH)^+ + H^+; K_1 = 10^{-3.7}\,M.\ pK_{s1} = 3.7$$

$$Hg(OH)^+ + H_2O \rightarrow Hg(OH)_2^0 + H^+; K_2 = 10^{-2.6}\,M.pK_{s2} = 2.6$$

At pH = 6.5 is $Hg(OH)_2^0$ the dominating form of mercury(II) according to the two equilibrium constants, that is, $[Hg(OH)_2^0] \approx 0.05$ mM.
The concentrations of $[Hg(OH)^+]$ and $[Hg^{2+}]$ are found from $[Hg(OH)_2^0] \approx 0.05$ mM and the equilibrium constants:

$$[Hg(OH)^+] = \frac{[Hg(OH)_2^0]\cdot 10^{-pH}}{K_2} = 10^{-8.2}\,M$$

$$[Hg^{2+}] = \frac{[Hg(OH)^+]\cdot 10^{-pH}}{K_1} = 10^{-11}\,M$$

2. The hydroxo-complexes are 100 times less toxic, that is, they have a LC_{50} value of 0.12 g/L = 0.55 mM.

The ratio between concentrations and LC_{50} values are found:

$$\frac{(10^{-8.2}\,\mathrm{M} + 0.05\,\mathrm{mM})}{0.55\,\mathrm{mM}} + \frac{10^{-11}\,\mathrm{M}}{0.0055\,\mathrm{mM}} = 0.09 < 1$$

The LC_{50} for daphnia is not exceeded.

A.10 Application of Double Logarithmic Diagrams to Determine the Conditional Constants for Complex Formation

Let us consider a complex formation according to the following reaction scheme:

$$Me^{n+} + L^{m-} = MeL^{(n-m)+} \qquad \log K = F$$

exemplified by (Glycinate = Gly):

$$Fe^{3+} + Gly^{-} = FeGly^{2+} \qquad \log K = 10.8$$

The conditional constant, $K_{cond} = [FeGly^{2+}]/[Fe'][Gly']$.

The following equations can be applied to find the conditional constant as function of pH:

$$[Fe'] = \left[Fe(III)_{total}\right] - \left[FeGly^{2+}\right] \tag{A.23}$$

$$[Fe'] = \left[Fe^{3+}\right] + \left[FeOH^{2+}\right] + \left[Fe(OH)_2^{+}\right] + \left[Fe(OH)_3\right] + \left[Fe(OH)_4^{+}\right] \tag{A.24}$$

$$[Fe'] = \left[Fe^{3+}\right]\left(1 + \frac{{}^*K_1}{[H^+]} + \frac{{}^*\beta_2}{[H^+]^2} + \frac{{}^*\beta_3}{[H^+]^3} + \frac{{}^*\beta_4}{[H^+]^4}\right) \tag{A.25}$$

$$[Fe'] = \frac{[Fe^{3+}]}{f_{Fe}} \tag{A.26}$$

$$\left[Gly'\right] = \left[Gly_{total}\right] - \left[FeGly^{2+}\right] \tag{A.27}$$

$$\left[Gly'\right] = \left[H_2Gly^{+}\right] + \left[Hgly\right] + \left[Gly^{-}\right] \tag{A.28}$$

$$[Gly'] = [Gly^-]\left(1 + \frac{[H^+]}{K_2} + \frac{[H^+]^2}{K_1K_2}\right) \quad (A.29)$$

$$[Gly'] = \frac{[Gly^-]}{f_{gly}} \quad (A.30)$$

Based on these equations, it is possible to calculate f_{Fe} and f_{Gly} as function of pH, provided of course that the various applied mass equations constants are known: log $^*K_1 = -3.05$, log $^*\beta_2 = -6.31$, log $^*\beta_3 = -13.8$ and log $^*\beta_4 = -22.7$. The pK values for H_2Gly^+ are 3.1 and 9.9.

Figure A.15 shows calculate f_{Fe}, f_{Gly} and the product of the two, $f_{Fe}\ f_{Gly}$, calculated as function of pH. The Equations A.25 through A.26 and A.29 through A.30 are applied. f_{Fe}, f_{Gly}, and the product $f_{Fe}\ f_{Gly}$ are found at a given pH and the conditional constant is calculated as $K_{cond} = K\ f_{Fe}\ f_{Gly}$, K_{cond} has a maximum around pH = 3.2. At this pH, log K_{cond} = log K + log $f_{Fe}\ f_{Gly}$ = 10.8 – 7.0 = 3.8.

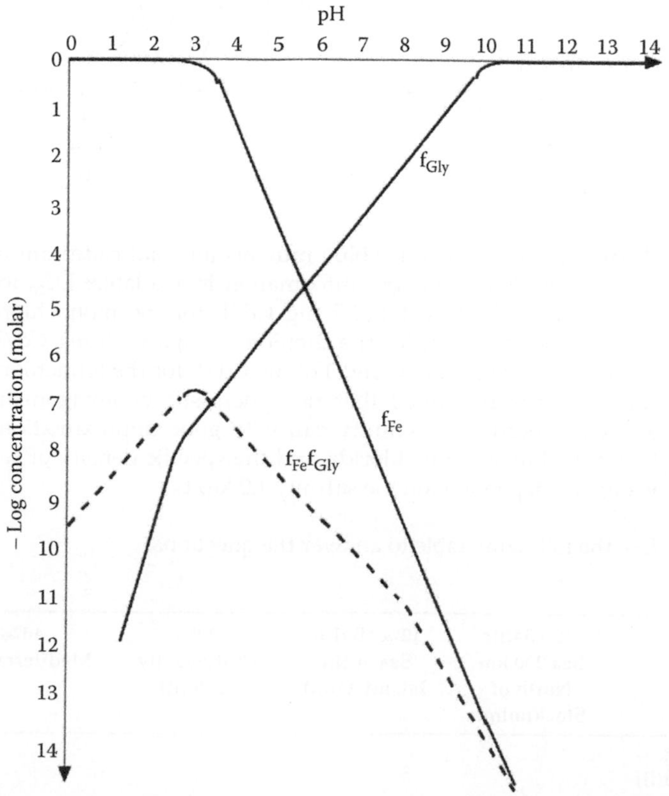

FIGURE A.15
f_{Fe} and f_{Gly} and the product $f_{fe}f_{Gly}$ are found as function of pH by the Equations 12.23 through 12.30.

Illustration A.6

1. By the application of the diagrams shown below, answer the following questions:
   What is the dominant form of cadmium(II) in water with different salinities? $M_{NaCl} = 58.44$ g/mol.

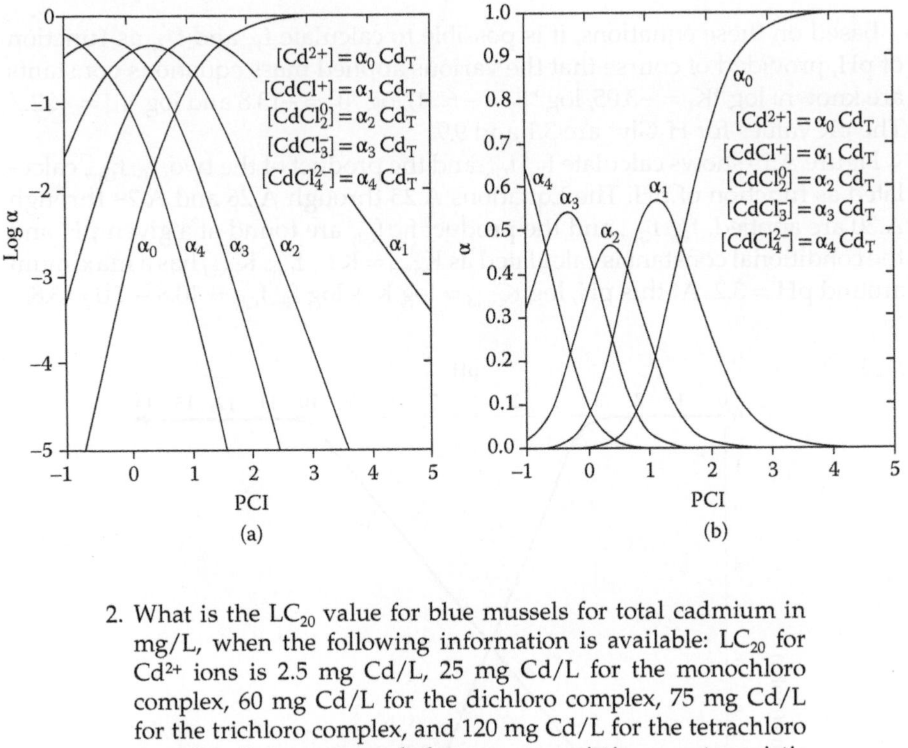

(a) (b)

2. What is the LC_{20} value for blue mussels for total cadmium in mg/L, when the following information is available: LC_{20} for Cd^{2+} ions is 2.5 mg Cd/L, 25 mg Cd/L for the monochloro complex, 60 mg Cd/L for the dichloro complex, 75 mg Cd/L for the trichloro complex, and 120 mg Cd/L for the tetrachloro complex. It is presumed that no synergistic or antagonistic effect will occur. The salinity can with good approximations be referred to sodium chloride and the specific density of sea water is independent on the salinity 1.0 kg/L.

   Use the following table to answer the questions:

Salinity	**5‰ (Baltic Sea 200 km North of Stockholm)**	**12‰ (Baltic Sea at the Island Møn)**	**20‰ (Kattegat by Anholt)**	**40‰ (The Mediterranean Sea)**
Dominating form of Cd(II)				
Approximate LC_{20} value in mg/L				

SOLUTION

1. Use the diagrams to answer the questions. The answer can be found in the table.
2. For the calculations of $LC_{20,Cd(II),pCl}$ for the four pCl values read on the diagram the α-values. Calculate $LC_{20,Cd(II),pCl}$ as:

$$\frac{1}{LC_{20,Cd(II),pCl}} = \sum_{n=0}^{n=4} \frac{\alpha_n}{LC_{20,Cd[Cl]_n^{2-n}}}$$

Salinity	5‰ (Baltic Sea 200 km North of Stockholm)	12‰ (Baltic Sea at the Island Møn)	20‰ (Kattegat by Anholt)	40‰(The Mediterranean Sea)
pCl	1.07	0.69	0.47	0.16
Dominating form of Cd(II)	$CdCl^+$	$CdCl^+$ and $CdCl_2^0$	$CdCl_2^0$	$CdCl_2^0$ and $CdCl_3^-$
Approximate LC_{20} value in mg/L	9.4 mg/L	17.6 mg/L	30.4 mg/L	45.9 mg/L

A.11 Redox Equilibria: Electron Activity and Nernst's Law

There is an analogy between acid–base and reduction–oxidation reactions. Acid–base reactions exchange protons. Acids are proton donors and bases proton acceptors. Redox processes reactions exchange electrons. Reductants are electron donors and oxidants are electron acceptors. As there are no free hydrogen ions, protons, there are not free electrons. It implies that every oxidation is accompanied by a reduction and vice versa. Like pH has been introduced as the proton activity, we may introduce an electron activity defined as:

$$pe = -\log\{e^-\} \tag{A.31}$$

where e^- is the electron and p as usually is an abbreviation of −log. As free electrons do not exist, pe should be considered a concept that is introduced to set up a parallel to pH and facilitate the calculations of redox equilibria. The relationship between pe and the redox potential, E, is

$$pe = \frac{F * E}{2.3RT} \tag{A.32}$$

Redox Equations:

$$E = E_o + \frac{2.3RT}{nF(\log\{ox\}/\{red\})}$$

or

$$pe = pe_o + \frac{1}{n(\log\{ox\}/\{red\})}$$

$$\Delta G_o = -E_o\, nF = -RT\ln K = 2.3RT\log K = 2.3RT\, n\, pe_o$$

Notice that when the composition, that is, {ox}/{red} is given, the redox potential E or pe can be found.

Freshwater in equilibrium with the atmosphere (oxygen partial pressure 0.21 atm) will have the following pe:

$$pe = 20.78 + 0.5\log(0.21)^{0.5}\{H^+\}^2$$

From the pe value the ratio between any redox pair in freshwater is determined.

FIGURE A.16
Summary of the most important equations to calculate redox potential, pe, and the equilibrium constant.

where F is Faraday's number = 96485 C/mol = the charge of 1 mol of electrons, R is the gas constant = 8.314 J/mol K = 0.082057 L atm/mol K. Nernst law:

$$E = E^\circ + \frac{-RT}{nF} - \log\frac{\{ox\}}{\{red\}} \tag{A.33}$$

may be rewritten by the use of pe°:

$$pe^\circ = \frac{F * E^\circ}{2.3\ RT} \tag{A.34}$$

to

$$pe = pe^\circ + \left(\frac{1}{n}\right)\log\frac{\{ox\}}{\{red\}} \tag{A.35}$$

This yields furthermore the following relationship between the free energy and pe, as $\Delta G = -E\ n$:

$$pe = \frac{-\Delta G}{n * 2.3RT} \qquad pe^\circ = \frac{\Delta G^\circ}{n * 2.3RT} \tag{A.36}$$

Notice that these equations are applied on the half reaction and that an oxidant takes up an electron, which is transferred to a reductant. A redox process, however, requires as pointed out above that another reductant is available to deliver the electron and thereby this reductant is changed to the corresponding oxidant.

Equation A.36 implies that the following relationship between pe and the equilibrium constant, K, for the half reaction, in which the electron takes part, is valid:

$$pe^{\circ} = \frac{1}{n}\log K \qquad (A.37)$$

As illustration of the application of these equations, the following reaction is considered:

$$Fe^{3+} + e^{-} = Fe^{2+} \qquad (A.38)$$

The standard redox potential corresponding to the {ox} = {red} = 1 for this process can be found in any table of standard redox potentials to be 0.77 V by room temperature. It means

$$E^{\circ} = 0.77;\ pe^{\circ} = \frac{F * 0.77}{2.3 * R * 298}(t = 25°C) \qquad (A.39)$$

$$\log K = n * pe^{\circ} = 1 * \frac{0.77}{0.059} = 13.0 \qquad (A.40)$$

where $\log K = \log(\{Fe^{2+}\}/\{Fe^{3+}\}\{e^{-}\})$

If for instance in an acidic solution $[Fe^{3+}] = 10^{-3}$ and $[Fe^{2+}] = 10^{-2}$, we get, applying concentrations instead of activities as a reasonable good approximation:

$$pe = pe^{\circ} + \left(\frac{1}{n}\right)\log\frac{\{ox\}}{\{red\}} = 13 + 1 * \log 0.1 = 12.0 \qquad (A.41)$$

Illustration A.7

Find the pe value for a natural aquatic system at pH = 7.0 and in equilibrium with the atmosphere (partial pressure of oxygen = 0.21 atm.). What is the ratio $\{Fe^{2+}\}$ to $\{Fe^{3+}\}$ in the water?

SOLUTION

The following process determines the redox potential:

$$0.5O_2 + 2H^{+} + 2e^{-} = H_2O \qquad (A.42)$$

In tables, it is possible to find that log K for this process is = 41.56, or the standard redox potential can be found and log K calculated based on E^{o}.

It means that $pe^o = 41.55/2 = 20.78$.

$$pe = pe^o + 0.5\log\left(\sqrt{pO_2}\,\{H^+\}^2\right) = 20.78 + 0.5\log\left(\sqrt{0.21}\;10^{-14.0}\right) = 20.78 + 0.5\,(-14.34) = 13.61$$

This pe value is determining all ratios between oxidants and reductants, because the oxygen concentration in water in equilibrium with the atmosphere will inevitably have a constant concentration determined by the partial pressure of 0.21 atm. Therefore,

$$13.61 = pe = pe^\circ + \frac{1}{n}\log\frac{[Fe^{3+}]}{[Fe^{2+}]} = 13.0 + \log 10^{0.61} \tag{A.43}$$

The ratio $\{Fe^{2+}\}$ to $\{Fe^{3+}\} \approx 4.0$.

A.12 pe as Master Variable

Equation A.35 can be used to set up a graphical presentation in a double logarithmic diagram of a redox process. How is, for instance, the ratio $\{Fe^{2+}\}$ to $\{Fe^{3+}\}$ changed with changing redox potential or pe? Obviously, at low pe the reductant will be dominating and at high pe the oxidant will be dominating. When $pe = pe^o$ the two forms will be equal. The double logarithmic diagram is therefore very similar to the double logarithmic diagram applied for acid–base reactions. A double logarithmic diagram for $\{Fe^{2+}\}$ and $\{Fe^{3+}\}$ is shown in Figure A.17.

A.13 Examples of Relevant Processes in the Aquatic Environment

An aqueous solution at a given pH and a given pe determines the partial pressure of hydrogen and oxygen according to the following redox processes:

$$2H^+ + 2e^- = H_2(g) \qquad \log K = 0 \tag{A.44}$$

(The process is reference to the redox potential scale)

$$2H_2O + 2e^- = H_2(g) + OH^- \qquad \log K = -28 \tag{A.45}$$

$$O_2(g) + 4H^+ + 4e^- = 2H_2O \qquad \log K = 83.1 = 4 \times 20.78 \tag{A.46}$$

$$O_2(g) + 4e^- + 2H_2O = 4OH^-$$

It implies that we obtain the following relationships between the partial pressure and pH and pe:

$$\log pH_2 = 0 - 2\,pH - 2\,pe \tag{A.47}$$

$$\log pO_2 = -83.1 + 4\,pH + 4\,pe \tag{A.48}$$

If on the other side the partial pressure of oxygen can be considered constant and in equilibrium with aqueous solution and pH is given, the pe value can be found from Equation A.48.

Manganese is present in aquatic systems as manganese dioxide (s) and manganese ions according to the following redox potential:

$$MnO_2(s) + 4H^+ + 2e^- \rightarrow Mn^{2+} + 2H_2O \tag{A.49}$$

The concentration of manganese ions is determined by the redox potential for the aqueous solution. If the redox potential is determined by the partial

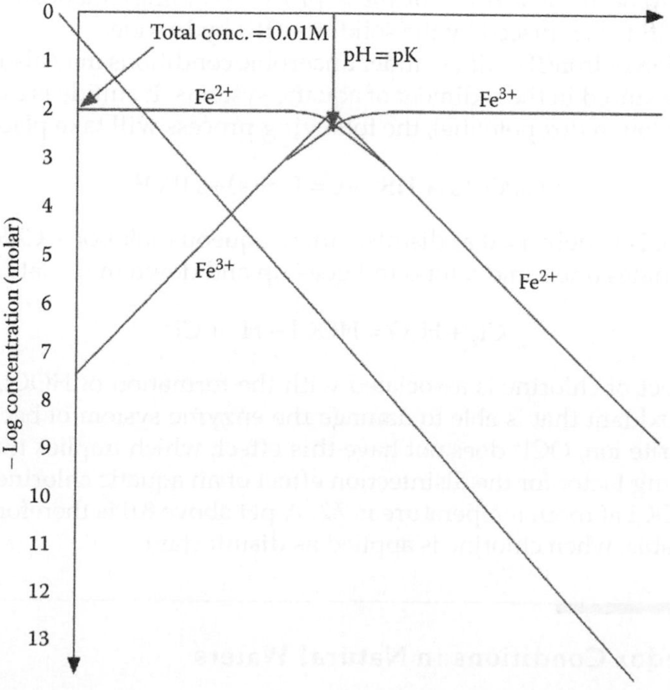

FIGURE A.17
(See color insert.) Double logarithmic diagram for the redox process: $Fe^{3+} + e^- = Fe^{2+}$.

pressure of oxygen in equilibrium with the aqueous solution, we obtain the following equation to determine the manganese ion concentration:

$$pe = 13.6 = 20.42 + 0.5 \cdot \log\left(\frac{[H^+]^4}{[Mn^{2+}]}\right) \quad (A.50)$$

$$\log[Mn^{2+}] = 2 \times (20.42 - 13.6) - 7 \times 4 = -14.4 \quad (A.51)$$

Iron(II) is often under anaerobic conditions present in sediment and soil as iron(II) sulfide. If the sediment or soil is exposed to air, the following oxidation process takes place:

$$2FeS_2(s) + 2H_2O + 7O_2 = 2FeSO_4 + 2H_2SO_4 \quad (A.52)$$

$$4FeSO_4 + O_2 + 2H_2SO_4 = 2Fe_2(SO_4)_3 + 2H_2O \quad (A.53)$$

$$Fe_2(SO_4)_3 + 6H_2O = 2Fe(OH)_3(s) + 3H_2SO_4 \quad (A.54)$$

These processes involve formation of sulfuric acid, and extreme low pH values may occur as a result of these processes. Simultaneously, the solid iron(II) sulfide is replaced by the solid iron(III) hydroxide.

Formation of iron(II) sulfide under anaerobic conditions may also mobilize phosphate stored in the sediment of aquatic systems. If sulfide is formed as a result of a low redox potential, the following process will take place:

$$FePO_4(s) + HS^- + e = FeS(s) + HPO_4^{2-} \quad (A.55)$$

Chlorine is widely used as disinfectant in aqueous solutions. Chlorine disproportionates (the same compound goes up and down in oxidation state):

$$Cl_2 + H_2O = HOCl + H^+ + Cl^- \quad (A.56)$$

The effect of chlorine is associated with the formation of HOCl, which is a strong oxidant that is able to damage the enzyme system of bacteria. The hypochlorite ion, OCl^- does not have this effect, which implies that pH is a determining factor for the disinfection effect of an aquatic chlorine solution. pK for HOCl at room temperature is 7.2. A pH above 8.0 is therefore not recommendable, when chlorine is applied as disinfectant.

A.14 Redox Conditions in Natural Waters

It is often convenient to consider the redox potential in natural aquatic systems, where it is presumed that pH = 7.0. A symbol pe^0 (w) is applied for these calculations. $pe^0(w)$ is analogous to pe^0, except that the activities of protons and hydroxide ions correspond to natural water at pH = 7.0. This implies that the following expression between pe^0 (w) and pe^0 is valid:

$$pe^0(w) = pe^0 + 0.5n \log K_w \tag{A.57}$$

where n is the number of moles of protons exchanged per mole of electrons.

$0.25\ O_2(g) + H^+ + e^- = 0.5\ H_2O$ has a pe^0 value of 13.75. The corresponding $pe^0(w)$ is therefore 20.75.

A table of $pe^0(w)$ values may be used to determine whether a system will tend to oxidize equimolar concentrations of any other system, which would require that it would have higher $pe^0(w)$. Sulfate/sulfide has for instance a $pe^0(w) = -3.5$, while CO_2/CH_2O has $pe^0(w) = -8.20$. Sulfate is accordingly able to oxidize CH_2O in natural water.

Illustration A.8

The air in immediate contact with a wetland has a partial pressure of carbon dioxide and methane on 100 and 250 Pa, respectively. There is equilibrium between the water in the wetland and the air above the wetland. pH in the wetland is 7.0. The temperature is 25°C.

1. Find pe and the redox potential for the water in the wetland, when $pe^0(W)$ for the carbon dioxide/methane redox equilibrium is −4.13.
2. What is the ratio iron(III) to iron(II) in the water? pe^0 for Fe^{3+}/Fe^{2+} is 13.0.
3. How much will it change the redox potential (pe) if pH is changed from 7.0 to 7.6?
4. At a later stage, it is found that pH is 7.2, and the redox potential is −0.23 V. Is methane produced under these conditions in detectable amounts?

SOLUTION

1. $$pe = pe^0(W) + \frac{1}{8}\log\left(\frac{pCO_2}{pCH_4}\right) = -4.13 + 0.125\log\left(\frac{100}{250}\right) = -4.18$$

   $$EH = 0.05895 * (-4.18) = -0.246$$

2. $$-4.18 = 13.0 + \log\frac{Fe^{3+}}{Fe^{2+}}; \frac{[Fe^{3+}]}{[Fe^{2+}]} = 10^{-17.2}$$

3. $$0.125\log\left(\frac{10^{-7.6}}{10^{-7.0}}\right)^8 = -0.6$$

4. The redox potential implies that 1/8 ($\log pCO_2/pCH_4$) >> 0.

Illustration A.9

In the lagoon of Venice, sulfate may be transformed to free sulfur in hot weather with no or little wind. pe for this redox reaction is 6.03.

1. What is pe under these conditions? pH is measured to be 7.5. A 10 mM concentration of sulfate is assumed.
2. What is the ratio $[Fe^{3+}]/[Fe^{2+}]$ under these conditions? pe^0 for Fe^{3+}/Fe^{2+} is 13.0.

SOLUTION

1. $pe = 6.03 + \log[SO_4^{2-}]^{1/6}[H^+]^{4/3}$

$= -4.3$

$-4.3 = 13.0 + \log\frac{[Fe(III)]}{[Fe(II)]}$

2. $\frac{[Fe(III)]}{[Fe(II)]} = 10^{-17.3}$

Figure A.18, named a redox stair case, can be used in the same manner to decide on which oxidants are able to oxidize which reductants. pe^0 values for various redox pair at four different pH are shown. The

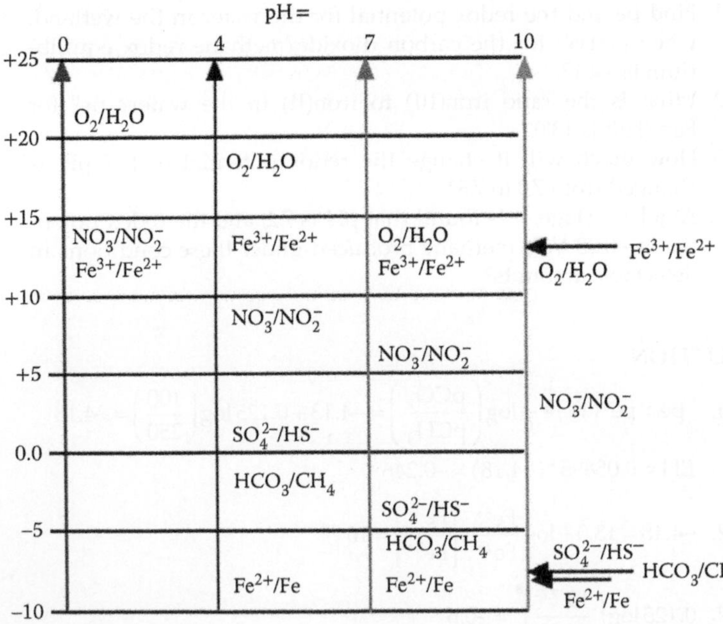

FIGURE A.18
pe^0 values for the most common redox pair in aquatic systems are shown for four different pH values.

TABLE A.5

Yields of kJ and ATPs per Mole of Electrons, Corresponding to 0.25 Mol of CH_2O Oxidized

Reaction	kJ/mol/e⁻	ATPs/mol/e⁻
$CH_2O + O_2 \rightarrow CO_2 + H_2O$	125	2.98
$CH_2O + 0.8\,NO_3^- + 0.8\,H^+ \rightarrow CO_2 + 0.4\,N_2 + 1.4\,H_2O$	119	2.83
$CH_2O + 2\,MnO_2 + H^+ \rightarrow CO_2 + 2\,Mn^{2+} + 3\,H_2O$	85	2.02
$CH_2O + 4\,FeOOH + 8\,H^+ \rightarrow CO_2 + 7\,H_2O + Fe^{2+}$	27	0.64
$CH_2O + 0.5\,SO_4^{2-} + 0.5\,H^+ \rightarrow CO_2 + 0.5\,HS^- + H_2O$	26	0.62
$CH_2O + 0.5\,CO_2 \rightarrow CO_2 + 0.5\,CH_4$	23	0.55

Note: The released energy is available to build ATP for various oxidation processes of organic matter at pH = 7.0 and 25°C.

oxidants placed higher in the figure can oxidize the reductants placed lower in the figure. At the same time, the figure indicates the pe value corresponding to the present at an equilibrium of a given redox pair. If we know for instance that oxygen is present at a normal concentration around 8–10 mg/L in an aquatic ecosystem, it will inevitably imply that pe must be around 13.75 and that the other redox pair must adjust their ratio according to this pe value. Notice that pe^0 for Fe^{3+}/Fe^{2+} is independent of pH because no protons participate in the redox reaction.

Finally, the redox stair case indicates which oxidant will be used for an oxidation under given circumstances. If pe is around 13–15 and pH = 7.0, oxygen will be applied as oxidant before iron(III), nitrate, and so on. When oxygen has been used Fe^{3+} will be used before nitrate and sulfate, and so on. This is a parallel to the titration of a mixture of acids with a base. The strongest acid will be neutralized first, then the second strongest acid, and so on.

It is interesting that the sequence of oxidants (oxygen is used as oxidant before iron(III), which is used before nitrate, sulphate, carbon dioxide) also is the sequence that gives the highest energy efficiency (most kJ for formation of ATP and thereby most exergy) (see also Jørgensen, 2012). This is shown in Table A.5.

A.15 Construction of pe–pH Diagrams

The equations for pe obtained from Equations A.47 and A.48

$$pe = -pH - 0.5 \log pH_2 \tag{A.58}$$

$$pe = 20.78 - pH + 0.25 \log pO_2 \tag{A.59}$$

can be plotted in a pH–pe diagram; see Figure A.19. The diagram gives the following important information: above the upper line, water is an effective

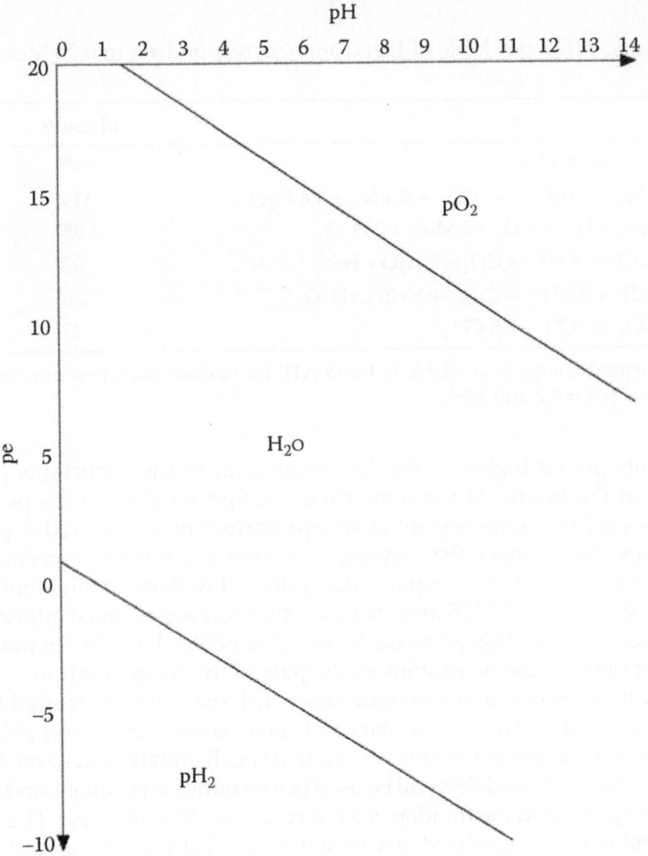

FIGURE A.19
The diagram represents the equilibria between water and oxygen (upper line) and between water and hydrogen (lower line).

reductant (forming oxygen) and below the lower line, water is an effective oxidant, producing hydrogen. Aquatic systems in equilibrium with the oxygen in the atmosphere will have a pe = f(pH) as the upper line. Under this normal aerobic condition, oxygen is an oxidant and hydrogen is a reductant.

pe–pH diagrams illustrate which components will prevail under given conditions, that is, at given pe and pH. It is, therefore, advantageous to construct pe–pH diagrams for the elements most commonly found in aquatic systems to be able to quickly determine which form of the considered element is stable under the prevailing conditions. At the same time, the pe–pH diagram gives a good overview of the possible processes where both redox processes and acid–base reactions can take place. Figure A.19 gives the information for the two elements oxygen and hydrogen. Figure A.20 gives the pe–pH diagram for the sulfur system.

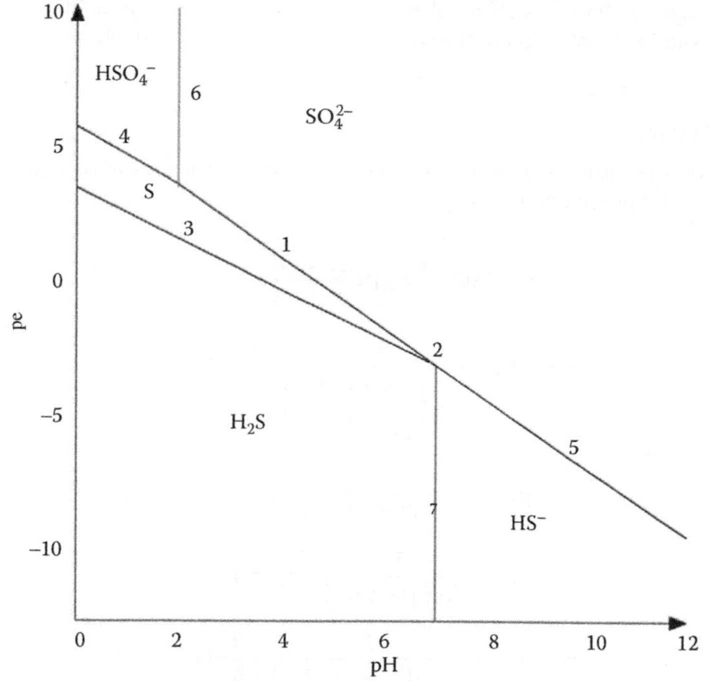

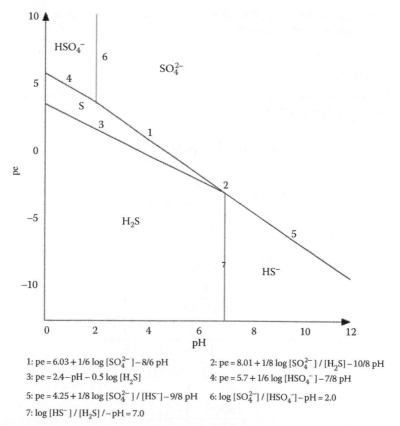

1: $pe = 6.03 + 1/6 \log [SO_4^{2-}] - 8/6\, pH$

2: $pe = 8.01 + 1/8 \log [SO_4^{2-}] / [H_2S] - 10/8\, pH$

3: $pe = 2.4 - pH - 0.5 \log [H_2S]$

4: $pe = 5.7 + 1/6 \log [HSO_4^-] - 7/8\, pH$

5: $pe = 4.25 + 1/8 \log [SO_4^{2-}] / [HS^-] - 9/8\, pH$

6: $\log [SO_4^{2-}] / [HSO_4^-] - pH = 2.0$

7: $\log [HS^-] / [H_2S] / - pH = 7.0$

FIGURE A.20
pe–pH diagram for the sulfate–sulfur–hydrogen sulfide system. If only one soluble species is included in the equation, the concentration 0.01 M is applied. If two soluble species are included, for instance equation number 2, they are both assumed to be 0.00 M.

Illustration A.10

Construct a diagram considering that sulfur can be found in natural aquatic ecosystem as sulfate, as sulfur (s), and as hydrogen sulfide (g) in equilibrium with aqueous solution of hydrogen sulfide. The following redox reactions can be found in the literature:

$$SO_4^{2-} + 8H^+ + 6e^- = S(s) + 4H_2O \qquad pe^0 = 6.03$$

$$SO_4^{2-} + 10H^+ + 8e^- = H_2S(aq) + 4H_2O \qquad pe^0 = 5.12$$

$$S(s) + 2H^+ + 2e^- = H_2S(aq) \qquad pe^0 = 2.40$$

$$HSO_4^- + 7H^+ + 6e^- = S(s) + 4\,H_2O \qquad pe^0 = 5.70$$

$$SO_4^{2-} + 9H^+ + 8e^- = HS^- + 4\,H_2O \qquad pe^0 = 4.25$$

Hydrogen sulfate has pK = 2.0 and hydrogen sulfide has pK = 7.0. The total concentration of the soluble sulfur species is 0.01 M.

SOLUTION

Based on the shown reactions, it is possible to obtain the following equations for the pe–pH diagram:

$$pe = 6.03 + \frac{1}{6}\log[SO_4^{2-}] - \frac{8}{6}pH$$

$$pe = 5.12 + \frac{1}{8\log\left(\frac{[SO_4^{2-}]}{[H_2S(aq)]}\right)} - \frac{10}{6}pH$$

$$pe = 2.40 - \frac{1}{2\log\,[H_2S]} - pH$$

$$pe = 5.70 + \frac{1}{6\log\,[HSO_4^{2-}]} - 876pH$$

$$pe = 4.25 + \frac{1}{8\log\left(\frac{[SO_4^{2-}]}{[H_2S(aq)]}\right)} - \frac{9}{8}pH$$

$$\log\left(\frac{\frac{[SO_4^{2-}]}{[HSO_4^{2-}]}}{[H_2S(aq)]}\right) - pH = -2.0$$

$$\log\left(\frac{[HS^-]}{[H_2S(aq)]}\right) - pH = -7.0$$

These equations can easily be plotted in a pe–pH diagram; see Figure A.20.

Figure A.21 shows another example, namely the pe–pH diagram for iron. It is constructed by the same method as Figures A.19 and A.20.

A.16 Redox Potential and Complex Formation

Figure A.22 shows how it is possible to calculate the pe° value for two different oxidation states of metal complexes. It is possible to go from Me^{2+} to MeL^{2+} by the processes (1) + (3) or by the processes (2) + (4) and these two

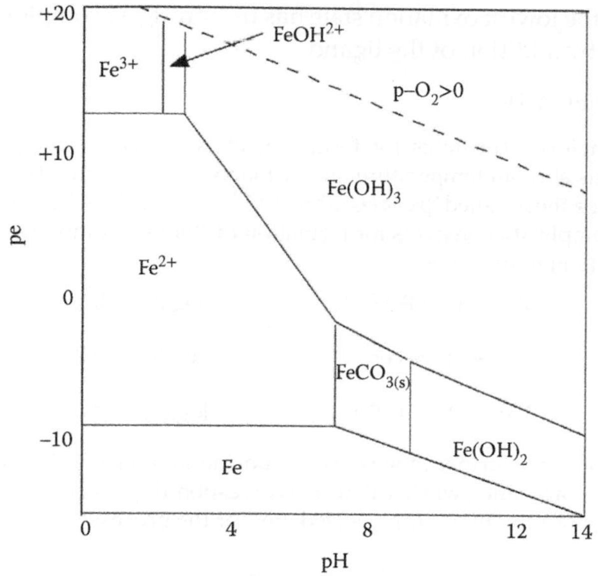

FIGURE A.21
pe–pH diagram for iron.

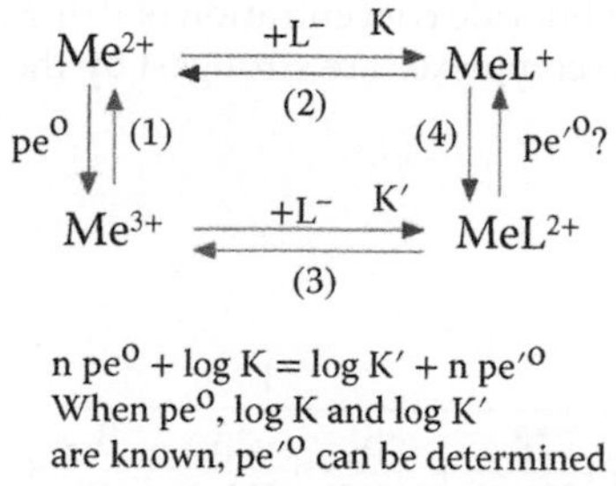

FIGURE A.22
It is possible to go from Me^{3+} to MeL^{2+} by the processes (1) + (3) or by the processes (2) + (4), and the two pathways must necessarily give the same result with respect to the equilibrium between Me^{3+} and MeL^{2+}. It implies that n pe^0 + log K = log K' + n pe'^0.

pathways must necessarily give the same result with respect to the overall equilibrium between Me^{2+} and MeL^{2+}. The equilibrium constants for the two pathways are therefore equal. The equilibrium constant for two successive processes are the product of the equilibrium constants for the two processes:

$$n\,pe^0 + \log K = \log K' + n\,pe'^0 \tag{A.60}$$

As seen from this equation, if the highest oxidation state has the strongest complex, pe^o will be reduced by addition of the corresponding ligand, and vice versa; if the lowest oxidation state has the strongest complex, the pe° will be increased by addition of the ligand.

Illustration A.11

The complexity constants for formation of Fe^{3+} hydroxo-complexes by hydrolysis at room temperature are for mono-, di-, tri-, and tetrahydroxo-complexes, the so called $^*\beta$ values, $10^{-2.2}$, $10^{-5.7}$, $10^{-15.6}$, and $10^{-21.6}$, respectively.

The complexity constants for formation of fluoride complexes for Fe^{3+} at room temperature are

$$Fe^{3+} + F^- = FeF^{2+} \qquad \log \beta_{1F} = 5.2$$

$$Fe^{3+} + 2F^- = FeF_2^+ \qquad \log \beta_{2F} = 9.2$$

$$Fe^{3+} + 3F^- = FeF_3^0 \qquad \log \beta_{3F} = 11.9$$

Iron(II) does not form complexes with fluoride, and the possible hydroxo-complexes are much weaker than the corresponding complexes for Fe^{3+}. And they can therefore be neglected. pe° for the process

$$Fe^{3+} + e^- = Fe^{2+}$$

is 13.00 at room temperature.

1. Find the pe value at equal concentrations of iron(II) and iron(III), pH = 5.0 and a fluoride concentration of 0.01 M.
2. Which iron(III) complexes are strongest by these conditions?

SOLUTION

1. $$\alpha_0 = \frac{[Fe^{3+}]}{Fe(III)}$$

$$= \frac{1}{\left(1 + \frac{^*\beta_1}{[H^+]} + \frac{^*\beta_2}{[H^+]^2} + \frac{^*\beta_3}{[H^+]^3} + \frac{^*\beta_4}{[H^+]^4} + \frac{\beta_{1F}}{[F^-]} + \frac{\beta_{2F}}{[F^-]^2} + \frac{\beta_{3F}}{[F^-]^3}\right)}$$

$$= \frac{1}{(1 + 10^{2.8} + 10^{4.3} + 10^{-0.6} + 10^{-1.6} + 10^{3.2} + 10^{5.2} + 10^{5.9})}$$

$$= \frac{1}{9.7 \times 10^5} = 1.03 \times 10^{-6}$$

$$pe = 13.00 + \log\frac{[Fe^{3+}]}{[Fe^{2+}]} = 13.00 + \log\frac{10^{-6}}{1} = 7.00$$

2. From the expression for $\alpha_0 = [Fe^{3+}]/Fe(III)$ can be seen that the dihydroxo, difluoride, and trifluoride complexes are contributing most to the formation of complexes.

References

Abedon, S.T., Breitenberger, C.A., Roden, E.E., Williams, J.B. 2008. Respiration. In: S.E. Jørgensen and B. Fath (Editors), *Encyclopedia of Ecology*. Academic Press, Oxford, UK, pp. 3010–3020.

Allen, R.G., Pereira, L.S., Raes, D., Smith, M. 1998. *Crop Evapotranspiration - Guidelines for Computing Crop Water Requirements*. FAO Irrigation and Drainage paper 56. FAO - Food and Agriculture Organization of the United Nations, Rome.

Allesina, S. and Bondavalli, C. 2003. Steady state of ecosystem flow networks: A comparison between balancing procedures. *Ecological Modelling*, 165:221–229.

Amthor, J.S. 2000. The McCree–de Wit–Penning de Vries–Thornley respiration paradigms: 30 years later. *Annals of Botany*, 86:1–20.

Arrigo, K.R. 2005. Marine microorganisms and global nutrient cycles. *Nature*, 437:349–355.

Artioli, Y. 2008. Enzymatic processes. In: S.E. Jørgensen and B. Fath (Editors), *Encyclopedia of Ecology*. Academic Press, Oxford, UK, pp. 1377–1383.

Azam, F., Fenchel, T., Field, J.G., Gray, J.S., Meyer-Reil, L.A., Thingstad, F. 1983. The ecological role of water-column microbes in the sea. *Marine Ecology Progress Series*, 10:257–263.

Baird, D., McGlade, J.M., Ulanowicz, R.E. 1991. The comparative ecology of six marine ecosystems. *Philosophical Transactions of the Royal Society of London. Series B: Biological Sciences*, 333:15–29.

Barausse, A. 2008. Light extinction. In: S.E. Jørgensen and B. Fath (Editors), *Encyclopedia of Ecology*. Academic Press, Oxford, UK.

Bateson, G. and Bateson, M.C. 1987. *Angels Fear: Towards an Epistemology of the Sacred*. Macmillan, New York.

Bateson, G. 1972. *Steps to an Ecology of Mind*. Ballantine, New York.

Bénard, H. 1901. Les tourbillons cellulaires dans une nappe liquide transportant de la chaleur par convection en régime permanent [Cell vortices in a liquid sheet transporting heat by convection in a steady regime]. *Annales De Chimie Et De Physique*, 23:62–144.

Bendoricchio, G. and Palmeri, L. 2005. Quo vadis ecosystem? *Ecological Modelling*, 184:69–81.

Benedict, F.G. 1938. *Vital Energetics. A Study in Comparative Basal Metabolism*. Publication No 503. Carnegie Institution of Washington, DC.

Bertalanffy, L.V. 1950. The theory of open systems in Physics and Biology. *Science*, 111:23–29.

Bertalanffy, L.V. 1968. *General System Theory*. Braziller, New York.

Bird, R.B., Stewart, W.E., Lightfoot, E.N. 2002. *Transport Phenomena* (2nd ed.). John Wiley & Sons, New York.

Blanchette, C.A., O'Donnell, M.J., Stewart, H.L. 2008. Waves as an ecological process. In: S.E. Jørgensen and B. Fath (Editors), *Encyclopedia of Ecology*. Academic Press, Oxford, UK, pp. 3764–3770.

Boltzmann, L. 1896. *Vorlesungen Uber Gastheorie [microform]/von Ludwig Boltzmann*. J.A. Barth, Leipzig, Germany.

Boltzmann, L. 1905. *The Second Law of Thermodynamics* (Populare Schriften. Essay No.3 [Address to Imperial Academy of Science in 1886]). Reprinted in English in: *Theoretical Physics and Philosophical Problems, Selected Writings* of L. Boltzmann. D. Riedel, Dordrecht, Netherlands.

Bowie, G.L., Mills, W.B., Porcella, D.B., Campbell, C.L., Pagenkopf, J.R., Rupp, G.L., Johnson, K.M., Chan, P.W.H., Gherini, S.A., Chamberlin, C.E. 1985. *Rates, Constants, and Kinetics Formulations in Surface Water Quality Modeling* (2nd ed.). EPA/600/3-85/040, June 1985. Environmental Research Laboratory, Office of Research and Development, U.S. EPA, Athens, GA.

Brillouin, L. 1962. *Science and Information Theory*. Academic Press, New York.

Brose, U. 2010. Body-mass constraints on foraging behavior determine population and food-web dynamics. *Functional Ecology*, 24:28–34.

Brown, J.H., Gillooly, J.F., Allen, A.P., Savage, V.M., West, G.B. 2004. Toward a metabolic theory of ecology. *Ecology*, 85:1771–1789.

Cano-Odena, A. and Vankelecom, I.F.J. 2008 Transport over membranes. In: S.E. Jørgensen (Editor), *Encyclopedia of Ecology*. Elsevier, Amsterdam, Netherlands.

Chapra, S.C. 1997. *Surface Water-Quality Modeling*. McGraw-Hill, New York, pp. 3583–3588.

Christensen, V. and Walters, C.J. 2004. Ecopath with Ecosim: Methods, capabilities and limitations. *Ecological Modelling*, 172:109–139.

Ciani, M., Comitini, F., Mannazzu, I. 2008. Fermentation. In: S.E. Jørgensen and B. Fath (Editors), *Encyclopedia of Ecology*. Academic Press, Oxford, UK, pp. 1548–1557.

Connell, D.W. 1997. *Basic Concepts of Environmental Chemistry*. Lewis Publication, Boca Raton, FL. 506 pp.

Cordell, D., Drangert, J.O., White, S. 2009. The story of phosphorus: Global food security and food for thought. *Global Change Biology*, 19:292–305.

Costanza, R., d'Arge, R., de Groot, R., Farber, S., Grasso, M., Hannon, B., Limburg, K., et al. 1997. The value of the world's ecosystem services and natural capital. *Nature*, 387:253–260.

Dagan, G., Fiori, A., Janković, I. 2008. Transport in porous media. In: S.E. Jørgensen (Editor), *Encyclopedia of Ecology*. Elsevier, Amsterdam, Netherlands, pp. 3576–3582.

Danckwerts, P.V. 1951. Significance of liquid-film coefficients in gas absorption. *Industrial and Engineering Chemistry*, 43:1460–1467.

Daskalov, G. 2002. Overfishing drives a trophic cascade in the Black Sea. *Marine Ecology Progress Series*, 225:53–63.

del Giorgio, P.A. and Williams, P.J.le.B. 2005. *Respiration in Aquatic Ecosystems*. Oxford University Press, Oxford, UK.

de Leiva Moreno, J.I., Agostini, V.N., Caddy, J.F., Carocci, F. 2000. Is the pelagic-demersal ratio from fishery landings a useful proxy for nutrient availability? A preliminary data exploration for the semi-enclosed seas around Europe. *ICES Journal of Marine Science*, 57:1091–1102.

Di Toro, D.M., 1985. A particle interaction model of reversible organic chemical sorption. *Chemosphere*, 14:1503–1538.

Doney, S.C., Fabry, V.J., Feely, R.A., Kleypas, J.A., 2009. Ocean acidification: The other CO_2 problem. *Annual Review of Marine Science*, 1:169–192.

Dunne, J.A. 2006. The network structure of food webs. In: M. Pascual and J.A. Dunne (Editors), *Ecological Networks: Linking Structure to Dynamics in Food Webs*. Oxford University Press, New York, pp. 27–86.

Dunne, J.A., Williams, J.A., Martinez, N.D. 2002. Food-web structure and network theory: The role of connectance and size. *Proceedings of the National Academy of Sciences of the USA*, 99:12917–12922.

EEA. 2001. *Eutrophication in Europe's Coastal Waters* No. 7. European Environment Agency, Copenhagen.

Eigen, M., 1971. Molecular self-organization and the early stages of evolution. *Quarterly Reviews of Biophysics*, 4:149.

EPA. 1996. *Soil Screening Guidance: Part 5: Chemical-Specific Parameters*. Technical Background Document No. EPA/540/R-95/128.

Eriksson, C. and Mortimer, D.C. 1975. Mercury uptake in rooted higher plants—laboratory studies. *Verhandlungen Des Internationalen Verein Limnologie*, 19:2087–2093

Evans, R. 1969. A proof that essergy is the only consistent measure of potential work. Ph.D. Thesis, Dartmouth College, Hanover, NH.

Fath, B.D., Patten, B.C., Choi, J.S. 2001. Complementarity of Ecological Goal Functions. *Journal of Theoretical Biology*, 208:493–506.

Fenchel, T. 2008. The microbial loop–25 years later. *Journal of Experimental Marine Biology and Ecology*, 366:99–103.

Fischer, H.B., List, E.J., Koh, R.C.Y., Imberger, J., Brooks, N.H. 1979. *Mixing in Inland and Coastal Waters*. Academic Press, New York.

Fogg, P.G.T. and Sangster, J.M. 2003. *Chemicals in the Atmosphere – Solubility, Sources and Reactivity*. John Wiley & Sons, Chichester, UK.

Galli, G., Barausse, A., Mazzoldi, C., Djakovac, T., Precali, R., Palmeri, L. Unpublished data. A trophic network model to study biomass oscillations of anchovy and sardine populations in the northern Adriatic pelagic ecosystem. Submitted.

Ganev, K. 2011. *Chapter 30: Air Pollution in Handbook of Ecological Models Used in Ecosystem and Environmental Management*. CRC, Boca Raton, FL. 620 pp.

Garcia, S.M., Zerbi, A., Aliaume, C., Do Chi, T., Lasserre, G. 2003. The ecosystem approach to fisheries. Issues, terminology, principles, institutional foundations, implementation and outlook. FAO Fisheries Technical Paper. No. 443, FAO, Rome.

Glansdorff, P. and Prigogine, I. 1971. *Thermodynamic Theory of Structure, Stability and Fluctuations*. Wiley-Interscience, New York.

Glazier, D.S. 2005. Beyond the '3/4-power-law': Variation in the intra- and interspecific scaling of metabolic rate in animals. *Biological Reviews*, 80:611–662.

Grant, P.R. 1986. *Ecology and Evolution of Darwin's finches*. Reprinted in 1999. Princeton University Press, New Jersey. 492 pp.

Hanski, I. 1999. *Metapopulation in Ecology*. Oxford University Press, Cambridge, UK. 422 pp.

Healey, F.P. 1980. Slope of the Monod equation as an indicator of advantage in nutrient competition. *Microbial Ecology*, 5:281–286.

Holling, C.S. 1959. Some characteristics of simple types of predation and parasitism. *The Canadian Entomologist*, 91:385–398.

Hundsdorfer, W.H. and Verwer, J.G. 2003. *Numerical Solution of Time-Dependent Advection-Diffusion-Reaction Equations*. Springer-Verlag, Berlin, Germany.

IPCC. 2007. *Fourth Assessment Report of the Intergovernmental Panel on Climate Change*. IPCC, Geneva, Switzerland.

Iqbal, M. 1983. *An Introduction to Solar Radiation*. Academic Press, New York.

Irmak, S. 2008. Evapotranspiration. In: S.E. Jørgensen and B. Fath (Editors), *Encyclopedia of Ecology*. Academic Press, Oxford, UK, pp. 1432–1438.

IUPAC. 1997. *Compendium of Chemical Terminology* (2nd ed., the "Gold Book"). Blackwell Scientific Publication, Oxford, UK.

Jackson, J.B.C., Kirby, M.X., Berger, W.H., Bjorndal, K.A., Botsford, L.W. Bourque, B.J., Bradbury, R.H., et al. 2001. Historical overfishing and the recent collapse of coastal ecosystems. *Science*, 293:629–638.

Jennings, S. and Kaiser, M.J. 1998. The effects of fishing on marine ecosystems. *Advances in Marine Biology*, 34:201–352

Jensen, M.E., Burman, R.D., Allen, R.G. 1990. *Evapotranspiration and Irrigation Water Requirements*. ASCE manual and Reports on engineering practice 70. ASCE, New York.

Jørgensen, S.E. 1994a. Review and comparison of goal functions in system ecology. *Vie Milieu*, 44:11–20.

Jørgensen, S.E. 1994b. *Fundamentals of Ecological Modelling* (2nd ed). *Developments in Environmental Modelling* 19. Elsevier, Amsterdam, Netherlands. 628 pp.

Jørgensen, S.E. 1995a. Exergy and ecological buffer capacities as measures of the ecosystem health. *Ecosystem Health* 1:150–160.

Jørgensen, S.E. 1995b. The growth rate of zooplankton at the edge of chaos: Ecological models. *Journal of Theoretical Biology*, 175:13–21.

Jørgensen, S.E. 2012. *Fundamentals of Systems Ecology*. CRC, Boca Raton, FL. 320 pp.

Jørgensen, S.E. and Bendoricchio, G. 2001. *Fundamentals of Ecological Modeling* (3rd ed.). Elsevier, Amsterdam, the Netherlands. 628 pp.

Jørgensen, S.E. and Fath Brian, D. 2004. Modelling the selective adaptation of Darwin's Finches. *Ecological Modelling*, 176:409–418.

Jørgensen, S.E. and B. Fath. 2011. *Fundamentals of Ecological Modelling* (4th ed.). Elsevier, Amsterdam, Netherlands. 400 pp.

Jørgensen, S.E., Ladegaard, N., Debeljak, M., Marques, J.C. 2005. Calculations of exergy for organisms. *Ecological Modelling*, 185:165–176.

Jørgensen, S.E. and Meyer, H.F. 1977. Ecological buffer capacity. *Ecological Modelling*, 3: 39–61.

Jørgensen, S.E., Patten, B.C., Straskraba, M. 1999. Ecosystem emerging: 3. Openness. *Ecological Modelling*, 117:41–64

Jørgensen, S.E. 1992a. Parameters, Ecological constraints and exergy. *Ecological Modelling*, 62:163–170.

Jørgensen, S.E. 1992b. Development of models able to account for changes in species composition. *Ecological Modelling*, 62:195–208.

Jørgensen, S.E. 1997. *Integration of Ecosystem Theories: A Pattern*. Academic Publishers, Dordrecht, The Netherlands.

Jørgensen, S.E. 2006. *Eco-Exergy as Sustainability*. WIT Press, Southampton, UK.

Jørgensen, S.E. 2008. Allometric principles. In: S.E. Jørgensen (editor), *Encyclopedia of Ecology*. Elsevier, Amsterdam, the Netherlands.

Jørgensen, S.E. and Fath, B.D. 2004. Application of thermodynamic principles in ecology. *Ecological Complexity*, 1:267–280.

Kadlec, R.H. and Wallace, S. 2008. *Treatment Wetlands* (2nd ed.). CRC Lewis Publishers, Boca Raton, FL.

Kadlec, R.H. and Knight, R.L. 1996. *Treatment Wetlands*. CRC Press, Boca Raton, FL.

Karickhoff, S.W. 1981. Semi-empirical estimation of sorption of hydrophobic pollutants on natural sediments and soils. *Chemosphere*, 10:833–846.

Karickoff, S.W., Brown, D.S., Scott, T.A. 1979. Sorption of hydrophobic pollutants on natural sediments. *Water Research*, 13:241–248.

Kie, J.G. 1999. Optimal foraging and risk of predation: Effects on behavior and social structure in ungulates. *Journal of Mammalogy*, 80:1114–1129.

Kubo, T., Tanaka, N., Hosoya, K. 2004. Target-selective ion-exchange media for highly hydrophilic compounds: A possible solution by use of the "interval immobilization technique. *Analytical and Bioanalytical Chemistry*, 378:84–88.

Lal, R., Kimble, JM., Follett, RF., Stewart, BA. 2001. *Assessment Methods for Soil Carbon*. Lewis Publication, Boca Raton, FL. 675 pp.

Lambers, H., Chapin III, F.S., Pons, T.L. 2008. *Plant Physiological Ecology* (2nd ed.). Springer, New York.

Lavery, T.J., Roudnew, B., Gill, P., Seymour, J., Seuront, L., Johnson, G., Mitchell, J.G., Smetacek, V. 2010. Iron defecation by sperm whales stimulates carbon export in the Southern Ocean. *Proceedings of the Royal Society B, Biological Sciences*, 277:3527–3531.

Levins, R. 1969. Some demographic and genetic consequences of environmental heterogeneity for biological control. *Bulletin of the Entomological Society of America*, 15:237–240.

Lewis, W.K. and Whitman W.G. 1924. Principles of gas absorption. *Industrial and Engineering Chemistry*, 16:1215–1220.

Li, Z., Wakao, S., Fischer, B.B., Niyogi, K. 2009. Sensing and responding to excess light. *Annual Reviews of Plant Biology*, 60:239–260.

Lignell, R. 1990. Excretion of organic carbon by phytoplankton: Its relation to algal biomass, primary productivity and bacterial secondary productivity in the Baltic Sea. *Marine Ecology Progress Series*, 68:85–99.

Lipton, B.H. 2005. *The Biology of Belief: Unleashing the Power of Consciousness, Matter & Miracles*. Hay House UK Limited, London, UK.

Liu, Y. and Chen, J. 2008. Phosphorus cycle. In: S.E. Jørgensen and B. Fath (Editors), *Encyclopedia of Ecology*. Academic Press, Oxford, pp. 2715–2724.

Lorch, W. 1987. *Handbook of Water Purification* (2nd ed.). John Wiley & Sons, New York.

Lotka, A. 1922. Contribution to the energetics of evolution. *Proceedings of the National Academy of Sciences USA*, 8:147–151.

Lovelock, J.E., 1979. *Gaia: A New Look at Life on Earth*. Oxford University Press, London, UK.

Margulis, L. and Lovelock, J.E. 1974. Biological modulation of the earth's atmosphere. *Icarus*, 21:471–489.

MacArthur, R.H. 1955. Fluctuations of animal populations and a measure of community stability. *Ecology*, 36:533–536.

Manahan, S.E. 2001. *Fundamentals of Environmental Chemistry*. Lewis Publishers, Boca Raton, FL.

Martell, S.J.D. 2008. Fisheries management. In: S.E. Jørgensen and B. Fath (Editors), *Encyclopedia of Ecology*. Academic Press, Oxford, UK, pp. 1572–1582.

Maturana, H. and Varela, F. 1980. *Autopoiesis and Cognition*. D. Reidel, Dordrecht Holland, Netherlands.

Maurer, B.A. 2008. Ecological process: Scale. In: S.E. Jørgensen (Editor), *Encyclopedia of Ecology*. Elsevier, Amsterdam, Netherlands.

McChesney, I.G. and Lincoln University. Centre for Resource, M. 1991. *The Brundtland Report and Sustainable Development in New Zealand/Ian G. McChesney*. Centre for Resource Management, [Lincoln] NZ.

McCutcheon, S.C. and Jørgensen, S.E. 2008. Phytoremediation. In: S.E. Jørgensen and B. Fath (Editors), *Encyclopedia of Ecology*. Academic Press, Oxford, UK, pp. 2751–2767.

Mitsch, W.J. and Gosselink, J.G. 2007. *Wetlands*. John Wiley & Sons, New York.

Morford, S.I., Houlton, B.Z., Dahlgren, R.A. 2011. Increased forest ecosystem carbon and nitrogen storage from nitrogen rich bedrock. *Nature*, 477:78–81.

Morowitz, H.J. 1968. *Energy Flow in Biology*. Ox Bow Press, Woodbridge, CT.

Newell, C.J., Rifai, H.S., Wilson, J.T., Connor, J.A., Aziz, J.A., Suarez, M.P. 2002. *Calculation and Use of First-Order Rate Constants for Monitored Natural Attenuation Studies*. EPA/540/S-02/500. United States Environmental Protection Agency, Cincinnati.

Nicolis, G. and Prigogine, I. 1977. *Self-Organization in Nonequilibrium Systems*. John Wiley & Sons, New York.

Nielsen, S.N. 1992a. Application of maximum exergy in structural dynamic models. Ph.D. Thesis, National Environmental Research Institute, Denmark, 51 pp.

Nielsen, S.N. 1992b. Strategies for structural-dynamical modelling. *Ecological Modelling*, 63:91–102.

Nixon, S.W. 1995. Coastal marine eutrophication: A definition, social causes, and future concerns. *Ophelia*, 41:199–219.

O'Malley, R. 2010. Ocean Productivity web site. http://www.science.oregonstate.edu/ocean.productivity/. Accessed 12th June 2012.

Odum, E.P. 1969. The strategy of ecosystem development. *Science*, 164:262–270.

Odum, E.P. 1983a. *Basic Ecology*. CBS College Publishing, Philadelphia, PA.

Odum, E.P. 1953. *Fundamentals of Ecology*. W. B. Saunders, Philadelphia, PA.

Odum, H.T. 1983b. *Systems Ecology*. John Wiley & Sons, New York.

Odum, H.T. 1988. Self-organization, transformity, and information. *Science*, 242:1132–1139.

Odum, H.T. 1996. *Environmental Accounting. EMERGY and Environmental Decision Making*. John Wiley & sons, New York.

Onsager, L. 1931. Reciprocal relations in irreversible processes. *Physical Review*, 37:405–426.

Patten, B.C., Straskraba, M., Jørgensen, S.E. 1997. Ecosystems emerging: 1. conservation. *Ecological Modelling*, 96:221–284.

Pauly, D. and Christensen, V. 1995. Primary production required to sustain global fisheries. *Nature*, 374:255–257.

Pauly, D., Christensen, V., Walters, C. 2000. Ecopath, Ecosim, and Ecospace as tools for evaluating ecosystem impact of fisheries. *ICES Journal of Marine Science*, 57:697–706.

Pauly, D., Christensen, V., Dalsgaard, J., Froese, R., Torres Jr, F., 1998. Fishing down marine food webs. *Science*, 279:860–863.

Peters, R.H. 1983. *The Ecological Implications of Body Size*. Cambridge University Press, Cambridge, UK.

Phillips, O.M. 1980. *The Dynamics of the Upper Ocean*. University Press, Cambridge, UK.

Postgate, J. 1998. *Nitrogen fixation*. (3rd ed.). Cambridge University Press, Cambridge, UK.

Prigogine, I. and Strangers, I. 1984. *Order Out of Chaos: Man's New Dialogue with Nature*. Bantham, New York.

Prigogine, I. 1967. *Thermodynamics of Irreversible Processes*. John Wiley & Sons, New York.

Prigogine, I. 1991. The arrow of time. In: C. Rossi and E. Tiezzi (Editors), *Ecological Physical Chemistry*. Elsevier, Amsterdam, the Netherlands, pp. 1–24.

Prigogine, I. 1996. Time, chaos and the laws of nature. In: P. Weingartner and G. Schurz (Editors), *Law and Prediction in the Light of Chaos Research*. Springer, Berlin, Germany, pp. 3–9.

Rascio, N. and La Rocca, N. 2008. Biological nitrogen fixation. In: S.E. Jørgensen and B. Fath (Editors), *Encyclopedia of Ecology*. Academic Press, Oxford, UK, pp. 412–420.
Redfield, A.C. 1958. The biological control of chemical factors in the environment. *American Scientist*, 46:205–221.
Reible, D.D. 1998. *Fundamentals of Environmental Engineering*. Lewis Publication, Boca Raton, FL. 526 pp.
Rose, K.A., Allen, J.I., Artioli, Y., Barange, M., Blackford, J., Carlotti, F., Cropp, R., et al. 2010. End-to-end models for the analysis of marine ecosystems: Challenges, issues, and next steps. *Marine and Coastal Fisheries: Dynamics, Management, and Ecosystem Science*, 2:115–130.
Rutheford, J.C. 1994. *River Mixing*. John Wiley & Sons, New York.
Salt, D.E., Blaylock, M., Kumar, N.P.B.A., Dhushenkov, V., Ensley, B.D., Chet, I., Raskin, I. 1995. Phytoremediation: A novel strategy for the removal of toxic metals from the environment using plants. *Biotechnology*, 13:468–474.
Sankaran, M., Hanan, N.P., Scholes, R.J., Ratnam, J., Augustine, D.J. Cade, B.S., Gignoux, J., et al. 2005. Determinants of woody cover in African savannas. *Nature*, 438:846–849.
Schneider, E.D. and Kay, J.J. 1994. Life as a manifestation of the second law of thermodynamics. *Mathematical and Computer Modelling*, 19:25–48.
Schrödinger, E. 1944. *What is life?* Cambridge University Press, Cambridge.
Shannon, C.E. and Weaver, W. 1963. *The Mathematical Theory of Communication*. University of Illinois, Urbana, IL.
Shertzer, K.W., Prager, M.H., Vaughan, D.S., Williams, E.H., 2008. Fishery models. In: S.E. Jørgensen and B. Fath (Editors), *Encyclopedia of Ecology*. Academic Press, Oxford, UK, pp. 1582–1593.
Shieh, J.H. and Fan, L.T. 1982. Estimation of energy (enthalpy) and exergy (availability) contents in structurally complicated materials. *Energy Sources*, 1–46.
Smith, L.S., Yamanaka, Y., Pahlow, M., Oschlies, A. 2009. Optimal uptake kinetics: Physiological acclimation explains the pattern of nitrate uptake by phytoplankton in the ocean. *Marine Ecology Progress Series*, 384:1–12.
Sterner, R.W. and Elser, J.J. 2002. Ecological stoichiometry: The biology of elements from molecules to the biosphere. Princeton University Press, Princeton, NJ.
Straskraba, M. 1979. Natural control mechanisms in models of aquatic ecosystems. *Ecology Modelling*, 6:305–322.
Suarez, M.P. and Rifai, H.S. 1999. Biodegradation rates for fuel hydrocarbons and chlorinated solvents in groundwater. *Bioremediation Journal*, 3:337–362.
Svirezhev, Y.M. and Svirejeva-Hopkins, A. 1998. Sustainable biosphere: Critical overview of basic concept of sustainability. *Ecological Modelling*, 106:47–61.
Svirezhev, and Yu, M., 1998. Thermodynamic orientors: How to use thermodynamic concepts in ecology? In: F. Müller and M. Leupelt (Editors), *Eco Targets, Goal Functions and Orientors*. Springer, Berlin-Heidelberg, Germany, pp. 102–122.
Szilárd, L. 1929. Über die Entropieverminderung in einem thermodynamischen System bei Eingriffen intelligenter Wesen [On the reduction of entropy in a thermodynamic system by the intervention of intelligent beings]. *Zeitschrift für Physik*, 53:840–856.
Tetko, I.V., Gasteiger, J., Todeschini, R., Mauri, A., Livingstone, D., Ertl, P., Palyulin, V.A., et al. 2005. Virtual computational chemistry laboratory - design and description. *Journal of Computer-Aided Molecular Design*, 19:453–463.

Tiezzi, E., 2003. *The Essence of Time* (125 pp) and *The End of Time* (202 pp). WIT Press, Southampton, UK.

Trapp, S. and Matthies, M. 1998. *Chemodynamics and Environmental Modeling*. Springer-Verlag, Berlin, Germany.

Tribus, M. and McIrvine, E.C. 1971. Energy and information. *Scientific American*, 225:179–188.

Ulanowicz, R.E. 2009. *A Third Window: Natural Life Beyond Newton and Darwin*. Templeton Foundation Press, West Conshohocken, PA.

Ulanowicz, R.E. 1997. *Ecology, the Ascendent Perspective*. Columbia University Press, New York.

Ulanowicz, R.E. and Abarca-Arenas, L.G. 1997. An informational synthesis of ecosystem structure and function. *Ecological Modelling*, 95:1–10.

Ulanowicz, R.E. 1986. *Growth and Development: Ecosystems Phenomenology*. Springer-Verlag, New York.

Ulanowicz, R.E. 2004. Quantitative methods for ecological network analysis. *Computational Biology and Chemistry*, 28:321–339.

Ulgiati, S. and Bianciardi, C. 1997. Describing states and dynamics in far from equilibrium systems: Needed a metric within a system state space. *Ecological Modelling*, 96:75–89.

Ursin, E. 1967. A mathematical model of some aspects of fish growth, respiration, and mortality. *Journal Fisheries Research Board of Canada*, 24(11):2355–2453.

van den Berg, N. and Ashmore, M. 2008. Nitrogen. In: S.E. Jørgensen and B. Fath (Editors), *Encyclopedia of Ecology*. Academic Press, Oxford, UK, pp. 2518–2526.

Verity, P.G., Smetacek, V., Smayda, T.J. 2002. Status, trends and the future of the marine pelagic ecosystem. *Environmental Conservation*, 29:207–237.

Vollenweider, R.A., Giovanardi, F., Montanari, G., Rinaldi, A. 1998. Characterization of the trophic conditions of marine coastal waters with special reference to the NW Adriatic sea: Proposal for a trophic scale, turbidity and generalized water quality index. *Environmetrics*, 9:329–357.

Wang, L. and D'Odorico, P. 2008. Decomposition and mineralization. In: S.E. Jørgensen and B. Fath (Editors), *Encyclopedia of Ecology*. Academic Press, Oxford, UK, pp. 838–844.

Weber, W.J.Jr. and DiGiano, F.A. 1996. *Process Dynamics in Environmental Systems*. Wiley Interscience, New York, NY.

Whitman, W.G. 1923. The two-film theory of gas absorption. *Chemical Metallurgical Engineering*, 29:146–148.

Widdison, P.E. and Burt, T.P. 2008. Nitrogen cycle. In: S.E. Jørgensen and B. Fath (Editors), *Encyclopedia of Ecology*. Academic Press, Oxford, UK, pp. 2526–2533.

Wiener, N. 1999. Cybernetics or control and communication in the animal and the machine. MIT Press, Cambridge, MA.

Woodward, G., Ebenman, B., Emmerson, M., Montoya, J.M., Olesen, J.M. Valido, A., Warren, PH. 2005. Body size in ecological networks. *Trends in Ecology and Evolution*, 20:402–409.

Zhang, J., Gurkan, Z., Jørgensen, S.E. 2010. Application of eco-energy for assessment of ecosystem health and development of structurally dynamic models. *Ecological Modelling*, 221:693–702.

Index

D